Thinking Swarms

Simon Ng • Jason Beaufort Scholz •
Hussein A. Abbass
Editors

Thinking Swarms

Springer

Editors
Simon Ng
Trusted Autonomous Systems (TAS)
Toowong, QLD, Australia

Jason Beaufort Scholz
RMIT University
Melbourne, VIC, Australia

Hussein A. Abbass
School of Systems and Computing
University of New South Wales
Canberra, ACT, Australia

ISBN 978-3-031-82789-1   ISBN 978-3-031-82790-7 (eBook)
https://doi.org/10.1007/978-3-031-82790-7

This work was supported by Trusted Autonomous Systems Defence CRC.

© The Editor(s) (if applicable) and The Author(s) 2025. This book is an open access publication.

Open Access This book is licensed under the terms of the Creative Commons Attribution 4.0 International License (http://creativecommons.org/licenses/by/4.0/), which permits use, sharing, adaptation, distribution and reproduction in any medium or format, as long as you give appropriate credit to the original author(s) and the source, provide a link to the Creative Commons license and indicate if changes were made.

The images or other third party material in this book are included in the book's Creative Commons license, unless indicated otherwise in a credit line to the material. If material is not included in the book's Creative Commons license and your intended use is not permitted by statutory regulation or exceeds the permitted use, you will need to obtain permission directly from the copyright holder.

The use of general descriptive names, registered names, trademarks, service marks, etc. in this publication does not imply, even in the absence of a specific statement, that such names are exempt from the relevant protective laws and regulations and therefore free for general use.

The publisher, the authors and the editors are safe to assume that the advice and information in this book are believed to be true and accurate at the date of publication. Neither the publisher nor the authors or the editors give a warranty, expressed or implied, with respect to the material contained herein or for any errors or omissions that may have been made. The publisher remains neutral with regard to jurisdictional claims in published maps and institutional affiliations.

This Springer imprint is published by the registered company Springer Nature Switzerland AG
The registered company address is: Gewerbestrasse 11, 6330 Cham, Switzerland

If disposing of this product, please recycle the paper.

Dedicated to the peoples of Australia, in recognition of indigenous and migrant contributions to an innovation ecosystem in which humans and machines can thrive together.

To our students, colleagues, mentees, and mentors—we learn as much from you as you do from us.

To our families for giving us love and time.

Foreword by Professor Calum Drummond

In the spirit of trusted autonomous systems and taking advantage of the human-machine interface, this foreword has been written with support from RMIT Val—a private secure generative AI tool.

In the realm of scientific exploration and technological innovation, few concepts capture the imagination and potential for transformative impact as vividly as "Thinking Swarms". This book, a culmination of dedicated research and cross-disciplinary collaboration, stands at the forefront of understanding and harnessing the power of collective intelligence, such as the Internet of Things, a principle beautifully illustrated by the natural world. Delving into the intricacies of swarm behaviour, the chapters draw inspiration from nature's finest examples—the mesmerising murmuration of starlings, the relentless coordination of ants, and the fluid dynamics of a fish shoal. These natural phenomena underscore a fundamental truth: the collective can achieve what the individual cannot. This principle is not only a spectacle of biological evolution but also a beacon guiding our quest to develop advanced, intelligent systems capable of addressing complex challenges.

The chapters within this book explore the multifaceted dimensions of thinking swarms, from foundational theories and behavioural insights to architectural frameworks and cognitive processes. Each section builds on the premise that while simple local interactions can produce complex global behaviours, there is an untapped potential in creating swarms that not only act collectively but also think both individually and as a unit.

RMIT University and the Trusted Autonomous Systems Defence Cooperative Research Centre (TAS DCRC) have been steadfast supporters of this endeavour, recognising the profound implications of swarm intelligence for defence and beyond. In the defence context, the application of swarm intelligence promises to revolutionise operational strategies, enhancing situational awareness, resilience, and adaptability. However, the path to realising these capabilities is fraught with challenges. Ethical, legal, regulatory, social, and technological barriers must be navigated with care and foresight. This book does not shy away from these issues; instead, it confronts them head-on, offering a comprehensive analysis and proposing pathways forward.

Innovation and the translation of research into practical, impactful applications remain critical. The true potential of swarm systems lies in our ability to bridge the gap between theoretical constructs and real-world implementation. As we advance, it is imperative that we foster a collaborative ecosystem where academia, industry, and government can work together to overcome obstacles and drive progress.

It is my pleasure to contribute this foreword to a work that embodies the spirit of innovation and the relentless pursuit of knowledge. I am intrigued by the possibilities that thinking swarms present. They are not merely a concept but a vision of a future where intelligent, autonomous systems operate in harmony with human intent, enhancing our capabilities and opening new frontiers. This book is a testament to the hard work and visionary thinking of its contributors, and I am confident that it will serve as a cornerstone for future research and development in this exciting field.

To the readers, I invite you to immerse yourselves in this exploration of thinking swarms. May it inspire you, challenge your perspectives, and ignite a passion for innovation that drives us all towards a brighter future.

Deputy Vice-Chancellor Research and Innovation and Vice-President, RMIT, Australia
Melbourne, July, 2024

Professor Calum Drummond AO

Foreword by Prof. Tim Barrett

In Defence strategic planning circles, there is always discussion on the changing character of warfare and what will be the next revolution in military affairs. In recent years this has been fuelled somewhat by images and intense commentary on the prime roles played by emerging drone capabilities employed in real-world scenarios, including land and sea domains in the Ukraine and the Red Sea. Heightened by the effect that these devices appear to have on strategic geopolitical outcomes, the debate has often focused solely on the number and type of drones employed on the battlefield. Indeed, the recent Defence Strategic Review (2023) promotes to some extent the development of trusted autonomous systems but does not necessarily prescribe their application.

Indeed, it is wrong to think of the drone itself as creating Defence's warfighting edge. It is a capability means to achieve warfighting ends. But to fully realise it's advantage, the capability also requires a disciplined and deliberate concept of operations that considers it's broader range of application. With the advent of digital transformation and artificial intelligence, the utility of the drone takes on a completely different edge with the ability to generate mass and agility by the aggregation of effect. This is not a simple thing to achieve and is not well understood. Despite some significant achievements in the commercial world demonstrating swarming techniques used in benign applications, its use in military applications is not yet fully settled.

This publication explores in depth the emerging science that underpins this edge. It demonstrates the importance of understanding concepts that will enable cooperative and collaborating drone systems to be used on the battlefield, for warfighting effect and to be used within the legal frameworks that satisfy our international obligations. The detail can inform the defence planner on what requirements need to be prioritised. For industry it can guide design and production, and for the operator in can provide assurance that the system has the warfighting edge and can be used in accordance with their rules of engagement. It is a worthy read.

Board Director, Trusted
Autonomous Systems
Defence CRC, Australia
Brisbane, July, 2024

AO, CSC, RAN (Rtd) VADM Tim Barrett

Preface

It has been said that "quantity has a quality all of its own". The murmuration of hundreds of thousands of starlings, a swarm of millions of ants, or a shoal of hundreds of millions of herring are well-known examples of swarms in nature and represent a specialised form of social organisation.

In nature, predators often have evolved exquisite and deep individual specialisations, and their prey have developed massive-scale organisational abilities to counter them. For example, the hawk has better eyesight, more lethal claws, higher speed, and better manoeuvrability than any single starling on which it would prey; yet, in a murmuration it is challenged. The hawk's superior ability to target and track an individual is made so much harder because the starlings all look the same and try to stay close together. Their closeness also creates a potentially fatal risk of collision for a hawk travelling at speed. The internal murmuration signal detection of the hawk travels, to even distant members, faster than the hawk can fly to reach them, allowing them to avoid the hawk without having to see it. Moreover, there is safety in numbers. Statistically speaking, the odds are better if you can hide deep in the crowd than if you are exposed and alone.

Other predators have, however, evolved the other way, with organisational abilities en masse, to allow them to overcome prey with individual specialisations. A documentary from BBC Earth (2009) illustrates this using a freshwater crab in a rainforest that is discovered by a raiding party of foraging "driver" ants. The ants start their attack by massing in numbers and swarming all over the crab. However, the ants' tiny jaws are ineffectual on the crab's hard and thick shell. Some of the ants start to probe possible weak points. They spray formic acid in the crab's eyes and mouth parts. One group of ants examines the legs and claws. It searches for vulnerable joints and cut its way into the leg. Large worker ants use their mandibles to cut away at the soft tissue and, as the tears widen, small worker ants are able to climb inside the crab's legs to reach inside to the muscles. One group of ants examines the crab's mouth, and another the eyes. They appear to work together, while larger ants hold back the crab's mouth and small ants attack the soft tissue behind it. Despite being dwarfed by the crab, there are so many ants attacking at once. Even a fully armoured crab does not stand a chance. This demonstrates the adhoc formation of subgroups or teams, separation of tasks, and separation of roles.

Further, in this documentary, these ants cooperate to build and sustain their colony at a massive and distributed scale, using a combination of tree leaves and their own bodies to form and hold together a bridge for the flow of ants across a large and shallow pond. This raises many more questions, such as how do they discern the need to instigate this, in which direction do they build, and for how long do they maintain their cooperative effort?

Nobody could have imagined, even a few years ago, how technology and the Internet would transform the way individuals work, communicate, entertain, and socialise. Martin Libicki wrote "The Mesh and The Net: Speculations on Armed Conflict in an Age of Free Silicon" in March 1994, introducing the concept of a battlefield mesh of many, small, cheap silicon sensors and effectors. He considered the implications for not only military platforms and organisations but also civilian information networks. This essay argued that a consequence of Moore's Law is organised intelligent devices (ubiquitous, tiny, and cheap). Later, in 1999, Kevin Ashton coined the term the "Internet of Things". Today arguably such a concept is supported by advances in machine learning: quantum computing, sensing and communication, and efficient means of energy conversion and storage. Yet, it is difficult to highlight examples of man-made swarming systems. Why is this so? Do they exist, but are waiting in secret? Or are their significant outstanding barriers, such as ethical, legal, regulatory, social, or technological ones? These chapters address these questions.

In general, we consider a "thinking swarm" as a collaborative problem-solving process in which a group of individuals work together in a decentralised and self-organised manner such that the collective intelligence of the group can be harnessed to solve complex problems more effectively than any individual could on their own. This includes the previous examples from biology but is broader in scope, to include:

1. Biological: mammals, insects, fish, bacteria, viruses, and cells, including neurons in the human brain
2. Human: brainstorming sessions, focus groups, protests, and flash mobs
3. Machine (Robotic): physical robots that work together to perform complex tasks
4. Machine (Computational): cyber robots including artificial intelligence, bots, and software viruses
5. Human–machine: combined thinking to solve complex problems, social media, robot shepherding, and botnets

Although models that include simple local social interactions consisting of rules such as repulsion, alignment, and attraction are sufficient to generate swarm morphologies, these illustrations from nature give a reason to assert that more is possible. This collection of chapters represents a small range of cross-disciplinary perspectives that we believe are critical to thought leadership for development of future technologies. We structured the book in the five parts described below.

Part I: Foundations of Thinking Swarm

The title *Thinking Swarms* was intended with two meanings. The first, as thinking about swarms: it is a collection of essays and papers that explore thoughts on the subject of swarms. The second, as an exploration of a more distilled idea: swarms don't simply adhere to the traditional notion of simple and identical automata interacting locally to achieve global effect; but rather, swarms that think both individually and collectively.

Chapter 1 is the result of a one-day workshop in which an audience of researchers and professionals from academia, industry, and government came together to talk about the topic of "thinking swarms". The defining characteristic of that discussion is captured here in the breadth of notions of swarm and of thinking, and how the two might be intertwined. A thinking swarm thus includes the surrender of agency to the collective but also the surrender of the collective to the individual. It is a system constituted by entities that think—with their own agency, purposes, and "lives" separate from that of the collective—their own distinct manifestations, but that also share information, goals and may be subordinated by the collective.

In this introductory part, we start with an exploration of traditional views of swarm robotics. Chapter 2 provides a scientometric analysis of how researchers in the open literature talk about autonomy, swarming, and teaming. This provides a clear baseline for terminology to complement the functional and ontological "models" put forward in Chap. 1. Chapter 3 presents an account of the various definitions of a swarm through swarm intelligence and swarm optimisation lenses. The chapter assists in the conceptualisations of swarms to frame subsequent chapters.

These first three chapters are rooted in a tradition that began with Beni and Wang [108], but Chap. 4 takes readers on a philosophical journey departing from tradition, which challenges assumptions for thinking about swarms from a biological perspective using the notion of enkapsis.

Finally, this part concludes with an exploration of how the popular media present the notion of swarms. Chapter 5 starts with concerns about perception, which are relevant given the current global interest in lethal autonomous weapons systems, and establishes a value-based lens to inform further discussion.

Part II: Thinking Swarm—Behaviour

When humans observe any autonomous system, the first thing they typically see beyond its static physical appearance is how it behaves dynamically in time and space, both in isolation and in response to stimuli. In Chap. 1, the authors discuss important concepts that describe how we might interpret the "embodied appearance" and "observed behaviour" we see, asking core questions: How does appearance influence our perception of a system, specifically, its "value-laden-ness"? How does it support essential requirements of "explicability", trust, and assurance? To these we might add the following: Is what we see actually representative of what is going on? And if not, can we go past the obvious in terms of appearance and see the hidden?

As users and regulators of a system, external observations of a system (whether how it looks or how it behaves) are often the only information we have about the system itself, its inherent capabilities, and its proclivities. As designers, appearance (static and dynamic) is the interface that relates our design to the wider social system within which the system operates.

How a system behaves is at the root of how we understand, predict, and ultimately trust that system. Inscrutable or chaotic behaviour undermines trust on the part of users but may enhance unpredictability for adversaries. Cooperation and proactive behaviour implies a degree of intelligence or intent that may not exist, producing unfounded expectations.

Chapter 6 examines the challenges of interpretation in terms of observing a system about which we have little inside knowledge, proposing a hermeneutic framework and an experimental approach to better understand how we interpret a system and how that interpretation engenders trust. However, are certain types of behaviour more or less important? What might the study of hermeneutics say about whether the ability of a system to emote or exhibit "helpful" behaviour enhances interpretability at the expense of effectiveness?

Chapter 7 offers us a route to identifying coherent behaviour and structure from noise. Finding coherency in chaos is a fundamental challenge when we observe any large-scale system, and it remains central to notions of interpretability.

Chapter 8 brings story analysis and mathematics together, applying techniques from information theory to better predict social behaviour. The authors suggest that observed behaviour can be transformed and stored in the information domain, allowing agents to reason in the information theoretic domain about behavioural events.

Finally, this part finishes with two chapters. Chapter 9 tackles the problem of how to frame behaviour from a regulatory perspective by examining the current state of regulation in terms of maritime autonomous systems and how swarming may challenge these approaches. Chapter 10 discusses the relationship between design choices and required law, which provides a complementary lens to the regulatory work in the preceding chapter. Unlike Chaps. 6 and 7, these two chapters start with a set of expected behaviours (lawfulness) and articulate the design requirements and policy choices needed to produce this behaviour in systems operating in real-world conditions.

Part III: Thinking Swarm—Topology and Architecture
Unlike observations about the appearance of a system, the underlying architectural topology and the mechanisms for reasoning (perception, planning, and reaction) are design choices that developers make.[1] They are the how if behaviour is the what. This is not to imply that one theme has primacy over another. We design systems

---

[1] In this discussion, we hold ourselves to artificially constructed or human-designed systems, no matter what design process is employed, rather than naturally occurring systems designed through evolutionary selection.

to exhibit desired behaviours, and we adjust our expected behaviours according to feasible engineering solutions.

In the introduction we ask the following questions: What control topology governs the system's or swarm's function? Is it defined, dynamic, or evolving? How does the system transform between topological models and maintain coherence of purpose and function? How do "thinking swarms" think? What are the cognitive strategy routines and processes and what do they mean in relation to the other three themes?

In this series of papers, the authors explore approaches to designing systems that exhibit desirable (or at least not undesirable) behavioural outcomes, providing insights into technical approaches, methods, and experimental results.

The part starts with Chap. 11, which specifies an architectural topology that provides a multi-agent system that has the ability to flexibly instantiate communities of agents (virtual and real) that can communicate and collaborate within a dynamic context. Their work builds on Pask's conversational skeleton by extending it through introducing situational awareness and sensemaking and narrative devices, such as "framing" and "analogy".

Chapter 12 proposes a control mechanism using only local interactions and a general communication topology to produce distinct patterns of behaviour from a multi-agent system, highlighting a way of achieving behavioural properties through local interactions and simple cognitive architectures.

Finally, Chap. 13 covers broad ground, including a broad review of the elements of a swarming system. To this end, it has relevance to many of the thematic layers of this text. However, Guihan's review of swarm architectures in the context of the challenges of underwater swarming robotics is of particularly note, not necessarily because it is comprehensive, but because it highlights the topological architectures arising from the conceptual assumptions and limitations imposed on engineers. The question that remains answers whether these assumptions are still valid as autonomy advances.

Part IV: Thinking Swarm—Cognition

Cognition sits at the heart of thinking. In the prior parts, we present definitions, forms and behaviours, and topologies. All are central to the ambition of seeing swarm robotics used effectively and ethically. However, the heart of any autonomous system, but especially thinking swarms, is the process in which it takes inputs, calibrates these against what it "knows" and its "intentions", and then acts.

Machine cognition, even in individual systems, is a subject that has been and will continue to be explored in depth in the academic literature. This part offers three perspectives on how swarms might think, and how that cognitive process can support intelligent collective and individual behaviour. Each of the three chapters brings a different perspective to the notion of how swarms think and learn. The first two chapters explicitly tackle the problem in the context of teaching behaviours to a swarm. The third chapter synthesises multiple approaches to machine cognition into a more complex whole (recalling the concept of enkapsis discussed in Chap. 4).

Chapter 14 tackles the challenge of instilling rich behaviours in a swarm using a genetic algorithm framework and exploiting simulation to allow the swarming agents to learn how to best communicate in a complex environment in which networks are disrupted by local effects through a process of behaviour testing, construction/evolution, and selection that uses a pattern matching rubric to enable selection in changing environments.

Chapter 15 proposes a curriculum to systematise the process of learning and the incorporation of expert knowledge in the act of design, focusing on three curriculum design approaches, namely, learner centred, teacher centred, and blended, using a series of case studies. Combined, Smith and Hussein examine behaviours from opposite sides of the conceptual fence.

Chapter 16 is, in many ways, a perfect ending for the book. Johnson et al. are field roboticists trying to solve the end-to-end swarming problem in the real world for an end user. Their chapter starts with notions of human intent and human and robotic perception and, through the application of a collection of techniques, ranging from machine learning to information entropy, fields a multirobot system that reacts dynamically (individually and collectively) to what humans are doing on the ground while coordinating efforts for global effect across the system as a whole.

Part V: Thinking Swarm—Way Forward

Chapter 17 concludes the book by drawing on threads from all chapters to ask questions that remain unanswered or, at the least, for which further research is needed. The chapter explores diverse questions, from social science and humanity ones to technological ones on modelling, simulation, and robotics.

Toowong, QLD, Australia	Simon Ng
Melbourne, VIC, Australia	Jason Beaufort Scholz
Canberra, ACT, Australia	Hussein A. Abbass
June 2024	

Acknowledgements

A book is a journey, with people, and with one's self. The journey of this book included many people and organisations we wish to acknowledge.

The editors wish to thank all authors for their contributions to this book. The multidisciplinary nature of the subject matter required a delicate balance between delving deep into individual disciplines and creating readable chapters about diverse disciplines. The authors were very patient in receiving critical feedback to balance and re-balance their contributions.

The editors send particular thanks to Dr S. Kate Conroy, who was instrumental in the early stages of this endeavour. Her energy and expertise drove the initial concept for the book, and her professional leadership informed the initial workshop and the engagement strategy with authors. We are indebted to her.

A special thanks goes to the Trusted Autonomous Systems Defence Cooperative Research Centre (TAS-DCRC) for funding this project to make the book public access.

Thanks also goes to the University of New South Wales in Canberra for the time taken by the third editor for this book project.

Contents

Part I Thinking Swarm: Foundations

1. **Thinking About Thinking Swarms** 3
   Jason Beaufort Scholz, Simon Ng, and Hussein A. Abbass

2. **Understanding the Roots of Swarm Intelligence in Defence to Find the Path Forward** 21
   Anton Klarin, Pi-Shen Seet, Janice Jones, Michael N. Johnstone, Helen Cripps, Jalleh Sharafizad, and Tony Marceddo

3. **Definitions, Sources and Categorisations for Thinking Swarms** 39
   Christine Boshuijzen-van Burken

4. **A Philosophical Analysis of "Thinking Swarms"** 49
   Christine Boshuijzen-van Burken

5. **Robots as Research and Drones as War** 61
   Tara Roberson and Christine Boshuijzen-van Burken

Part II Thinking Swarms: Behaviour

6. **Swarm Hermeneutics** 77
   Gavin Mount and David Beesley

7. **Identifying and Predicting Hidden Coordinated Behaviour Using Synthetic Language Narrative Models** 95
   Andrew D. Back and Beth Cardier

8. **Intelligent Swarming Narratives** 127
   Beth Cardier, Andrew D. Back, Jessica Korte, and Pauline Pounds

9. **Regulating Maritime Autonomous Swarms** 151
   Rachel Horne

| 10 | An Approach to the Legal Review of Autonomous Swarms 171
Damian Copeland, Philip Sammons, and Lauren Sanders |

Part III Thinking Swarm: Topology and Architecture

| 11 | Cognitive Architecture of Aware System of Systems 189
Peter Bernus, Beth Cardier, Ovidiu Noran, and Glen Smith |

| 12 | Trochoids 207
Jerome Moses Monsingh, Hoam Chung, and Arpita Sinha |

| 13 | The Barriers and Opportunities of Effective Underwater
Autonomous Swarms.. 243
Damien Guihen |

Part IV Thinking Swarms: Cognition

| 14 | Introducing Lifelong Learning in Swarm Robotics..................... 271
Phillip Smith, Aldeida Aleti, Asad I. Khan, Vincent C. S. Lee,
and Robert Hunjet |

| 15 | Learner-Centred, Teacher-Centred and Blended Curriculum
Design in Swarm Systems.. 291
Aya Hussein, Sondoss Elsawah, Eleni Petraki,
and Hussein A. Abbass |

| 16 | Hyper-Teaming 313
David Johnson and Felix Kong |

Part V Thinking Swarm: Way Forward

| 17 | Future Directions 329
Simon Ng |

References... 339

Index... 381

Editors and Contributors

About the Editors

Simon Ng is the Chief Technology Officer at Trusted Autonomous Systems (TAS). Graduating from Monash University with a PhD in 1998, he completed a Post-Doctoral Fellowship at CSIRO before joining DSTG, where he developed techniques for military operations experimentation, and applied systems methods to surveillance and response, space operations and autonomous aerial systems. He was a DSTG Group Leader for the Joint Systems Analysis and Aerial Autonomous Systems Groups, and the Associate Director of the Defence Science Institute. He was Australia's National Lead on The Technical Cooperation Program Technical Panel "UAS Integration into the Battlespace" and is a Graduate of the Australian Institute of Company Directors. His research interests include autonomy and artificial intelligence, robotics, systems science, and systems engineering.

Jason Beaufort Scholz is an Innovation Professor, with the Research and Innovation Capability department of RMIT University, Melbourne, Australia. He was the 2020 recipient of the McNeil Prize for outstanding contributions to the Royal Australian Navy and is a Graduate Member of the Australian Institute of Company Directors. He was CEO of TAS-DCRC in Australia from 2018 to 2022. He was the Chief Scientist and Engineer for TAS-DCRC and holds 34 years of research, development, and leadership experience with the Defence Science and Technology Group. His research interests include autonomous systems, artificial intelligence, human-machine command, control and communications, and research translation for impact.

Hussein A. Abbass is a Professor with the School of Systems and Computing, University of New South Wales, Canberra. He is a Fellow of IEEE, a Fellow of the Australian Computer Society, a Fellow of the UK Operational Research Society, a Fellow of the Australian Institute of Managers and Leaders, and a Graduate Member of the Australian Institute of Company Directors. He was the National President

(2016–2019) for the Australian Society for Operations Research, the Vice President for Technical Activities (2016–2019) for the IEEE Computational Intelligence Society, a Distinguished Lecturer (2022–2024) for the IEEE Computational Intelligence Society, and the Founding Editor-in-Chief of the IEEE Transactions on Artificial Intelligence (2020–2024). His current research focuses on trusted quantum-enabled human-AI-swarm teaming systems and distributed and trusted machine learning and machine education systems and algorithms.

Contributors

Hussein A. Abbass School of Systems and Computing, University of New South Wales, Canberra, ACT, Australia. h.abbass@unsw.edu.au, linked.in/husseinabbassoz/. Hussein Abbass is a full professor with the School of Systems and Computing, University of New South Wales Canberra. He is a fellow of IEEE, a fellow of the Australian Computer Society, a fellow of the UK Operational Research Society, a fellow of the Australian Institute of Managers and Leaders, and a graduate member of the Australian Institute of Company Directors. He was the national president (2016–2019) for the Australian Society for Operations Research, the vice-president for technical activities (2016–2019) for the IEEE Computational Intelligence Society, and an ExCom and AdCom member (2016–2019) of the IEEE Computational Intelligence Society. Prof Abbass is a distinguished lecturer for the IEEE Computational Intelligence Society and the founding editor-in-chief of *IEEE Transactions on Artificial Intelligence*. He is an associate editor of several IEEE journals, including the *IEEE Transactions on Cybernetics*. After 10 years in industry and academia, he joined UNSW Canberra in 2000 and has been a full professor since 2007. His current research focuses on trusted quantum-enabled human-AI-swarm teaming systems, and distributed and trusted machine learning and machine education systems and algorithms.

Aldeida Aleti Monash University, VIC, Australia. Aldeida.Aleti@monash.edu, linked.in/aldeida-aleti. Aldeida Aleti is a professor at Monash University, specialising in the intersection of AI and software engineering. Her research focuses on developing AI methods that automate various software development tasks, including software design and testing. Using her expertise, Prof Aleti has led and contributed to multiple teams working on AI and software engineering projects, amounting to a cumulative value exceeding $15 million. She takes pride and joy in mentoring and advising over 16 PhD students and five research fellows, contributing to their academic and professional growth. In addition to her academic role, Prof Aleti holds the position of associate dean for engagement and impact in the Faculty of IT at Monash University.

Andrew D. Back School of Electrical Engineering and Computer Science, University of Queensland, QLD, Australia. a.back@uq.edu.au, linked.in/andrewdback. Andrew Back has a background in machine learning, AI, and signal processing. Adjunct A/Professor Back is developing a groundbreaking paradigm for AI using the concept of emergent, beyond-human, synthetic languages. This approach demonstrates significant potential for social awareness, and understanding the meaning of systems. This has implications for security, intelligence, trust, and AI comprehension by rapidly classifying sequential behaviours with minimal data. He has held research fellowships in Australia, the United States, and Japan. A/Prof Back has made significant contributions to industry, founding a private computational finance company providing software to international banks and companies operating on the New York Stock Exchange and Chicago Board Options Exchange. A/Prof Back (h-index 24, 7,200 citations) has published in numerous journals and authored several book chapters on machine learning and AI.

David Beesley School of Media and Communication, RMIT University, VIC, Australia. david.beesley@rmit.edu.au, linked.in/david-beesley-99a79934. David Beesley is an affiliate of the ARC Centre of Excellence for Automated Decision-Making & Society at RMIT University. Dr Beesley is a media professional and documentary filmmaker and the technical services and facilities manager for the School of Media and Communication and School of Design (College of Design & Social Context) at RMIT University. He completed his PhD thesis "Head in the Clouds: documenting the rise of personal drone culture", which was a project-based longitudinal ethnographic documentary looking at the significance of drone cultures in Melbourne, Australia. Dr Beesley's research areas include communication technologies and digital media studies, drones, drone cultures, swarms, ethnography, resilient hermeneutics, and the changing role of the human in human machine teaming.

Peter Bernus Institute for Integrated and Intelligent Systems, School of Information Communication and Technology, Griffith University, QLD, Australia. P.Bernus@griffith.edu.au, linked.in/peter-bernus-3b8720. Peter Bernus has worked internationally on aspects of enterprise integration as a researcher, consultant, and project leader for industry, government, and Australian defence. A/Prof Bernus expertise includes systems engineering, artificial intelligence, enterprise engineering, virtual enterprises, and company networks. He is currently working on intelligent cognitive architectures for cyber-physical systems, the harmonisation of systems engineering and enterprise architecture standards in ISO and IEC as Standards Australia's representative, covering enterprise, systems and software, and smart manufacturing. A/Prof Bernus chaired the IFIP-IFAC Task Force for Architectures for Enterprise Integration, a peer-selected body of over 50 industry professionals and academics that developed GERAM, the generalised enterprise reference architecture and methodology, and ISO 15704:2019, in line with A/Prof Bernus's work. With Professors Günter Schmidt, Jacek Blazevicz, and Michael Schaw, he is series editor for the International Handbooks on Information Systems for Springer Verlag. He is also a member of the editorial boards of several international journals and conferences, a member of INCOSE and the Systems Engineering Society of Australia, and a former foundation chair of the International Federation for Information Processing (IFIP) Working Group (WG) 5.12 and International Federation of Automatic Control (IFAC) Technical Committee (TC) 5.3 committees on enterprise integration and interoperability.

Christine Boshuijzen-van Burken The University of New South Wales, Canberra, ACT, Australia. c.vanburken@unsw.edu.au, linked.in/christine-boshuijzen-van-burken-60366578. Christine Boshuijzen-van Burken is a senior research associate at the University of New South Wales. Dr Boshuijzen-van Burken holds a PhD in ethics of technology from Eindhoven University of Technology in the Netherlands. Her current research focuses are the ethics of autonomous systems and AI. Her research interests include ethics of technology, military ethics, refugee and humanitarian logistics, and reformational philosophy. She is the associate professor by special appointment in Christian Philosophy at Eindhoven University of Technology.

Beth Cardier Institute for Integrated and Intelligent Systems, School of Information Communication and Technology, Griffith University, QLD, Australia. b.cardier@griffith.edu.au, linked.in/bethcardier. Beth Cardier analyses narrative to understand the level of information above the text, where contextual and causal information is structured. Dr Cardier recently completed an Advance Queensland fellowship at the Trusted Autonomous Systems Defence CRC and Griffith University, exploring how machines can adapt to open world situations. This research included collaboration with industry partner Downer on risk modelling, and separately, with Canadian Aviation Electronics (CAE) under a grant from the Queensland Defence Science Alliance, the development of new visualisations for an augmented reality prototype. She has collaborated with Disney Research, foremost context logician Keith Devlin at Stanford, and the Virginia Modeling Analytics and Simulation Center, which developed a prototype immersive modelling environment using her methods. She has an interdisciplinary PhD in creative writing/information systems from the University of Melbourne and is an adjunct principal research fellow at Griffith University, as well as an assistant Professor at the Eastern Virginia Medical School.

Hoam Chung Department of Mechanical and Aerospace Engineering, Monash University, VIC, Australia. hoam.chung@monash.edu, linked.in/hoam-chung-77a792b6/?ppe=1. Hoam Chung is currently a lecturer in the Mechanical and Aerospace Engineering department at Monash University. Dr Chung's research interests include model predictive control theory and application, autonomous miniature flyers, semi-infinite optimisation problems, real-time system design, and the software/hardware implementation of autonomous unmanned systems (Unmanned Aerial Vehicles - UAVs, Unmanned Ground Vehicles - UGVs, and Unmanned Underwater Vehicles UUVs), rotorcraft dynamics, and robotics. From

2000 to 2009, he worked on the Berkeley AeRobot (BEAR) project as a PhD student and later a post-doctoral researcher. He received his BA and MS degrees in precision mechanical engineering from Hanyang University, Korea, in 1997 and 1999, respectively.

Damian Copeland Directorate of Operations and International Law, Defence Legal, Department of Defence, ACT, Australia TC Beirne School of Law, University of Queensland, QLD, Australia. damian.copeland@uq.edu.au, linked.in/damian-copeland-4901252111. Damian Copeland is director of operations and international law within Australia's Department of Defence. He is responsible for the conduct of Article 36 legal reviews for the Australian Defence Force. He is also an adjunct associate professor in the "Law and Future of War" team in the School of Law at the University of Queensland. His PhD considers the challenges in conducting an Article 36 legal review of an autonomous weapon system and proposes a "functional approach" to legal review that considers weapons law and targeting rules engaged by the autonomous functionality. The views in his chapter are his own and do not represent those of the Australian government nor of the Australian Defence Force.

Helen Cripps School of Business and Law, Edith Cowan University, WA, Australia. h.cripps@ecu.edu.au, linked.in/hdcripps/. Helen Cripps is an honorary senior lecturer at Edith Cowan University's School of Business and Law (ECU's SBL). Dr Cripps she has conducted research in maritime, social media, technology adoption, tourism, entrepreneurship, and innovation across Australia, Croatia, Finland, Norway, Sweden, the United Kingdom, and the United States. Her previous roles include economic analysis and policy development for the Western Australian government, community development coordination for the City of Vincent, and managing business relationships at ECU. Dr Cripps is a sought-after speaker and has received multiple awards for technology, innovative teaching, and industry engagement. Her current projects include developing an AI startup for geospatial technology in biodiversity management, creating business-related educational content, and branding boutique spirit brands. She is also working on the Apple Heritage and Produce Trail, inspired by her father, John Cripps, breeder of the pink lady apple. In addition, Dr Cripps is researching the pink lady apple brand's success and the economic effects of agritourism.

Sondoss Elsawah Capability Systems Centre, University of New South Wales, Canberra, Australia. s.elsawah@unsw.edu.au, linked.in/sondoss-elsawah-5a471a187/. Sondoss Elsawah is an associate professor of Systems and Computing, and the director of the Capability Systems Centre at the University of New South Wales. A/Prof Elsawah led the systems engineering discipline at UNSW Canberra. Her research programme focuses on understanding the behaviour of large complex problems and systemic risks that arise from the interactions between social, technological, and environmental systems. She is an expert on the application of systems methodologies to support design, decision-making, and education. She has published widely, including journal articles, conference papers, book chapters, and technical reports. Building on her academic achievements, A/Prof Elsawah has been recognised as a thought leader in her scientific fields in Australia and internationally. For example, she was elected as a distinguished fellow of the Modelling and Simulation Society of Australia and New Zealand. Sondoss is a senior editor of the *Journal of Environmental Modelling and Software* (A*) and the *Journal of Group Decision and Negotiation* (A*).

Damien Guihen AMC Search, Australian Maritime College, University of Tasmania, TAS, Australia. damien@guihen.com. Damien Guihen is a senior scientist at AMC Search, a division of the Australian Maritime College, University of Tasmania. Dr Guihen is interested in the application of technology to help understand biological and physical processes and has a PhD in physical oceanography from the National University of Ireland. His subsequent research has involved the deployment of robotic platforms in Europe, Australia, and the Antarctic. His current work involves the development of new methodologies and tools for applying marine robotics to questions of environmental observation. He also develops and delivers courses in the operation and analysis of autonomous maritime systems.

David Johnson Mission Systems, NSW, Australia. david.johnson@missionsystems.com.au, linked.in/david-johnson-53aa9411/. David Johnson is director and co-founder of Mission Systems. Dr Johnson has worked in radar, perception, and autonomous systems since 2001, beginning his professional career at Roke Manor Research, then Siemens' UK R&D facility. During his time at Roke Manor, he worked on a number of civil and defence projects relating to radar and avionics systems, focusing on small target detection in clutter. In 2006, Dr Johnson moved to Australia to undertake a PhD at the Australian Centre for Field Robotics. Following the completion of his PhD, he was employed with the Rio Tinto Centre for Mine Automation. From late 2014 to the founding of Mission Systems in 2017, and as defence research lead for the Australian Centre for Field Robotics (ACFR), he built his own research group and obtained industry funding for increasingly significant projects in the areas of distributed sensing and multi-agent autonomy.

Michael N. Johnstone School of Science, Edith Cowan University, WA, Australia. m.johnstone@ecu.edu.au. Michael N. Johnstone teaches network security and mobile application development at ECU's School of Science. A/Prof Johnstone has published in high-impact journals, including *IEEE Transactions on Information Forensics and Security*, *Sensors*, and *Computer Networks*, and has been featured in The Conversation and on ABC Radio. As a member of the Centre for Securing Digital Futures at ECU, his research focuses on secure development methodologies, wireless sensor networks, and IoT device security, specifically, in critical infrastructure. Having over 30 years of ICT experience, he provides cybersecurity consultancy for private industry, government, and research organisations. Before academia, he held roles such as programmer, systems analyst, project manager, and network manager. A/Prof Johnstone also serves on various A-ranked cyber-related conference committees and was the theme lead for Network Forensics-Response to Emerging Threats in the federally funded Cyber Security Cooperative Research Centre. He is a member of the Australian Computer Society and the Association for Computing Machinery.

Janice Jones Flinders University, SA, Australia. janice.jones@flinders.edu.au. Janice Jones (Jane) is an associate professor of Human Resource Management at Flinders University's College of Business, Government & Law. She has a PhD (Flinders University) and M. Commerce (UNSW). Her research focuses on the interfaces between innovation, technology, and human resource management. She has been the chief or partner investigator on competitive research projects in Australian Category 1, 2, 3, and 4 grants over the past five years. This includes two Department of Defence Strategic Policy Grants Program projects. She has published in high-impact journals, including *Technological Forecasting and Social Change*, the *Journal of Technology Transfer*, and *Industrial Marketing Management*. Her research projects have led to major submissions to the Australian government and publications in high-impact journals and media outlets such as The Conversation. She has extensive experience in collaborating with public and private sector organisations across levels of government, as well as peak bodies and nongovernment organisations.

Rachel Horne Formerly of Trusted Autonomous Systems Defence Cooperative Research Centre, Australia. DrRachelHorne@Outlook.com.au, rachel-horne-46074120. Rachel Horne is a leading expert in the regulation of autonomous systems, with significant professional and academic experience in this field. In her prior role as general manager—law, regulation, and assurance at Trusted Autonomous Systems Defence Cooperative Research Centre, Dr Horne delivered world-leading initiatives aimed at improving regulatory pathways for autonomous systems. These included delivery of the first Australian-centric technical standard for autonomous vessels, delivery of the largest commercial demonstration of autonomous vessels in Australia to date, and delivery of the platform www.rasgateway.com.au. Dr Horne's career has spanned defence, commercial, and government spheres, in a range of legal, regulatory, policy, and leadership roles. Dr Horne holds a PhD from Queensland University of Technology, a Graduate Certificate in Maritime Studies from the Australian Maritime College, and a Master of Law, Graduate Diploma of Legal Practice, and a Bachelor of Law with Honours from the Australian National University.

Robert Hunjet Defence Science and Technology Group, SA, Australia. robert.hunjet@defence.gov.au, linked.in/robert-hunjet-b32258ba. Robert Hunjet is the programme leader for the Australian Defence Science and Universities Network at the Defence Science and Technology Group. In this role Dr Hunjet drives engagement between academia, industry, and the defence innovation, science, and technology ecosystem. He received his PhD from the University of Adelaide, Australia, in 2014 and is an adjunct associate professor with the Trusted Autonomy Group in the School of Systems and Computing, University of New South Wales Canberra. He is the Australian lead for the AUKUS Pillar II AI and Autonomy Working Group and sits on numerous boards and panels, including the steering committee for the Centre of Advanced Defence Research in Robotics and Autonomous Systems and the Governance Panel of the Defence AI Research Network.

Aya Hussein Faculty of Science & Technology, University of Canberra, ACT, Australia. aya.hussein@canberra.edu.au, linked.in/aya-hussein-20b463125. Aya Hussein is a lecturer in AI and machine learning in the Faculty of Science & Technology, University of Canberra. Dr Hussein was a post-doctoral fellow at the University of New South Wales Canberra from 2020 to 2023. Her research focuses on multi-agent systems, swarm robotics, reinforcement learning, and human-machine interaction. She is also interested in designing the teaming and knowledge-exchange settings between humans and machines. Dr Hussein serves as an associate editor at the *IEEE Transactions on Artificial Intelligence*. She is an associate fellow of the Higher Education Academy. Dr Hussein received her bachelor and master degrees from Cairo University and her PhD from UNSW.

Asad I. Khan Sensor Analytics, Australia. asad.khan@sensoranalytics.com.au, linked.in/aikd101. Asad I Khan has over 25 years experience in the IT sector as a senior enterprise systems manager, researcher, and university lecturer. Dr Khan received his bachelor's degree from the University of Engineering and Technology, Pakistan, followed by a master's with distinction and PhD (in engineering) from Heriot-Watt University, UK. He was appointed a lecturer at Heriot-Watt and then assumed a senior IT management role with Monash University in Australia. He later served as a tenured-track faculty member at the same place in parallel computing. His work on bio-inspired computing methods led to several large research grants and a national HPC award. He has over 100 refereed publications including two research monographs and several book chapters, three of which in a Wiley's The PROSE award honourable mention title. He currently operates Sensor Analytics Australia, specialising in machine leaning algorithms for industrial and surveillance uses.

Anton Klarin School of Management and Marketing, Curtin University, WA, Australia. anton.klarin@curtin.edu.au, linked.in/anton-klarin-6566b716/. Anton Klarin received his PhD from the University of New South Wales and is a senior lecturer in Innovation, Entrepreneurship, Strategy and International Business at Curtin University's School of Management and Marketing. Dr Klarin's research encompasses, and has been published on, the topics of strategic choices of emerging market firms (e.g. in the *Journal of World Business* and *Journal of Management Inquiry*), institutional environments in emerging markets, and interdisciplinary research using scientometric methods (in *Technological Forecasting and Social Change*, *Systems Research and Behavioral Science*, the *Journal of Business Research*, *International*

Business Review, and other high-impact outlets). He has also published a policy document on the Internet of things and received an Emerald Literati Award for Excellence.

Felix Kong The Robotics Institute, School of Mechanical and Mechatronic Engineering, The University of Technology Sydney, NSW, Australia. Felix.Kong@uts.edu.au, linked.in/fkong. Felix Kong is a distinguished lecturer at the University of Technology Sydney with over five years of dedicated experience in robotics and algorithmic research. Since January 2023, Dr Kong has been advancing the field of dynamic motion planning and control for multilegged robots, multirobot coordination for comprehensive information gathering across diverse environments, and long-range ship-routing algorithms influenced by stochastic weather forecasts. Prior to this role, Dr Kong was a post-doctoral researcher at the same institution, where significant contributions were made in optimisation and control of a novel helicopter swashplate, path planning for underwater gliders, and short-range ocean flow field prediction using machine learning. In addition, Dr Kong holds the position of chief scientist at Crest Robotics, leveraging expertise to lead pioneering projects in robotic innovation. Dr Kong's work continues to push the boundaries of robotics and automation, contributing valuable insights to academia and industry.

Jessica Korte School of Computer Science, Queensland University of Technology, QLD, Australia. jessica.korte@qut.edu.au, linked.in/jessica-korte-60642832. Jessica Korte is passionate about the ways good technology can improve lives. To ensure technology is "good", Dr Korte advocates involving end users in design processes. Her philosophy for technology design is the needs of people who are disempowered or disabled by society should be considered first, everyone else will then benefit from technology that maximises usability. This philosophy held true in her recently completed TAS-DCRC fellowship Project, the Auslan Communication Technologies Pipeline project, which looked to foreground the visual-gestural language expertise of Deaf signers in the creation of technologies for the recognition, production, and processing of Australian Sign Language communication.

Vincent C.S. Lee Monash University, VIC, Australia. vincent.cs.lee@monash.edu, linked.in/vincentcslee. Vincent CS Lee is Associate Professor Department DS & AI, FIT, Monash University. Australia. A/Prof Lee received a PhD in adaptive signal processing from The University of Newcastle (NSW) in 1992, and an MSc (Computer System Engineering) and Bachelor of Electrical Engineering from the National University of Singapore in 1984. A/Prof Lee received specialised avionic engineering training at the Royal Air Force College, UK, sponsored by MOD UK & Singapore. He is FIEAust, CPEng, and Senior Member of IEEE. A/Prof Lee's research is multi-disciplinary, spanning digital health, signal/information processing, financial engineering, educational data mining, explainable AI, deep ML, computer vision, and multi-agent autonomous systems. A/Prof Lee has published more than 200 Q1 papers as well as peer-reviewed international conference proceedings. A/Prof Lee has been general chair/co-chair of technical/steering committees and an invited keynote speaker for IEEE/ACM flagship conferences. He is associate editor for SCI Q1 *Journal of Intelligent Manufacturing Systems*.

Tony Marceddo Edith Cowan University, WA, Australia. a.marceddo@ecu.edu.au, linked.in/anthony-marceddo-7a442126/. Tony Marceddo is an honorary associate professor at ECU's Centre for Securing Digital Futures, focusing on cybersecurity, artificial intelligence, autonomous systems, information warfare, digital citizenship, and human behaviours. Previously, he was general manager of Vault Cloud, the first Australian cloud service certified by the Australian government for handling protected-level information. Having over 30 years of leadership in defence, intelligence, space, cyber, and communications, A/Prof Marceddo held key roles, including general manager of Australian Intelligence and Cyber Security at Northrop Grumman, integrating M5 Network Security and delivering critical Defence projects. At Raytheon, A/Prof Marceddo managed intelligence and cyber business, overseeing contracts, including the Joint Defence Facility Pine Gap and Canberra Deep Space Communication Complex. His more than 25 years in Australian defence intelligence earned him a US Intelligence Medallion and an Australian Intelligence Community Medallion.

Jerome Moses Monsingh Robert Bosch Centre for Cyber Physical Systems, Indian Institute of Science, Bangalore, India. jerome.monsingh.c2022@iitbombay.org, linked.in/jerome-moses-67658736/. Jerome Moses is currently a project scientist at the Robert Bosch Centre for Cyber Physical Systems, Indian Institute of Science, Bangalore, India. He previously held a teaching position at the National Institute of Technology, Calicut. He earned his doctoral degree from the Indian Institute of Technology Bombay and Monash University, Melbourne, in 2022. His research interests encompass cooperative control, reinforcement learning, control theory, and robotics.

Gavin Mount School of Humanities, University of New South Wales Canberra, ACT, Australia. g.mount@unsw.edu.au, linked.in/gavin-mount-763a9026. Gavin Mount is a Nexus fellow and senior lecturer at UNSW Canberra. His primary areas of research expertise have been in the areas of ethnic conflict and nationalism. More recently, he has been publishing on political influence using microtargeting, hybrid warfare, and the use of simulations in decision-centric warfare. He has extensive experience providing education to defence personnel at the Australian Defence Force Academy (ADFA) and the Australian Command and Staff College. The recipient of several teaching awards, he teaches in the subjects of great power politics and conflict transformation and is currently leading an international team on the pedagogies of war crimes. He has published works on nationalism (CUP) and hybrid warfare (ANU Press) and currently holds several editorial positions including commissioning editor of Australian Outlook since 2016.

Simon Ng Defence Cooperative Research Centre for Trusted Autonomous Systems, QLD, Australia. simon.ng@tasdcrc.com.au, linked.in/simon-ng-8a425b274. Simon

Ng is chief technology officer at Trusted Autonomous Systems Defence Cooperative Research Centre (TAS-DCRC). Graduating from Monash University with a PhD in 1998, Dr Ng completed a post-doctoral fellowship at the Commonwealth Scientific and Industrial Research Organisation (CSIRO) before joining Defence Science and Technology Group (DSTG), where he developed techniques for military operations experimentation, and applied systems methods to surveillance and response, space operations, and autonomous aerial systems. Dr Ng was previously DSTG group leader for both the Joint Systems Analysis and Aerial Autonomous Systems Groups, and associate director of the Defence Science Institute. Dr Ng was Australia's national lead on The Technical Cooperation Program Technical Panel "UAS Integration into the Battlespace", and is a graduate of the Australian Institute of Company Directors. His research interests include autonomy and artificial intelligence, robotics, systems science, and systems engineering.

Ovidiu Noran Institute for Integrated and Intelligent Systems, School of Information Communication and Technology, Griffith University, QLD, Australia. O.Noran@griffith.edu.au, linked.in/onoran. Ovidiu Noran is a senior lecturer and has worked as an engineer and architecture consultant in companies based in Europe and Australia. Dr Noran has taught enterprise architecture, database design, systems analysis and design, and agile business analysis at Griffith University. His research interests include artificial intelligence, business analysis, serious games, software engineering, and enterprise architecture, and he prefers action research. Dr Noran has a PhD in enterprise architecture, a BEng (Hons) degree in HVAC, and a Master of ICT and of Learning and Teaching in Higher Education. He is a member of Engineers Australia, the Association of Enterprise Architects, the International Institute of Business Analysts and of Standards Committees ISO/IEC SC7/WG42, ISO TC184 SC5/WG1, and Standards Australia IT-015.

Eleni Petraki Faculty of Education, University of Canberra, ACT, Australia. eleni.petraki@canberra.edu.au, linked.in/eleni-petraki-3960851a/. Eleni Petraki is an associate professor at the Faculty of Education, University of Canberra. She is also the associate dean of research for higher degrees in the Faculty of Education, a role to which she was appointed because of her two university Excellence for HDR Supervision awards. Having over 30 years of experience in tertiary education, specialising in linguistics and language teacher education, she has taught internationally in Europe, the United States, Asia, and Australia. Her research focuses on employing interaction analysis in various settings, curriculum design, human-AI teaming, and development of AI literacy. Her most recent work explores the application of curriculum design principles for human-AI education and the integration of AI and education technologies in educational settings.

Pauline Pounds Robotics Design Lab, EECS, University of Queensland, QLD, Australia. pauline.pounds@uq.edu.au. Pauline Pounds specialises in UAV propulsion, dynamics, sensing, and control. Prof Pounds' recent work has been in bipedal locomotion using control moment gyroscopes. Her BE in systems engineering

(2002) and PhD in robotics (2008) are both from the Australian National University. She is a 2013 ARC DECRA and 2020 TAS-DCRC fellow, the 2015 Queensland Science Young Tall Poppy, and the 2020 ATSE Batterham Medal winner. She is president of the Australian Robotics and Automation Association as of 2022, was IEEE RAS RA-L senior editor in aerial and field robotics 2021–2023, and CTO of Olaeris Inc from 2010 to 2017.

Tara Roberson ARC Centre of Excellence for Engineered Quantum Systems, University of Queensland, QLD, Australia. tara.roberson@qut.edu.au, linked.in/tara-roberson. Tara Roberson is an experienced science communicator with a background in research management and stakeholder engagement. Dr Roberson has worked in a range of areas, with an emphasis on responsible development of emerging technology, including quantum technologies and autonomous systems. She is an associate investigator in the ARC Centre of Excellence for Engineered Quantum Systems and holds a PhD in science communication from the Australian National University.

Philip Sammons Contour Advisory, ACT, Australia University of New South Wales, NSW, Australia. philip@sammons.com.au, linked.in/phil123123. Philip is a mechatronic systems engineer specialising in robotics and ICT systems integration. Having a strong background in designing and developing innovative solutions involving simulation, software integration, and control, he excels in developing and delivering solutions to complex problems that involve emerging and disruptive technology. Philip is passionate about staying updated with the latest advancements in robotics and artificial intelligence. His master degree thesis "Development of a System in the Loop Simulator for the Control of Multiple Autonomous Unmanned Aircraft" involved the use of simulation to demonstrate the control of drone swarms with a Bayesian stochastic control algorithm.

Lauren Sanders TC Beirne School of Law, University of Queensland, QLD, Australia. International Weapons Review, QLD, Australia. l.sanders@uq.edu.au, linked.in/lauren-sanders-a126a597. Lauren is an adjunct associate professor at the Law and Future of War team with the University of Queensland. Her research areas focus on novel and emerging military technologies, export controls, and the legal review of autonomous weapons and military AI. A/Prof Sanders teaching (including as an adjunct senior lecturer at the University of Adelaide) focuses on international humanitarian and criminal law, and counter-terrorism law, building on her PhD research into domestic accountability for core international crimes and universal jurisdiction. She is a part-time special counsel with the Operations and International Law Directorate within Australia's Department of Defence and a managing director of a legal firm that advises defence industry on how to incorporate international law into capability design. The views in her chapter are her own and do not represent those of the Australian government nor of the Australian Defence Force.

Jason Beaufort Scholz Research and Innovation, RMIT University, VIC, Australia. jason.scholz@rmit.edu.au, linked.in//jason-scholz-3139a878/. Jason Scholz is an innovation professor at RMIT University and commercialisation strategist for Consunet. Prof Scholz is the 2020 recipient of the McNeil Prize, honouring significant contributions to Royal Australian Navy capability. He is former CEO, and chief scientist of the Trusted Autonomous Systems Defence Cooperative Research Centre, where he oversaw one of the most successful programmes for translation of research to industry for defence impact. Prof Scholz has over 30 years experience in defence science and technology, specialising in the decision sciences, artificial intelligence, communications, and control. He has led the successful research, development, and translation of new telecommunications, control systems, logistics, and decision support technologies with embedded AI into defence, which has had significant operational effects and created over A$200 million in savings.

Jalleh Sharafizad School of Business and Law, Edith Cowan University, WA, Australia. j.sharafizad@ecu.edu.au, linked.in/jalleh-sharafizad-4970baa/. Jalleh Sharafizad is a senior lecturer in Entrepreneurship and Innovation at ECU's SBL. Dr Sharafizad has achieved notable success in the fields of entrepreneurship, small business, and regional entrepreneurship development. Her extensive publications include contributions to prestigious international journals, such as *Entrepreneurship and Regional Development*, the *Journal of Business and Industrial Marketing*, the *International Journal of Hospitality Management*, the *Journal of Hospitality and Tourism Management*, *Human Resource Management*, *Studies in Higher Education Journal*, and *Education + Training*. Dr Sharafizad has been awarded the University of Antwerp's Staff Exchange Erasmus programme award and Edith Cowan University's Athena SWAN Advancement Scheme award. She has effectively and successfully led multiple research projects funded by both Edith Cowan University and the Western Australian government.

Pi-Shen Seet School of Business and Law, Edith Cowan University, WA, Australia. p.seet@ecu.edu.au, linked.in/pishenseet/. Pi-Shen Seet is a professor of Entrepreneurship and Innovation at Edith Cowan University's School of Business and Law. Holding a PhD from the University of Cambridge and a Master of Defence Studies from the University of Canberra, Prof Seet is a graduate of the Australian Army Command and Staff College, SAF Command and Staff College, and the New Zealand Grade II Staff Course. He has extensive command and staff experience with the SAF and has collaborated closely with the Australian and New Zealand defence forces. His research explores the intersection of innovation and societal impact, securing around A$400,000 in competitive research grants, including projects under the Department of Defence Strategic Policy Grants Program. His work has been published in high-impact journals, including *Technological Forecasting and Social Change* and the *Journal of Technology Transfer*. He has contributed significantly to federal government submissions and been published on The Conversation and ABC News.

Arpita Sinha Systems and Control Engineering, Indian Institute of Technology, Bombay, India. arpita.sinha@iitb.ac.in, linked.in/drarpitasinha. Arpita Sinha received her PhD in aerospace engineering from the Indian Institute of Science, Bangalore, India, in 2007. Prof Sinha is currently a professor in the Systems and Control Engineering department at the Indian Institute of Technology (IIT). Before joining IIT Bombay, she was a post-doctoral researcher at Cranfield University, UK. Her research interests include the guidance and control of multiple autonomous vehicles, path planning, consensus algorithms, and distributed decision-making systems.

Glen Smith Institute for Integrated and Intelligent Systems, School of Information Communication and Technology, Griffith University, QLD, Australia. Glen.Smith@griffith.edu.au. Glen Smith had a career in applied psychology (cognition, psychometrics, experimental design and analysis) from 1979 until retiring in 2017. Dr Smith is currently an adjunct professor at the Institute for Integrated and Intelligent Systems, Griffith University. He was previously a reader in the Psychology Department (The University of Queensland), a senior lecturer (University of Melbourne's Psychology department), and a senior research officer at the Australian Council for Educational Research. Dr Smith moved to applied research at Defence Science and Technology Organisation (DSTO) to manage a multidisciplinary team developing C2 support technology and processes (in the last period he was the head of group). Dr Smith was educated at the University of Adelaide (majoring in Pure Mathematics, with first class honours in Computing Science, and in Psychology) and was awarded a PhD in psychology on mathematical modelling of decision-making. This mix of mathematics, computing, and psychology helped him cross discipline boundaries. Dr Smith has published 65 refereed papers and 2,100 citations (h-index 23) and attracted external competitive funding for his research. From 1984 to 2016 Glen was a registered psychologist.

Phillip Smith Defence Science and Technology Group, SA, Australia. phillip.smith14@defence.gov.au, linked.in/phillip-smith-235102178. Phillip Smith is a senior researcher at Defence Science and Technology Group (DSTG). Dr Smith received his bachelor degrees in Robotics & Mechatronics (Hons) and Computer Science in 2016 from Swinburne University. Following this, he received his PhD from Monash University in 2020, with his thesis "Creative Swarms for Wireless Network Assistance in Restricted Environments". Dr Smith's continued robotic research at DSTG specialises in multi-agent robotic behaviours, including swarming robots and collaborative heterogeneous teaming robotics. His research in the tactical and scalable autonomy team (at DSTG) focuses on combining theoretical and practical approaches to research. Dr Smith has taken part in numerous field trials and Defence exercises.

List of Figures

Fig. 1.1	Age and industry groups of participants............................	5
Fig. 1.2	Participants' experiences with swarm systems, robotics and artificial intelligence...	6
Fig. 2.1	Swarms and automation literature map...............................	26
Fig. 2.2	Swarms and automation growth of the literature in the last 20 years..	27
Fig. 5.1	Leximancer concept map..	66
Fig. 6.1	Boyd's OODA loop, simplified and complex [435]	80
Fig. 6.2	Hermeneutic framework	81
Fig. 6.3	Gatwick airport, UK (adobestock.com)	85
Fig. 6.4	Murmuration of starlings (adobestock.com)	86
Fig. 6.5	Drone lightshow (adobestock.com)	87
Fig. 6.6	Building an authentic scenario-based simulation	91
Fig. 7.1	The first stage of the narrative prediction model is to obtain an information topology language model by symbolisation of sensory narrative inputs.	114
Fig. 7.2	In the proposed model, narrative prediction is achieved by transforming the input narrative to an information space.	114
Fig. 7.3	A conceptual view of the information topology search process.....	115
Fig. 7.4	A representation of some unobserved behavioural properties for a coordinated group of agents within a physical or virtual swarm	120
Fig. 7.5	The symbolic ranked probabilities of the behaviour of some coordinated group of agents.	120
Fig. 7.6	A representation of some coordinated behavioural signals (blue) and other unrelated behaviours, all of which are not directly observable, but occur within a swarm of agents	121
Fig. 7.7	The observed swarm behaviour using a simple synthetic language representation.	122

Fig. 7.8	A simple synthetic language representation of some coordinated behaviour (blue) and other unrelated behaviours, all of which are not directly observable but occur within a swarm of agents	122
Fig. 7.9	Applying the synthetic language ICA (SLICA) approach enables the hidden agent behaviour to be extracted from the observed swarm behaviour	123
Fig. 7.10	The proposed synthetic language ICA (SLICA) approach is demonstrated to extract the hidden agent behaviour from within the observed global swarm behaviour (compare with Fig. 7.5)	123
Fig. 8.1	Narrative characterises an unexpected piece of information using a "swarm" of reference frameworks	132
Fig. 8.2	Increasingly complex system-level narrative structures	135
Fig. 8.3	A combination of system-level structures, precedence, and nesting	137
Fig. 8.4	Simple thresholding of scene identifications (left) misses a cluster of low-confidence identifications of a threat that could have been identified as meaningful through mutual proximity (right)	140
Fig. 8.5	Diagram of collective sensing mechanisms in Pounds' whisker drones	142
Fig. 9.1	Categorisation of vessels in Australian maritime regulatory framework by R Horne	158
Fig. 11.1	National Institute of Standards and Technology's standards landscape for smart manufacturing (redrawn after [542])	191
Fig. 11.2	ISO15704/Generalised Enterprise Reference Architecture and MethodologyGERAM's modelling framework	192
Fig. 11.3	Example of life cycle relationship between two entities	192
Fig. 11.4	Chain of life cycle relationships	193
Fig. 11.5	A long chain of life cycle relationships	193
Fig. 11.6	Models populating the modelling framework during the creation versus during the operation of an entity	194
Fig. 11.7	Conversation theoretic process pattern (called a "P-individual")	195
Fig. 11.8	Multiple ways of physically manifesting P-Individuals	196
Fig. 11.9	Sharing situation awareness through conversation	197
Fig. 11.10	Internal levels of knowledge	198
Fig. 11.11	Recursive process-pattern for aware entities	199
Fig. 12.1	Representative examples of a trochoid. (**a**) Epitrochoid (when $d < r$). (**b**) Hypotrochoid (when $d > r$)	211
Fig. 12.2	(**a**) Representation of the eigenvalues of (**a**) $-\mathscr{L}$, (**b**) $-k_p \mathscr{L} \otimes R(\theta)$ and (**c**) A	217

Fig. 12.3	Eigenvalues of A satisfying conditions (**a**) $\lambda_{3a} = \lambda_{\overline{3b}-1}$ (**b**) $\lambda_{3a} = \lambda_{3\overline{a}-1}$ (**c**) $\lambda_{3b} = \lambda_{3\overline{b}-1}$	221
Fig. 12.4	Representation of the eigenvalues of $-\mathscr{L}$ and the choice of μ_a and μ_b	228
Fig. 12.5	Geometrical interpretation of placing two eigenvalues	232
Fig. 12.6	Interaction topology for four agents. The arrow in the edge indicates the direction of information flow	236
Fig. 12.7	Trajectories of four agents. The snapshot shows the agent trajectories at $t = 2{,}500$ seconds with random starting positions and velocities. Circles denote the starting positions of the agents, and the squares denote the snapshots of the vehicles at 2,500 seconds	236
Fig. 12.8	(**a**) Eigenvalues of $-\mathscr{L}$. (**b**) Eigenvalues of $-g_p I - k_p \mathscr{L} \otimes R(\theta)$. The angle of rotation is chosen to be 2.28488 rad (**c**) Eigenvalues of Λ. The gains g_p and k_p are chosen such that there are two pairs of eigenvalues on the imaginary axis	237
Fig. 12.9	Patterns when $\frac{|\lambda_{2b-1}|}{|\lambda_{2a-1}|}$ is rational ((**a**)–(**d**)) and irrational ((**e**)–(**f**))	238
Fig. 12.10	Experimental set-up	240
Fig. 12.11	Simulation results for double integrators: (**a**) Case 6 and (**b**) Case 4	240
Fig. 12.12	Experimental results for double integrators: (**a**) Case 6 and (**b**) Case 4	240
Fig. 13.1	A simplified view of the layers of design in an AUV swarm, in which three vehicles are shown, and the dashed line indicates the potential for more	249
Fig. 13.2	A 3D perspective of underwater swarm considerations for describing the design of swarming architecture	250
Fig. 13.3	Modes of communication and control of multiple vehicles.	251
Fig. 13.4	A perspective on where platforms come from, with associated challenges and benefits	255
Fig. 13.5	A flow diagram of an idealised workflow from objective development, through planning, monitoring and analysis.	256
Fig. 13.6	Typical limiting factors for swarm function over different temporal and spatial scales	258
Fig. 13.7	The enabling developments that will allow swarms to scale over time and space	263
Fig. 13.8	A suggested description of autonomous underwater swarm complexity	265
Fig. 14.1	Example of HGN training on two patterns (A and B), then presented with a B pattern with noise	275
Fig. 14.2	Manually designed training environments for R-HGN.	279
Fig. 14.3	Proposed lifelong learning cycle.	279

Fig. 14.4	Example of a behaviour being added to the statistics table.	280
Fig. 14.5	Example of a statistics table update.	281
Fig. 14.6	Example of group learning in R-HGN statistics table.	282
Fig. 14.7	Fitness difference between the three LML swarms and static swarm.	285
Fig. 14.8	Task-specific improvement (fitness in Step 5 is greater than Step 1) of the three LMLs using ten-task moving-window-average	285
Fig. 14.9	Repertoire size and behaviour usage for LML repertoire in each task	286
Fig. 15.1	Steps of the Dick and Carey model	298
Fig. 15.2	Skill analysis in the swarm decision-making case study	298
Fig. 15.3	Learner-centred, teacher-centred and blended curriculum design.	302
Fig. 15.4	Decomposition of the unmanned aerial vehicle behaviour in the search and rescue case study	303
Fig. 15.5	Three lessons of increasing complexity in the curriculum for autonomous learning in the air combat task.	305
Fig. 15.6	Learning from a teacher with a limited teaching budget.	310
Fig. 16.1	Hyper-teaming functional network diagram	315
Fig. 16.2	Hyper-teaming framework	319
Fig. 16.3	Left: human intent prediction at mission-start. Right: evolution of human intent prediction over time	322
Fig. 16.4	Left: human intent prediction at mission-start. Right: evolution of human intent prediction over time	322
Fig. 16.5	A snapshot of Vulcan flight data from Year 2	324

List of Tables

Table 1.1	Identified concepts related to the embodiment and behaviour of robotic systems and, specifically, robotic collectives	7
Table 1.2	Identified concepts related to the control and cognition of robotic systems and, specifically, robotic collectives	8
Table 1.3	Categories of swarm behaviours	12
Table 1.4	Categories of swarm functions	13
Table 1.4	(continued)	14
Table 2.1	Top trending, top impact terms and indicative journals	28
Table 2.1	(continued)	30
Table 2.2	Top 20 countries that published swarms and automation journal articles	31
Table 3.1	Comparison of swarm optimisation algorithms: the basic information and the metaphor. Updated from [537]	44
Table 4.1	Aspects in Dooyeweerd [437]	55
Table 5.1	Overview of media articles collected from Google News	65
Table 6.1	Theory and application of hermeneutic frameworks	83
Table 12.1	Overview of pattern formation in the literature	213
Table 12.2	Simulation results	239
Table 14.1	Computation time and memory usage of swarms in the final task	287

Part I
Thinking Swarm: Foundations

Chapter 1
Thinking About Thinking Swarms

Jason Beaufort Scholz, Simon Ng, and Hussein A. Abbass

We have long been fascinated by the motion of bird flocks and insect swarms, admiring their fluid dynamics and purposeful interactions. It seems that nature has grasped the concept of producing emergent global properties from local rules and interactions to allow large-scale systems of agents to act cohesively [51]: think birds, herds, schools of fish and even crowds on the subway. Yet, the disparate ideas of what a swarm is (or is not) still hamper discussion, especially among stakeholders who are less interested in swarming as a discipline and more interested in swarming as a means to an end. To provide a common basis for the collection of papers provided in this volume, we conducted a workshop intended to tease out concepts and ideas relevant to the general idea of "thinking swarms". This chapter reports on the findings and presents a taxonomy that emerged from the workshop.

1.1 Introduction

The study of swarms has led to important algorithms widely used today in optimisation [293, 537] and computation. However, application of swarming principles, generally taken to be local interactions resulting in large-scale global behaviours

---

J. B. Scholz (✉)
Research and Innovation, RMIT University, Melbourne, VIC, Australia
e-mail: jason.scholz@rmit.edu.au

S. Ng
Trusted Autonomous Systems (TAS), Toowong, QLD, Australia
e-mail: simon.ng@tasdcrc.com.au

H. A. Abbass
School of Systems and Computing, University of New South Wales, Canberra, ACT, Australia
e-mail: h.abbass@unsw.edu.au

and properties in a multi-agent system, has not translated into significant practical robotic applications, although considerable effort is being made to exploit swarming principles to enhance the function of robotic systems [706], demonstrating the elegant expression of nature in artifice.

Swarming, as a term, is fraught. To the layperson, swarming is often used to describe the visual appearance of a system. Swarming tactics on the battlefield, for instance, may or may not actually employ swarming techniques as defined by Beni [105] and others. A detailed RAND study that examined the military history of swarm tactics [263] offers a definition using observable behaviours ("swarming occurs when several units conduct a convergent attack on a target from multiple axes") rather than technical foundations. This is not a criticism. Ultimately, the avenue for arriving at some military benefit is less important than the benefit itself, and this historical case study almost certainly provides evidence that the technical principles of swarming, specifically, local interactions determining global behaviour, apply equally well to Mongol hordes as they do to herds of bison.

We decided to conduct a workshop with prospective authors and thought leaders from academia, government and industry to improve our understanding of swarms, to think about swarms and to speculate on a future where swarms can think. We used the insights from this workshop to structure our approach to this book, but the workshop also provided a basis for building a collective wisdom and a shared language, creating a community of sorts, within the authorship group for this monograph.

The data presented in this chapter was collected during a four-part workshop used to inform the organisation of this book and to provide a basis for a shared conceptualisation among contributing authors. The study obtained approval from the Human Ethics Panel at the University of New South Wales (approval number HC220304). Contributions to these workshops were voluntary. Participants were unidentifiable in the collected data. Participants were given 2 weeks to withdraw from the study after the workshop. No participant withdrew. Mentimeter software was used to collect data from the participants. Participants were also given butcher's papers and stickers, of which data were photographed at the end of each session. Figure 1.1 depicts the age and industry distribution of participants. while Fig. 1.2 shows the distribution of participants' experience with swarm systems, robotics and artificial intelligence. It is important to note that participants were given the choice to answer, which resulted in slightly different numbers of total participants answering each question. In total, 43 participants actively contributed to the online data collection system, and nearly 50 participants overall at the workshop contributed to the ground discussions.

Fig. 1.1 Age and industry groups of participants

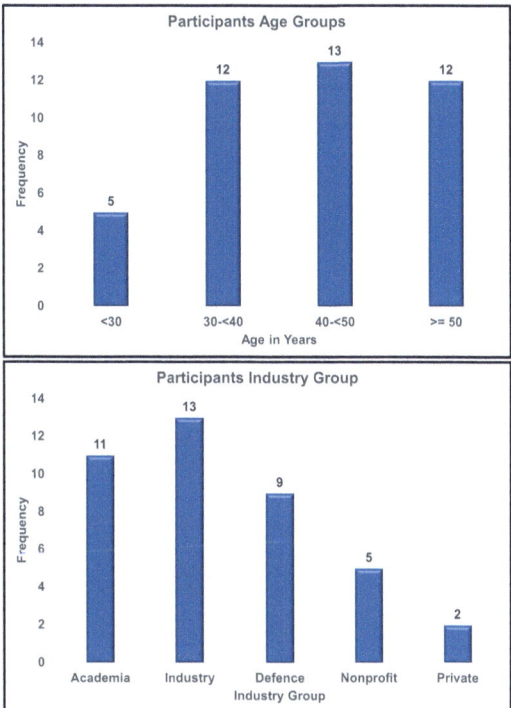

1.2 Construct Analysis

We provided little formal basis of definition but instead aimed at shared language to unify a diverse community. The analysis of the results of this workshop remains ongoing, but what has become clear is that when asked to explore the concept of swarms, teams and individuated agents, the workshop participants painted a rich picture encompassed within four categories, themselves clustered as either form or function. Each category contained a large set of constructs, written here as pairs of concepts.

Kelly [713] saw constructs as dimensions along which individuals interpret a system or situation. Here, we take the concepts from the workshop and combine them as pairs that define either end of a singular construct dimension. The list that follows is not exhaustive or exclusive. Personal constructs are just that, personal, and the existence of a pair of concepts in tension as a construct dimension does not preclude one or both of those same concepts existing as a pole in another construct.

Concepts can be relevant across categories or constructs within the same category. Such instances are indicated in bold. Where they relate to different categories, they should be interpreted within that categorical context. As an example, the concept "dangerous" appears in both **appearance** and **behaviour**. In the former case, this construct relates to how the system looks and, in the latter case, to how

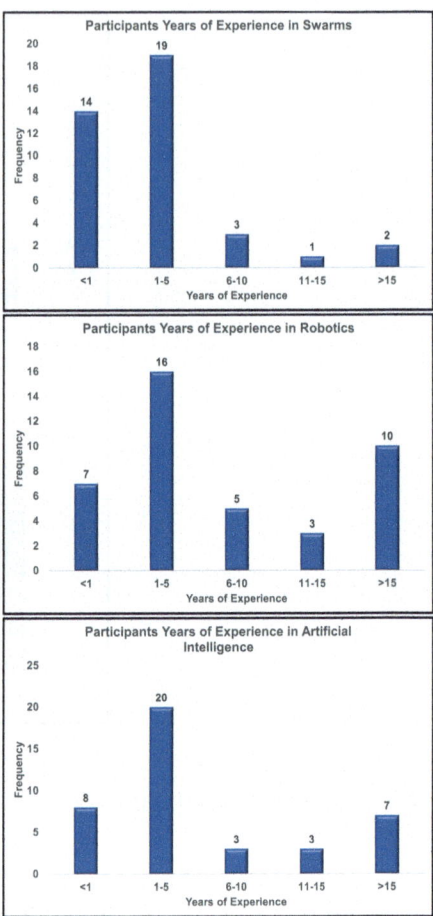

Fig. 1.2 Participants' experiences with swarm systems, robotics and artificial intelligence

it behaves. A simple example might be a system that bristles with threatening red lights as distinct from a system that acts erratically.

In some categories, the repetition of a concept (such as selfish) in more than one construct reflects multiple valid meanings.

1.2.1 Form

Form constitutes the external "observables" of any robotic system: its appearance and its behaviour.

Appearance, a system's basic appearance, is often the first thing we see when we examine a robotic system. Constructs within this category include artificial, natural; dangerous, safe; and emotional, clinical. Many of the concepts are value-laden:

"thinking swarms" could appear benign, dangerous, safe, strong or natural, fragile and so on. The value imposition matters when we consider that robots, swarms and "thinking swarms" exist as sociotechnical (rather than simply technical) systems. Therefore, appearance is the first layer of social interpretation.

The second layer is the system's overt **behaviour**: robots are as robots do. Constructs from this category include random, coherent; competitive, cooperative; helpful, selfish; and anticipatory, reflexive. In some instances, value-laden concepts are present in this category too. These appear to be associated with a particular instance of a system, for example, a collection of armed drones, rather than with the concept of swarm. These concepts are also highly interpretive. Selfish is a judgement on an observed behaviour. What appears selfish to one person may be sensible or self-preserving to another, or it may be motivated by selfless actions that are not explicable to an observer.

Embodiment and behaviour contribute to our narrative interpretation of a system, but may mislead us: what we think we see is not necessarily what is actually happening.

Table 1.1 provides a list of the constructs participants identified relating to embodiment and behaviour (external observables).

Table 1.1 Identified concepts related to the embodiment and behaviour of robotic systems and, specifically, robotic collectives

Form	Construct		
Physical appearance what it looks like	Artificial	....	Natural
	Organic	....	Mechanical
	Human-like	....	Inhuman
	Fragile	....	Robust/strong
	Dangerous	....	Safe
	Amorphous	....	Structured
	Homogeneous	....	Heterogenous
	Emotional	....	Clinical
	Team	....	Swarm
	One	....	Many
Observable behaviour how it moves or acts	Independent	....	Collective
	Reactive	....	Proactive
	Emotional	....	Clinical
	Competitive	....	Cooperative
	Inscrutable	....	Understandable
	Adversarial	....	Friendly
	Dangerous	....	Benign
	Selfish	....	Helpful
	Chaotic/random	....	Ordered/coherent
	Unlawful	....	Lawful
	Irresponsible	....	Responsible

1.2.2 Function

Function describes how a system actually operates. Participants highlighted a range of important topological concepts that informed their consideration of how a system might work. Individuation and individuality matter in a thinking swarm; agents in a "thinking" swarm were considered to have individual existences, with the potential for individuated goals and intentions. The distribution of control centralisation was a common construct: centralised or decentralised, top-down or bottom-up and master and slave. Other related concepts were also highlighted as significant: the degree of order inherent in the underlying system, implying topological evolution or dynamics that might be applied within the system (chaotic.......coherent), and the relationship between agents (might evolve or change dynamically, from strict hierarchy to holon to local) (Table 1.2).

Constructs expressing a view of the system's cognitive qualities were also discussed. Some constructs were common across the cohort: dumb, smart; programmed, adaptive; deterministic, learning; reactive, proactive; and global, local. Others were less common, but just as interesting. Example concepts from these

Table 1.2 Identified concepts related to the control and cognition of robotic systems and, specifically, robotic collectives

Function	Construct		
Control topology how nodes relate	Individual	....	Aggregated
	Decentralised	....	Centralised
	Local	....	Global
	Fractionated	....	Integrated
	Coupled	....	Decoupled
	Hierarchical	....	Holonic/collective
	Separable	....	Inseparable
	Emergent/flexible	....	Predefined/fixed
Cognitive features how the system thinks	Planned	....	Reflexive
	Proactive	....	Reactive
	Local knowledge	....	Global knowledge
	Pre-programmed	....	Learning/adaptive
	Individual goals	....	Shared goals
	Collective	....	Independent
	Human-in-the-loop	....	Human-on-the-loop
	Automated	....	Autonomous
	Teaming	....	Swarming
	Creative/ideating	....	Programmed
	Deterministic	....	Statistical
	Controlled	....	Commanded
	Passive	....	Agentic
	Ungoverned/unregulated	....	Regulated

constructs include ideating, agentic and self-learning. Are these cognitive qualities that we should strive for and how do they relate to problems of regulation, planning, strategy and order, which also feature in the conceptual space drawn out in discussion?

1.2.3 Design Considerations

In system design, we might consider four important questions emerging from the interpretative ideas presented earlier.

- What is the collective **physical appearance** of the system? How does it influence our perception of the system, namely, its "value-laden-ness"? What does it mean for value-sensitive design?
- What observable **observed behaviour** does the system exhibit? How does this support "explicability", trust and assurance?
- What **control topology** governs the system's function? Is it defined, dynamic or even evolving? How does the system transform between topological models and maintain coherence of purpose and function?
- How do "thinking swarms" think? What are the **cognitive features**, routines and processes, and what do they mean in relation to the other three themes?

The constructs can be used to describe a particular system, and here we use the examples from Sect. 1.1 to illustrate this. This approach provides us with a way of determining where and how a particular system fits in relation to other systems, but also provides a conceptual template to inform comparisons of particular systems.

In the preface, we used starling murmuration as an example of a natural swarming system that has distinct advantages in terms of protecting starlings from predators through the selfish herd effect [699]. Wave patterns [763] in the murmuration can confuse predators, and the scale makes it hard for predatory birds to follow individual targets.[1] Virtual murmuration using simple rules suggests that local information (coordination with approximately seven nearest neighbours) and enough space to manoeuvre is all that is needed to create large-scale coherency [95, 374, 858]. Starlings are, for all intents and purposes, identical.

The forms a murmuration manifested within the constructs that emerged from the workshop include the following:

- **Physical appearance**: We can see concepts immediately linked to this system, including natural, organic, biological, cohesive, amorphous, homogenous and swarm.
- **Observed behaviour**: Observed dynamic properties include flocking, flowing, collective, reactive, reflexive, inscrutable and coherent.

---

[1] It should be noted that starling murmuration has other potential benefits. The primary one is considered to be attracting starlings to large-scale roosts overnight.

- **Control topology**: From research and mathematical studies/simulations, we know control in a murmuration is likely, for example, individual, decentralised, localised, decoupled and emergent.
- **Cognitive features**: We know that each node has the following cognitive properties: local knowledge, individual goals, agentic, reactive, autonomous and programmed. It is not clear whether it can be said that the murmuration has its own cognitive strategy.

A second case study referred to the tactics and capabilities that African "driver" ants use to attack large, seemingly impervious, prey, such as the freshwater crab. We know that ants communicate through pheromone release and physical contact signalling [424]. The use of pheromones by ants is complex, given that different pheromones are used to send different types of messages related to not only the source of food but also whether there is enough food present, whether the source has "run out" or to signal the presence of predators. Multimodal signalling is common, implying significant complexity in communication and significant scope for communication over large spatiotemporal distances.

This is distinctly different in form and function to the murmuration of starlings. Using constructs from the workshop, we see concepts among ants that include the following:

- **Physical appearance**: natural, inhuman, biological, cohesive, structured, homogeneous and team
- **Observed behaviour**: independent,[2] proactive, cooperative, friendly, helpful and, or ordered/coherent.
- **Control topology**: individual, decentralised, global, many, holonic, integrated and emergent
- **Cognitive features**: planned, global knowledge,[3] individual goals, shared goals,[4] independent, teaming, agentic and proactive

A comparison of these concepts provides qualitative insight into how the two systems differ. In physical appearance, we see similarities, and the two systems appear like a single cohesive whole.

However, the nature of the two systems differs when we examine their behaviour. The ants behave with a greater degree of independence, overtly cooperate and act proactively. We cannot interpret underlying causes for the observed differences or claim that each system will always exhibit such behavioural characteristics. However, the behavioural differences are salient because they alter our interpretation

---

[2] Ants are homogeneous in appearance, but work independently, doing different things, though contributing to shared social objectives.

[3] Ants exhibit global and local knowledge. Global knowledge is captured by the pheromone signals, which transmit information across the population.

[4] Each ant is aware of the shared goal of "feed the colony" but has its own local goals driven by local information.

of the underlying mechanisms governing them, allowing us to attribute a degree of intelligence to one system that appears missing from the other.

However, is this difference real? What might the underlying cause of the difference be? One possibility lays in control topology [95]. Ants use long-range, integrated pheromone networks, allowing for unified and coherent mechanisms of control. Ants also seem to work as local coherent cells, in which they cooperate with immediate neighbours to achieve short-term goals, implying holonic structures that might only last a brief while but serve local functions using cooperation. Starlings, however, use much simpler rules of interaction, which are highly reflexive, to maintain their coherency without requiring long-range interactions or local cooperation.

Cognitive strategies also underlie some of the observed behavioural differences. Ants use global knowledge explicitly, whereas starlings leverage local knowledge through short-range interactions with their nearest neighbours only. The ants are agentic, acting to enhance the outcome for the colony, and proactive, seeking to deliberately spread information in nuanced ways. However, they still share local goals to solve shorter-term problems. Starlings do not exhibit these specific behaviours, at least in the process of generating murmurations.[5]

This comparison reveals that thinking swarms can operate in distinctly different, and often blended, ways that have similarities and differences in appearance deriving from combinations of underlying mechanisms that also overlap but do not fully correlate.

1.3 Functional Analysis

A workshop was conducted with the authors to examine the functions of "thinking swarms".

As a starting point, we used the categorisation proposed by Schranz et al. [706] of "swarm behaviours". This is reproduced in Table 1.3, including brief explanations of the meaning of each subcategory included. The categories in italics in the table were augmented by Schranz et al. [706] on the framework originally proposed by Brambilla et al. [145].

In the same spirit of augmentation, we acknowledge that swarm behaviours are but one of a broader range of functions, so the group was charged with proposing changes to the aforementioned model and adding new categories to complete the functions more comprehensively. This activity was performed interactively as a "swarm" effort by the group using sticky notes attached to a paper.

---

[5] It is noted that we do not have direct access to the control or cognitive mechanisms of either species. Much of this discussion is drawn from observational data or stimulation models, or both, that provide surrogate evidence.

Table 1.3 Categories of swarm behaviours

Swarm behaviours	Subcategory	Description
Spatial organisation	Aggregation	Move close to one another for interaction
	Pattern formation	Form a shape, includes chain formation for communication
	Self-assembly	Connect together to make a structure
	Object clustering and assembly	Manipulate distributed objects, e.g. to build something
Navigation	Collective exploration	Navigate through environment to map, search, monitor, communicate
	Coordinated motion	Move in formation to navigate a space
	Collective transport	Move objects too large or heavy for an individual
	Collective localisation	Find position and orientation to establish a local coordinate system
Decision-making	Consensus	Agree/converge on a single choice
	Task allocation	Dynamic task assignment to individuals to maximise performance of whole
	Collective fault detection	Determine individuals that deviate due to fault
	Collective perception	Better perceive objects by resource allocation
	Synchronisation	Synchronous action by aligning frequency and phase of cycles in the swarm
	Group size regulation	Form groups of desired size, split into multiple groups if too many
Miscellaneous	Self-healing	Fault recovery due to individual deficiencies, minimise impact on the whole
	Self-reproduction	Create new robots or replicate a pattern, eliminate need for a human engineer to create more robots
	Human-swarm interaction	Control robots in a swarm or receive information from them

The summary result of this work is presented in Table 1.4. To explain, in addition to behaviours, four new functions were identified, including subcategories. Here, we interpret behaviours as actions that change awareness of the world. Behaviours might therefore reasonably include communications, but it was decided to separate this function because of its overall importance. Behaviours, in turn, are driven by the difference between purposes and awareness. That is, if there is no difference between purposes and awareness, then arguably one is satisfied, and there is no required behaviour. It is only when an error or difference exists between these that

Table 1.4 Categories of swarm functions

Swarm function	Subcategory	Description
Governance	Ethical framework	Human responsibility, human competency (selecting, training, ...)
	Legal framework	LOAC (including proportionality), ROE; regulatory safety compliance; policing
	Technical compliance	Software controls (e.g. policy, ROE, LOAC); security (including anti-tamper, integrity, e.g. through-life history of individuals in swarm); explainable behaviours
	Sociotechnical compliance	Human-swarm interaction cognitive design; mission tasking validation; systemic verification; de-briefing: mission fulfilment, actual effect
	Managed levels	Individuals (goals, meta-cognition, geofence); swarms (common goal, contour size regulation); mixed swarms with tasked individuals (mixed collective and individual goals); multi-swarms (multiple-common goals, cooperate and compete)
Communications	Interaction levels	Human-swarm; inter-swarm; intra-swarm; swarm-individual/other machine(s)
	Interaction modes	Data ferrying; leader selection; mass synchronisation
	Bound action	Contracts, agreements, consensus (including coalition formation); roles; delegation; task assignment
	Environments	Denied; degraded; intermittent; limited
Purposes	Goal management	Selection, evaluation; agent systems (e.g. beliefs, desires, intentions model)
	Ends	Project (overwhelm, overpower, outnumber); protect (shield, block)
	Ways	Distract, deceive; hunt, track, pursuit; role-play, role-hide; probe for response; task cluster, concentrate effort
	Means	Self-destruct; attack; use numbers; assemble into a structure
	Performance assessment	Mission/effects assessment; mission assurance (tolerance of loss)/survival
Behaviours	Resource management	optimisation; learning to swarm; tasking; planning/courses of action; decision choice
	Move	Spatial (re-/dis-) organisation (aggregation, pattern formation, self-(dis-) assembly (e.g. for mobility), self-culling, area coverage, synchronised, serialised, self-heal, self-reproduce); navigation (collective exploration, coordinated motion, collective localisation)

(continued)

Table 1.4 (continued)

Swarm function	Subcategory	Description
	Carry	Collective transport (e.g. leader-follower); object clustering and (re-/dis-) assembly; maintenance, recover lost; supply chains
	See	Heterogeneity/homogeneity (sensors, appearance)
	Communicate	Using movement, using environment (e.g. pheromones); using spectrum
	Effect	Act; destroy; self-destroy; safe modes; effectors
Awareness	Impact assessment	Fusion; mission aware; systemic threats
	Situation assessment	Fusion; context; orient
	Object assessment	Fusion; distributed detection/collective perception and tracking; self-perception (including faults); observe
	Sense	Sense; sensors

some behaviour is needed. Governance was recognised as an additional function because it was considered important by the group to make it explicit.

The terms used to describe the lowest level categories were drawn from the group.

Under **governance**, the ethical framework identifies the centrality of humans to 'command' swarms and retain responsibility and competency in their use. The legal framework supports the need for swarms to comply with laws of armed conflict and rules of engagement for military use, as well as regulatory safety and the ability for effective policing. Technical compliance recognises the ability for some governance controls to be embedded in the technologies within code, in security features and in explainable behaviours. Sociotechnical compliance recognises governance embedded in the interface and interaction between humans and technical systems including the cognitive design, in the validation of mission tasking, in systemic verification of a holistic system and in the ability to debrief or account for swarm actions. Managed levels acknowledge that governance may be facilitated through certain "natural" constraints at the levels of individual (individual purposes), swarm (collective purpose), swarm with individuals (mixed) and multiple-swarms (multiple-mixed).

Under **communications**, interaction levels identify the types of interaction between human and swarm, across swarms, between members within a swarm and between a swarm and an individual. Interaction modes identify the ways that swarms can communicate, which may include physically ferrying data, through dynamic leader selection, and mass synchronisation. Bound action includes means by which action is assigned through contracts, agreements and consensus seeking and includes consideration of roles, delegation and task assignment. Communication environments acknowledge that media for communication can be constrained by being denied, degraded, intermittent and limited.

Under **purposes**, goal management includes goal selection, goal evaluation and achievement through agency under frameworks such as the belief-desire-intension (BDI) model. Ends, ways and means is a military approach to framing purpose at three levels of abstraction. Ends include strategic consideration to project and protect. Ways may include distract, deceive, hunt, track, pursuit, role-play, role-hide, probing for response, task clustering and concentration of effort. Means may include self-destruction, attack, use of numbers and assembly into a structure. Performance assessment recognises the importance of assessment of mission achievement and mission assurance (including notions such as tolerance of losses).

Under **behaviours**, resource management covers optimisation, learning to swarm, tasking, planning including course of action development and analysis and decision choices. The categories to move, carry, see, communicate and effect embrace the literal functionality of swarms. Move for swarms then includes spatial organisation, assembly, self-culling, coverage of an area or volume, synchronised movement, serialised movement, self-healing, self-culling, self-reproduction and navigation through collective exploration, motion and localisation. Carry is a logistic function covering collective ability to transport something, object clustering and reassembly, maintenance, recovery from loss and supply chains. See includes heterogeneous and homogeneous means for sensing of any conceivable kind. Communicate includes through movement, through the environment and through the spectrum (e.g. electromagnetic, acoustic, particles, ...). Effect includes to act, destroy, self-destroy, safe modes and effectors of any conceivable type.

Under **awareness**, the language of information fusion [704] was used to express three levels of abstraction of situation awareness applied to machine swarms. At the highest level, impact assessment includes swarm ability to assess threats and opportunities and relate to the strategy behind the mission according to the situation at hand. Situation assessment includes orientation to the context of a situation and comprehension of its meaning according to objects. Object assessment includes the perception of objects in volumes of space and time through distributed/collective detection and object tracking. Sense covers what is observable and sensors.

1.4 Infrastructure Analysis

The infrastructure workshop attempted to identify the infrastructure that needs to be in place for swarm systems to operate. Although the aim of swarm systems is to be fully autonomous, distributed and decentralised computational units, the requirements driven by practical real-world applications impose constraints where swarm systems need to integrate with existing infrastructure and new infrastructure may need to be developed for swarm systems.

The word "infrastructure" was interpreted through three lenses. The first was the physical infrastructure. The second was policies, processes and procedures. The third was from a human-machine teaming perspective. Each lens brought the physical, workflow and human dimensions to identify the infrastructure from an engineering, policy and human interaction perspective.

We asked the audience to describe a scenario within their group. Mining was used as the common domain, and individual groups were left to decide what tasks they saw for a swarm in mining. Participants were then asked to discuss the infrastructure challenges they saw. Mentimeter was used to collect answers to specific questions from every individual. The groups were then asked to refine their discussion and focus on one of the three lenses discussed above. It is important to say that the number of participants contributing to each question varied for each of the six questions asked. The number of contributors to each question was 34, 34, 28, 32, 31 and 31, respectively.

When participants were asked about the type of tasks swarm systems could support in the mining industry, 102 responses were received from 34 participants. The analysis of these answers revealed eight groups of tasks presented as follows:

1. **Environmental monitoring:** Air quality monitoring including oxygen and toxic monitoring, discovering leaks, gas detection and source localisation, detection of soil differences, weather monitoring, monitoring rock movements, safety check for changes in geological structures, collecting soil samples and radiation monitoring
2. **Survey and mapping:** Aerial and underground mapping and survey, footage capturing, exploration, resource discovery and localisation, mine design support, navigating continuously evolving environments, mapping objects from different angles and searching inaccessible or dangerous areas
3. **Communication:** Providing communication backbones and relays and forming underground mesh networks
4. **Logistic and construction support:** Navigation in confined spaces, equipment inspection, transportation of personnel and material movements, resource extraction, supporting infrastructure construction (wiring, paved roads, etc), obstacle removal and general logistic and construction support
5. **Command and control:** Situation awareness, change detection, coordination of activities, risk monitoring, security patrolling and monitoring operations and search and rescue operations
6. **Medical support:** Health monitoring of humans, rescuing humans, injury evacuations, supplying medicine and medical materials to employees and accident detection systems
7. **Uncrewed vehicle operations:** Monitoring the operations of uncrewed vehicles and providing safety assessment support
8. **Mining specific tasks:** Ore localisation, ore body characterisation, mineral extraction, ore transportation, area illumination, explosives and extraction operations and post-blast assessment

When participants were asked about the types of challenges when operating swarm systems, 106 responses were received from 34 participants. The analysis of these answers revealed seven groups of challenges, which are presented as follows:

1. **Environment:** The operating environment is harsh because of many factors including lightening/darkness conditions, humidity, temperature, noise, dust, gasses from explosives, water underground obscuring sensors, explosive risks and lack of access to power sources.
2. **Regulatory and ethical:** Regulations are still evolving, especially in terms of the use of uncrewed aerial vehicles. Participants mentioned the challenges associated with obtaining beyond line-of-sight autonomous operations, possible disapproval from unions, challenges in switching between modes of operations such as full-autonomous to cooperative modes and back, difficulties in accessing safety data about operations of swarms in mines and the ethical aspects that are discussed in detail in a dedicated question later.
3. **Human factors:** The impact of swarm operations on humans is multidimensional including a possible negative impact on the morale of workers, cultural inertia or misunderstanding of swarm operations, as well as the risks of interaction between swarms and humans and possible human injuries, and human perception and ignorance of swarm operations. Other factors concerned convincing management of the utility of swarm operations in mines and the level of investors' confidence in swarm operations.
4. **Communication:** Although swarms could offer communication support, they need to communicate among themselves. This raises challenges relating to GPS-denied environments, communication latency in harsh environments, electromagnetic and environmental interferences and the reliability of swarm operations under communication failures.
5. **Cybersecurity:** Participants questioned the risks associated with possible cybersecurity hazards in swarm operations. Not many details were discussed on this point. We can extrapolate issues associated with intruders and denial of service attacks and the possible negative consequences from unwelcome cyber-attacks on the swarm.
6. **Vehicle issues:** The main issue that was raised about the vehicles used in swarm operations is their endurance, particularly with respect to battery life and lack of power sources underground.
7. **Workforce:** A number of workforce issues were raised ranging from training the workforce to become accustomed to swarm operations, the lack of qualified swarm operators and the high salaries of roboticists.

The third question focused on the infrastructure requirements for swarm operations. A total of 97 responses were received from 28 participants. The analysis of these answers revealed six groups of infrastructure requirements, which are presented as follows:

1. **Land:** The primary point made was the need for launching and recovery sites, including capabilities for particular types of uncrewed vehicles, such as allowing for the vertical take-off of vehicles or runways for fixed-wing ones. Sites for emergency landing and equipping the land site with sensors and lights were also discussed.

2. **Equipment:** Discussions mainly focused on power and communication support equipment. These include power generation and management systems, power charging stations, Wi-Fi beacons and repeaters and network access points, light sources and passive radars.
3. **Workforce:** Access to a workforce that has the right skill sets and training were highlighted. This included both technical operators of swarm and those specialised in the legal and regulatory issues of swarm operations. Education, training and skilling the workforce were raised as important challenges.
4. **Logistic support:** Participants identified the need for the availability of spare parts, self-cleaning stations and general logistic support.
5. **Software:** Swarm operations require software support, including data analysis systems, information technology systems, human-machine interface design and friendly human-robot interaction systems.
6. **Processes and procedures:** This point is covered in detail in a later question. In this question, participants identified the need for human-robot teaming processes, procurement processes appropriate for swarm operations and good risk-management processes.

The three remaining questions were more focused and identified challenges related to communication networks, computer networks and ethical challenges. For each of these dimensions, 28, 30, 31 and 31 data points were collected from 28, 32, 31 and 31 participants, respectively. Some participants did not contribute to the questions. A summary of the issues in each category is presented as follows:

1. **Communication network challenges:** These include bandwidth, congestion, signal attenuation, data loss, denied communication and weather.
2. **Computer network challenges:** Many answers overlapped with the previous communication challenges. Additional points made included back-door hacking, memory and storage limitations, data management and data analysis.
3. **Ethical challenges:** These include community acceptance of the technology, the need to deliver real values and not just innovation for the sake of being innovative, job losses, human injury and safety, human cognitive load, invasion of privacy, cybersecurity, damage to environment, unions, accountability and the responsibility of swarm actions, adequate governance and the ethical issues related to the problem domain, in which mining has its own ethical challenges.
4. **Regulatory challenges:** These included the certification and assurance of swarm systems, safety qualification approvals, the complexity of intersecting and overlapping and sometimes contradictory, regulatory frameworks, the lack of precedent, slow adoption, the lag between regulation and innovation, open world risk, risk frameworks and mitigation strategies, the lack of policies and procedures for swarm operations and workplace safety.

This analysis highlights that swarm operations have a long way to go. Although the engineering community is focusing on designing reliable algorithms and systems for swarm operations, the deployment of these systems in the real-world needs to consider the infrastructure requirements, including the availability of regulatory, ethical and risk management frameworks to operate the technology and the implications of these on the workforce, processes, procedures and policies.

1.5 Conclusion

This chapter reported on a workshop that the editors conducted to lay out the foundations for this book. The workshop achieved multiple objectives. It brought people together and formed a "thinking swarms" community. Most of the people who contributed to this book attended the workshop. The workshop identified the constructs, taxonomy and ontology discussed in this chapter. Consequently, these findings formed the logic for structuring this book around five parts spanning chapters on foundations, behaviour, topology, cognition and the way forward.

Acknowledgments The workshop and human study reported in this chapter were approved by the Human Research Ethics UNSW Low-Risk Committee (protocol number HC220304). Informed consent was obtained from each participant. The authors wish to express their appreciation to all attendees in the workshop.

Open Access This chapter is licensed under the terms of the Creative Commons Attribution 4.0 International License (http://creativecommons.org/licenses/by/4.0/), which permits use, sharing, adaptation, distribution and reproduction in any medium or format, as long as you give appropriate credit to the original author(s) and the source, provide a link to the Creative Commons license and indicate if changes were made.

The images or other third party material in this chapter are included in the chapter's Creative Commons license, unless indicated otherwise in a credit line to the material. If material is not included in the chapter's Creative Commons license and your intended use is not permitted by statutory regulation or exceeds the permitted use, you will need to obtain permission directly from the copyright holder.

Chapter 2
Understanding the Roots of Swarm Intelligence in Defence to Find the Path Forward

A Scientometric Study of Autonomous Systems

Anton Klarin, Pi-Shen Seet, Janice Jones, Michael N. Johnstone, Helen Cripps, Jalleh Sharafizad, and Tony Marceddo

Swarm intelligence, inspired by the decentralised, adaptive and self-synchronising behaviours of natural swarms, is a pivotal component of autonomous systems, enhancing efficiency, robustness and scalability. The research in this area is nascent and interdisciplinary. To drive this important research forward, it is necessary to adopt a systems perspective on what is available in the current literature. This chapter offers a comprehensive systems perspective of the integration of swarm intelligence within the broader domain of automation, emphasising its application in the defence sector. A systems perspective of an interdisciplinary field is afforded through scientometrics. Using VOSviewer algorithms, we analysed 1706 publications from Scopus relating to swarms and automation, identifying five main research clusters: particle swarm optimisation and algorithms, autonomous robotics, stability and control of autonomous agents, communication and infrastructure and machine learning. The findings reveal the significant role of the Internet of Things (IoT) in enabling swarm intelligence, the foundational impact of autonomous

---

A. Klarin (✉)
School of Management and Marketing, Curtin University, Bentley, WA, Australia
e-mail: anton.klarin@curtin.edu.au

P.-S. Seet · H. Cripps · J. Sharafizad
School of Business and Law, Edith Cowan University, Joondalup, WA, Australia
e-mail: p.seet@ecu.edu.au; h.cripps@ecu.edu.au; j.sharafizad@ecu.edu.au

J. Jones
Flinders University, Bedford Park, SA, Australia
e-mail: janice.jones@flinders.edu.au

M. N. Johnstone
School of Science, Edith Cowan University, Joondalup, WA, Australia
e-mail: m.johnstone@ecu.edu.au

T. Marceddo
Edith Cowan University, Joondalup, WA, Australia
e-mail: a.marceddo@ecu.edu.au

© The Author(s) 2025
S. Ng et al. (eds.), *Thinking Swarms*, https://doi.org/10.1007/978-3-031-82790-7_2

technologies and the prevalence of unmanned aerial vehicles in military research. In addition, the study underscores the necessity for addressing ethical and legal implications, particularly in military contexts, to ensure sustainable and trustworthy deployment of these technologies. The holistic systems approach highlights the interconnectedness of swarm technologies across various disciplines and suggests directions for future research to address the complexity and ethical considerations. This research contributes to both the academic understanding and practical applications of swarm intelligence, advocating for interdisciplinary efforts and robust governance frameworks to harness the full potential of swarm thinking in defence and beyond.

2.1 Introduction

This chapter examines the state-of-the-art literature on the role of swarm technologies within the wider automation literature in an attempt to better understand the roots of swarm intelligence research and practice. Swarms are generally considered to be sets of distributed agents that self-synchronise their actions. Thus, designed swarms fall under the wider umbrella of autonomous systems because of their inherent ability to self-synchronise [18]. This study aims to provide an overarching analysis of foundational swarm research, namely, in automation and autonomous systems, focusing on the defence/military sector via a scientometric review of the academic research within the past decade. Swarm intelligence holds profound significance within the broader spectrum of autonomous systems, because it excels in addressing intricate problems through decentralised, adaptable and self-organised systems. By leveraging the principles of collaboration, emergence and adaptation observed in natural swarms, the incorporation of swarm intelligence fosters the creation of innovative autonomous solutions characterised by heightened efficiency, robustness and scalability. This paradigm shift creates unprecedented opportunities in the realm of defence/military applications, enabling the effective tackling of multifaceted challenges across domains such as autonomous systems, smart grid management, data clustering and task allocation [668]. For the purposes of this study, the roots of swarm intelligence were considered a broad family of autonomous systems. Autonomous systems research is inherently interdisciplinary, so we present an overarching interdisciplinary perspective on swarm intelligence within the broader autonomous systems field.

To analyse swarms within automation, an understanding of the complex systems that underpin these technologies in the wider environment is necessary. The origins of swarm intelligence can be traced back to the foundational research of entomologist and biologist EO Wilson, who studied the collaborative behaviours of social insects in the mid-twentieth century [834]. The term "swarm intelligence" gained significance with the influential work of Eric Bonabeau, Marco Dorigo and Guy Theraulaz in the late 1980s and early 1990s [134]. The ecosystem of the phenomena is best understood through the lens of a systems research perspective, which provides an interdisciplinary approach to study complex systems in society, nature and designed systems [190]. Given that swarms are natural phenomena (cf. insects, birds and fish), they are amenable to examination and analysis by this approach. Further, given the multiplicity, complexity and interdisciplinarity of swarm intelligence, a whole-of-systems approach allows for links and interrelationships between subsystems and constructs, and chains of causality between constructs, to surface [431, 524]. This practice of analysing the whole rather than individual subsystems is referred to as holism [115, 614] and is ideal in identifying the wider environment within which swarm intelligence operates, which is what we aimed to do in this study. Further, we draw the reader's attention to a highly relevant definitional study of "thinking swarms" according to systems thinking in this handbook [164], as well as a philosophical analysis of "thinking swarms" from the systems perspective [165].

There have been several literature review studies of swarms in the past [146, 635] that have provided reviews of algorithms underpinning the functioning of swarms or types of swarms. However, there has not been a study that has provided an analysis of swarms as a part of autonomous systems, as well as a discussion of swarms in a military application context. Therefore, this study offers a unique overview of what the literature discusses in terms of swarms as part of autonomous systems in the military sector. We also identify a number of gaps in the literature that form directions for future research and implications for practice.

In the conduct of literature reviews, scientometric methods most often use software that positions and clusters terms, themes and research directions according to algorithms, thus offering unbiased, transparent and replicable results. These features provide robust and reliable results in mapping the literature. Scientometric methods, therefore, allow researchers to gain a bird's-eye perspective on the scholarship, in this case, automation in defence, in which all published academic research on the topic is arranged under one map with distinct research streams [245, 468, 665]. Through this approach, the study of the themes and research streams uncovers those areas in which research is abundant or, conversely, limited, allowing a researcher to derive a deeper understanding of the subject's breadth and limitations and, as a result, suggest avenues for more tailored research directions [35, 308, 687].

2.2 Scientometric Literature Review Approach

A common saying is "we cannot see the forest for the trees", and therefore, a big picture approach should be applied before we engage in any in-depth study of a topic area. By providing a map of the scholarship and the clusters within, we can see the entirety and complexity of the topic (often) from an interdisciplinary perspective. Scientometric mapping with its taxonomies and typologies offered by review studies plays an important role in many research disciplines and topics [143] and can provide a bird's-eye view of the automation domain that is essential to understanding the breadth of the interdisciplinary literature.

In conducting this scientometric review of the literature, this study was informed by Petticrew and Roberts [652], Siddaway et al. [738] and Tranfield et al. [789] who described processes for conducting systematic literature reviews, as well as following procedures for conducting robust scientometric studies [470]. These processes/steps include (1) identification of a research field and a research objective, (2) identification of a review range, (3) establishing search criteria and data extraction, (4) data set screening for exclusion and inclusion, (5) results analysis and their interpretation and (6) discussion of the results.

In the first step, in recognition of the rapidly expanding use of swarms in automation, we derived the objective of this study, which was to identify the discussions of automation and swarms to provide a broad systems analysis of the interdisciplinary literature. In the second step, the study presents a large overarching systems perspective of the existing research using a substantial body of swarms and automation publications from the largest structured and extractable academic database, Scopus [356]. In the third and fourth steps, we extracted all available journal article publications including reviews, notes, editorials, letters and surveys on swarm* AND automation or autonomous as of 21 October 2022 from Scopus, which resulted in 2247 publications. We excluded all studies that did not relate to artificial swarm designs, including biology, marine research, ecology, pharmacology, medical sciences and other studies that discussed swarms in the natural environment or outside the scope of intelligent swarms, leaving a data set of 1706 publications. The Scopus database has cited references back to the 1970s, and, as discussed, given that the term "swarm intelligence" only emerged from the 1980s onwards, it is likely that the key "roots" of the research would have been captured.

In Steps 5 and 6, VOSviewer software, which is capable of mapping large maps into distance-based clusters using a co-occurrence matrix, in which items that have high levels of similarities are algorithmically located close to each other (for more details, see [262]), was chosen for an unbiased outlook on the research. Using the clustering technique by [262], a set of items that are closely related are assigned to colour-coded clusters. Each item can only occur in one cluster. We used the default settings of the software to generate the co-occurrence clusters because these are considered to be the most optimal parameters [262, 472]. The default settings demonstrate terms that have a minimum of 10 occurrences, meaning that terms that appear in fewer than 10 publications are not extracted into the clustering analysis. This study combines bibliometric author, publication, source, institution, keyword and country-based analyses together with content-based analysis that is possible through an extraction and linkages of commonly occurring noun phrases to provide an overarching analysis of automation in the defence literature [473].

2.3 Swarms in Automation Scholarship

The algorithmic positioning and clustering produced five distinct clusters corresponding to research themes: (1) particle swarm optimisation and other algorithms (green), (2) autonomous robotics including swarms (red), (3) stability and control of autonomous agents (blue), (4) communication and infrastructure of autonomous systems (yellow) and (5) machine learning (lilac). As per the aforementioned clustering technique of Waltman et al. [820], any terms evident in the map can only appear in one cluster, and the size of each term indicates its occurrence in the selected literature under investigation. To provide the state of the art of swarms and automation scholarship, it is necessary to discuss each cluster using the themes that are present within them. The results of the thematic representations built into a visualised map that was generated by the VOSviewer software are presented in Fig. 2.1. Furthermore, Fig. 2.2 shows the cumulative growth of the literature over the last 20 years, indicating an increasing interest in this topic year on year.

In addition to the visual representation of the swarms and automation scholarship presented in Fig. 2.1, this study offers two tables that provide bibliometric (using citation information) thematic and author results. Therefore, Table 2.1 offers the top trending themes that appeared in the most recent articles and the themes that are prevalent in the documents that received the highest normalised citation counts. Table 2.2 displays the top 20 countries arranged by publication volume, including the total number of citations for each country, the average publication year and an average number of citations per publication. To contextualise swarms within automation scholarship, it is necessary to briefly outline each cluster using the themes that are present within, which provides an overarching understanding of the field to gauge the basic dynamics of the existing research on swarms in automation (Fig. 2.1).

2.3.1 Green Cluster: Particle Swarm Optimisation and Other Algorithms

Much of the existing literature covers types of swarms in natural science and the resultant algorithms that take their inspiration from their biological foundations. By far, the most common algorithm for swarms is particle swarm optimisation (PSO), which is featured in the green cluster and inspired by bird flocking and fish schools. The basic premise of PSO is that each particle (one device/item/agent) has a position and velocity, and the best position of the swarm is dependent on the global best where all agents converge into an optimal position [456]. The second most common algorithm technique in swarming is ant colony optimisation (ACO), in which agents choose the most optimal path to reach a goal according to the most efficient path

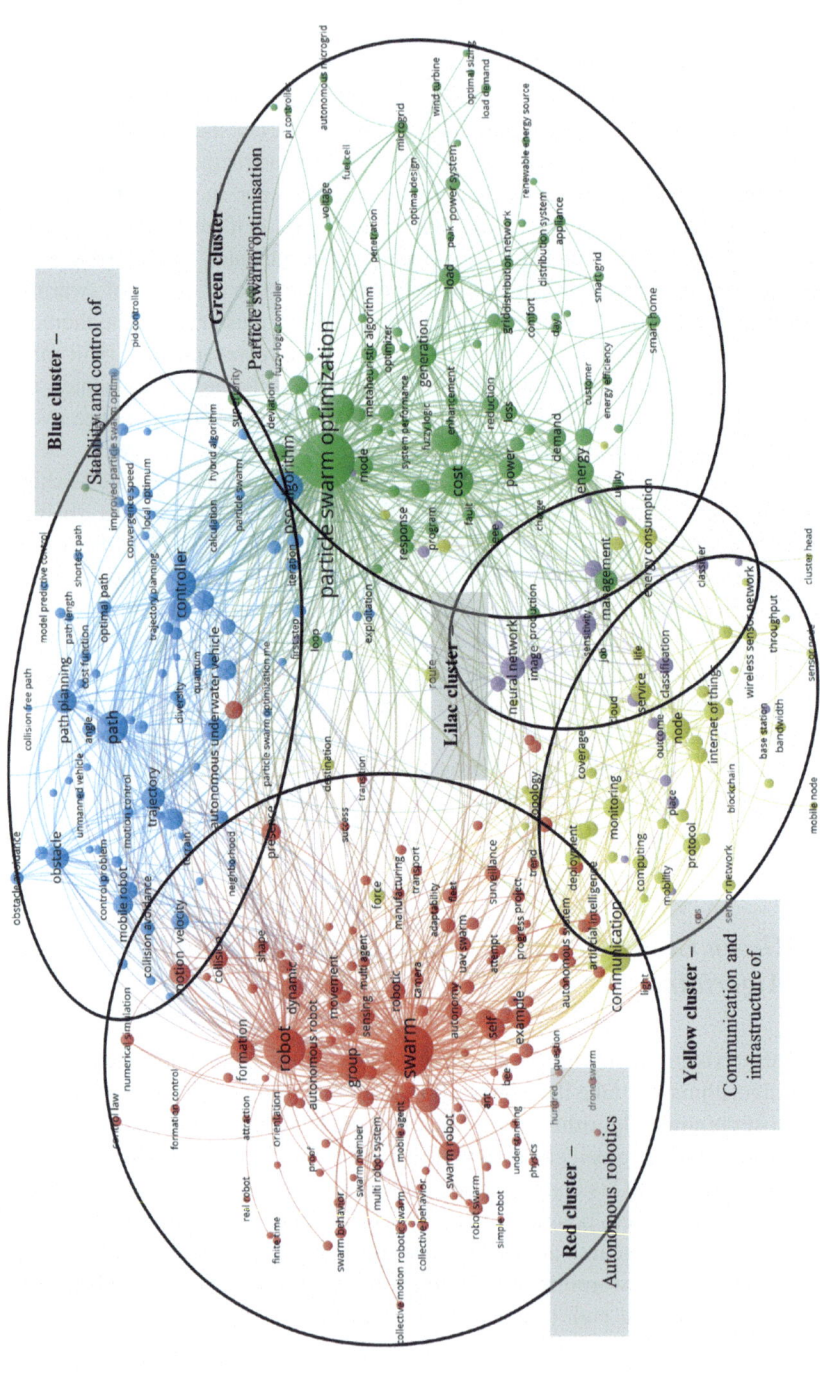

Fig. 2.1 Swarms and automation literature map

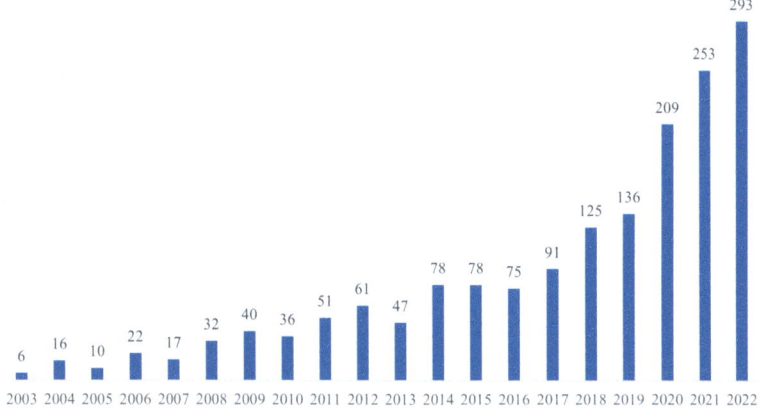

Fig. 2.2 Swarms and automation growth of the literature in the last 20 years

laid out by agents previously, similar to ants using pheromone concentration levels to find the most efficient path to the goal [250]. Since the 2000s, a number of further swarm algorithms have been developed including firefly, bat, artificial bee colony, glow-worm, whale and others (see, e.g. [729]). It is important to note that hybrid algorithms that combine swarm algorithms with other swarm and nonswarm algorithms, including grey wolf optimiser and simulated annealing, are common in the literature (e.g. Shami et al. [722]). We do note that a large part of the literature has not discussed swarms and their applications but PSO as a solution for a large number of applications. For example, Daixin et al. [221] showed that connecting systems through PSO results in energy-saving of individual items in a hospital environment. Therefore, the green cluster offers a variety of algorithmic solutions for a wide range of needs including energy generation.

2.3.2 Red Cluster: Autonomous Robotics, Including Swarms

The red cluster discusses potential uses of swarms including rescue, surveillance and navigation. Schranz et al. [707] offered a taxonomy of swarm behaviours that consists of (1) spatial organisation (aggregation, pattern formation, self-assembly and object clustering and assembly), (2) navigation (collective exploration, transport and localisation as well as coordinated motion), (3) decision-making (consensus, task allocation, fault detection, perception, synchronisation and group size regulation) and (4) miscellaneous behaviours (self-healing, self-reproduction and human-swarm interaction). Cheraghi et al. [195] suggest fields in which swarm applications are already in use. This includes agriculture, in which the SAGA experimental system consists of drones that monitor and perform weeding of fields, and industrial, in which Firebots are able to build structures as well as the FlyZoo

Table 2.1 Top trending[a], top impact terms[b] and indicative journals[c]

Cluster	Top trending terms	Top impact terms	Indicative disciplines	Indicative journals with no. of articles
Green	Grey wolf optimiser	Wind turbine	Engineering	*IEEE Access* (72)
Particle swarm optimisation and other algorithms	Salp swarm algorithm	Optimal design	Computer science	*Applied Soft Computing Journal* (22)
	Whale optimisation algo	Optimal sizing	Energy studies	*Neural Computing and App's* (15)
	Diesel generator	Diesel generator	Mathematics	*Energies* (20)
	Metaheuristic algorithm	Charge	Decision sciences	*Expert Systems with Applications* (19)
	Renewable energy system	Multi-objective OP		
	Superiority	Renewable energy system		
	Multi-objective OP	Autonomous operation		
	PI controller	Autonomous microgrid		
	Appliance	System performance		
Red	Drone/UAV swarm	Autonomous UAV	Decision sciences	*IEEE Access* (72)
Autonomous robotics including swarms	Potential application	Cyber-physical system	Innovation and engineering management	*Swarm Intelligence* (16)
	Artificial intelligence	Biology	Defence studies	*Autonomous Robots* (15)
	Foundation	Real robot	Robotics and industrial electronics	*Swarm and Evolutionary Computation* (12)
	Human operator	World	Engineering	*Robotica* (10)
	Individual agent	Manoeuvre		
	Guidance	Collective behaviour		
	Cyber-physical system	Emergence		

		Collision	Community		
		Manoeuvre	Surveillance		
		Perception	Engineering		
Blue		(Shortest) path length	Optimality	Decision sciences	*Engineering IEEE Access* (72)
Stability and control of autonomous agents		Collision-free path	Global search	Computer science	*Journal of Intelligent and Robotic Systems Th'ry & App's* (24)
		Autonomous navigation	Underwater environment	Economics and business	
		Model predictive control	Cost function	Mathematics	*Robotics and Autonomous Systems* (22)
		Collision avoidance	Control problem		*Energies* (20)
		Dynamic obstacle	Quantum		*IEEE Robotics and Auto'n Letters* (25)
		Hybrid algorithm	Local optima		
		Complex environment	Automated guided vehicle		
		Path planning	Search space		
		Improved PSO	Exploitation		
Yellow		Blockchain	Blockchain	Energy	*IEEE Internet of Things Journal* (11)
Communication infrastructure of autonomous systems		Smart city	Deployment	Computer science	*IEEE Access* (72)
		Real-time application	Collection	Mathematics	*Swarm Intelligence* (16)

(continued)

Table 2.1 (continued)

Cluster	Top trending terms	Top impact terms	Indicative disciplines	Indicative journals with no. of articles
	Internet of Things	Smart city	Engineering	*IEEE Robotics and Automation Letters* (25)
	Firefly algorithm	Internet of Things	Decision sciences	*Energies* (20)
	Attack	Cloud computing		
	Perf'ce metric/analysis	Monitoring		
	Ocean	Extensive simulation		
	Cloud	Real time application		
	Energy consumption	Delay		
Lilac	Deep learning	Outcome	Computer science	*IEEE Access* (72)
Machine learning	Machine learning	Deep learning	Mathematics	*Applied Soft Computing J'l* (22)
	Outcome	Place	Decision sciences	*Neural Computing and App's* (15)
	Dataset	Machine learning	Engineering	*Mathematical Problems in Engineering* (18)
	Support vector machine	Sensitivity		*Expert Systems with Applications* (19)
	Feature extraction	Classifier		
	Classifier	Better performance		
	Software testing	Dataset		
	Neural network	Feature extraction		
	Person	Person		

[a] Top trending terms represent the most recent average publication period sorted by recentness
[b] Top impact terms represent the highest normalised average citation counts in descending order
[c] Indicative journals in brackets show the total number of articles from the entire data set, not per cluster

Table 2.2 Top 20 countries that published swarms and automation journal articles

Country	Articles	Citations	Av. pub. year	Av. citation
China	511	8499	2018.18	16.63
United States	250	14,779	2014.43	59.12
India	211	2807	2018.58	13.30
United Kingdom	108	3919	2016.29	36.29
Australia	72	4808	2016.85	66.78
South Korea	67	873	2018.47	13.03
Iran	66	1625	2017.73	24.62
Italy	63	912	2017.14	14.48
Canada	58	1822	2017.71	31.41
Germany	58	1696	2016.17	29.24
Japan	50	504	2014.34	10.08
Saudi Arabia	49	794	2019.63	16.20
France	43	944	2016.26	21.95
Turkey	40	2,289	2016.85	57.23
Egypt	32	199	2020.09	6.22
Malaysia	30	780	2017.77	26.00
Pakistan	29	445	2019.28	15.34
Taiwan	28	789	2015.79	28.18
Spain	28	481	2016.82	17.18
Brazil	26	547	2017.46	21.04

Future Hotel, which is fully automated and hazardous zones, in which snake robots are able to crawl into areas that are unreachable for humans. Other areas that are engaged in research include military, medical, astronomy and household applications [195].

2.3.3 Blue Cluster: Stability and Control of Autonomous Agents

The blue cluster describes the stability and control of swarms. For example, Sharma et al. [729] compared the optimal performance algorithms on the 3D path planning of UAVs and found several issues such as the following: (i) PSO-based agents can become trapped, (ii) ACO-based systems suffer from computational complexity, (iii) firefly algorithms are excellent in exploration but lacking in exploitation, (iv) bat algorithms require tuning, (v) the artificial bee colony (ABC) has a slow convergence speed and (vi) the convergence speed of bacterial foraging solutions slows down when the number of targets increases. Therefore, the authors suggest research into hybrid algorithms to resolve these issues in controlling UAV swarms.

2.3.4 Yellow Cluster: Communication and Infrastructure of Autonomous Systems

The IoT allows for communication between devices through communication systems and is thus an integral part of automation and swarming. The yellow cluster is concerned with the communication of devices among each other and thus the inevitable ventures into the IoT and sensor network discussions. For example, Sharma and Tripathi [730] offered a review of swarm intelligence algorithms in IoT optimisation, including ABC, ACO and social spider optimisations, in resolving real-world issues.

2.3.5 Lilac Cluster: Machine Learning

The lilac cluster describes machine learning applications. A deep learning model reliant on PSO has shown itself to be a functional vehicle-classification model, which is relevant for security, traffic analysis and self-driving applications [39]. The literature offers a variety of hybrid machine learning algorithms based on swarm algorithms to resolve issues and create efficiencies in real-world simulated environments (see, e.g. [40, 139]). Relative to the other clusters, this is the smallest one, and this may be because swarm intelligence-based approaches to machine learning have traditionally been understudied [121].

2.4 Discussion

2.4.1 Contributions to the Literature on Swarm Intelligence

Swarm intelligence emerges as a compelling subject of research for defence or military applications. Key considerations centre on the concept of "coordination without control". The advantageous outcome of decentralised control is the scalability of the swarm system, which is capable of exhibiting complex emergent behaviours according to a simple set of instructions and interactions with the environment. In addition, it demonstrates robustness in adapting to changes and handling failures within itself and its operational environment. The mapping exercise helped to draw out three key findings about the role of swarm research and technologies within the wider automation literature.

First, as shown in the analysis of the yellow cluster and Fig. 2.1, swarm intelligence is increasingly driven by the IoT. As the chapter by Scholz et al., Chap. 1 [705] finds that swarm systems' effectiveness is determined by their need to integrate with existing infrastructure and new infrastructure. Artificial intelligence (AI and autonomous) systems are leading the digital revolution that is transforming modern defence forces and industrial applications. Such technologies provide strategic, operational and tactical advantages in the new battlespace and in industry, and these advantages are enabled by the IoT (Seet et al. [715]). For example, to facilitate swarm intelligence and sensors that allow for seamless communication and cooperation, cyber capabilities and security are of immense importance given that potentially vulnerable devices are connected through the IoT (Baig et al. [92]). Interestingly, no chapters in this monograph deal with the importance of the IoT to swarm intelligence. This is an important area for further research because it is through the IoT that an adversary's cyber-attack, resulting in the control of devices/weapons, could lead to irreversible consequences.

Second, we find that research on autonomous or unmanned technologies provide important foundation on which swarm intelligence is built and from which it continues to be developed. In our data set of journal articles on swarms and automation, we extracted over 70 studies that have depicted potential uses of swarms in a military context. The literature suggests that UAVs are the preferred technology in studying tactical military advantages. In a recent study, Lappas et al. [500] offered algorithms that allow UAVs and unmanned ground vehicles (UGVs) to gain surveillance and monitoring advantages through real-time situational awareness, thus allowing for effective decision-making and a reduction in casualties and human exposure. Weinberg [829] proposed using high-powered radio frequencies to disable hostile UAV swarms. Valianti et al. [807] successfully simulated the rotary-wing and fixed-wing swarm of UAVs to track-and-jam rogue drones. Huang et al. [398] stress-tested a flocking algorithm that is common in UAV clusters to find that data spoofing is capable of delaying network traffic arrival by almost 50% and therefore shows vulnerabilities in the current flocking systems. Many applications and solutions are not available in the academic literature but are instead provided in reports and other grey literature. A review study by Schranz et al. [707] referred to some military swarm applications, including the OFFSET system, which is capable of using over 100 swarm tactics in deploying reconnaissance missions in cities using UAVs and UGVs. Further, a swarm of Perdix drones has been developed by the Pentagon. These are capable of surveillance and targeted assassinations [398]. In terms of future research, the chapter by Cardier et al. [174] in this handbook builds on autonomous technology research by proposing a new approach to autonomous sensing and decision-making. Further, Horne [384], in her chapter on maritime autonomous systems, builds on other research developing aquatic missions using swarms that are also on the agenda of the military sector. For example, CARACaS, developed by NASA, was adopted by the US Navy to use a swarm of boats to perform a variety of tasks, including intercepting enemy vessels and escorting and protecting naval assets [393]. Overall, there are very limited practical applications of research on swarm intelligence; therefore, this monograph is instrumental in bringing the state of the art of swarms as well as directions for future research.

Third, our study points to the need for further research into the technical challenges and fundamental questions relating to swarm intelligence, specifically the problem of complexity. This is not due to the technologies themselves but the unforeseen consequences of the convergence of these technologies. As noted in the chapter by Guihen [342], there are multiple layers of complexity that need to be further addressed, and in the context of underwater swarm technology, these include the degree of fleet heterogeneity, the distributed estimate of state, task replanning and allocation. This reinforces the views of [121], who argue that as technology progresses, it leads to the generation of massive amounts of data, resulting in increased complexity and computational challenges for real-world applications. In addition, machine learning, specifically, deep learning, as well as many real-world applications, poses intricate optimisation problems that require thorough attention to achieve more accurate analyses [349]. However, similar to Darwish et al. [668], considering the rapid evolution of machine learning, which is still relatively new compared with other identified areas, we anticipate that machine learning may ultimately bring about similar advancements in the capabilities of swarm intelligence. These advancements could either build on the capabilities provided by machine learning, specifically deep learning, or introduce entirely new ones.

Finally, only a handful of publications in the current swarms and automation literature have discussed any ethical or legal implications of using swarms and related technologies [34, 205, 403], let alone the need for compliance in ensuring ethical technologies and their use. Thus, it is imperative that research clearly identifies the path towards ethicality in swarm and wider autonomous technologies, especially in the military context, to ensure sustainable development, future use and trust in technologies.

2.4.2 Contributions to Practice of Swarm Intelligence

The future of combat will involve extensive use of swarm technologies, and current research is providing conceptual models and solutions for eventual implementation in practice. The proliferation of technologies such as UAVs, UGVs and AI necessitates machines to surpass human capabilities in capacity and processing speed. Swarm intelligence, in this context, is poised to redefine the nature of future warfare by providing essential combat information, mitigating infrastructure damage and minimising the loss of human life. Our scientometric study identifies a few application-oriented factors related to the role of swarm research and technologies within the wider automation literature.

First, building on the findings about the importance of IoT technologies to swarm intelligence as indicated earlier, it is necessary to develop a system of systems (SoS) approach to govern and manage the implementation of swarm intelligence. One such example is hypersonic unmanned combat aerial vehicles (UCAVs) that will be able to provide an advantage in severe confrontation environments. Luo et al. [549] suggested the need for a dedicated ad hoc network that is capable of high levels of mobility, dynamic topology, vast geographic coverage with low delay and robust design. Therefore, the most pressing issues are the antenna design, spread spectrum modulation selection, channel code design and MAC/routing protocol optimisation [549]. Even before the defence sector can obtain hypersonic UCAVs, it is necessary to develop a SoS that will allow UAVs to effectively perform their missions. Therefore, an effective governance of SoSs will provide a strategic advantage for the military [458].

Second, it remains unclear whether it is possible to continue to separate the professional behaviour of the operators of these machines with those of AI-enabled autonomous systems, particularly given the collective nature of swarm intelligence. There are challenges ethically, legally and relating to governance, especially in terms of clearly delineating the potentially harmful impact of such machines in society. Despite this, in the short- to medium-term, there is a need to elevate individuals' professionalism among all relevant stakeholders in achieving safe and ethical outcomes as perceived by society [469]. This is perhaps an issue highlighted currently in the civilian area, particularly in terms of the unregulated use of UAVs. Chapters in this monograph focus on theoretical frameworks to enhance trust in digital architecture (e.g. Cardier et al. [174] 2024) or improve the trustworthiness of swarm technology or data it produces [82]. However, as our study and the chapter by Johnson et al. [434] on hyper-teaming show, the community of the Trusted Autonomous Systems CRC is largely driven and supported by the Australian Department of Defence and related stakeholders. Further, although the military entrusts much of the "thinking" to the machines, it is important to still acknowledge that the responsibility for the actions of AI technologies falls on the hands of the professionals in charge of these technologies [469]. Further research and practice need to consider the importance of instilling ethics in the design, development and use of these autonomous technologies to prevent ethical issues and ensure long-term trust in the technology (see, e.g. Boshuijzen-van Burken [161]). However, the definitions of autonomous systems and thinking swarms, as discussed in Chap. 1, open the question of whether this may be a false or unattainable objective. Specifically, for a system to achieve genuine autonomy, it might require a degree of "freedom" to make choices in situations in which there are no predefined instructions or clear answers. This presents a significant challenge, underscoring the need for further exploration of the tension between control and autonomy. It raises the question of whether swarms should transition from predictable, deterministic methods to unpredictable, probabilistic approaches to enhance their resilience.

We recognise that complexity, as indicated earlier, does play a role here in that many of those who develop AI are not necessarily part of one single profession. We see this trend of multidisciplinary knowledge as an enabler that will act to reduce complexity, bearing in mind our initial comments about using holistic approaches to problem-solving. At the national and international level, institutional efforts should be accelerated to facilitate research and development into swarm technology, to agree on ethical norms to provide frameworks in which swarm intelligence can be deployed safely and effectively and to provide defendable forensic methods that can be used for incident capture, analysis and attribution in case of loss or harm, which is a risk for any new technology. As discussed in the chapter by Roberson and Boshuijzen-van Burken [164], AI and swarm technology have an image problem in terms of trust, because of the data-driven nature of the algorithms leading to a black-box mentality. Confidence in such methods is bolstered by the use of explainable AI or XAI (Yang et al. [850]).

2.5 Conclusion

This chapter provided an analysis of foundational swarm research, namely, in automation and autonomous systems, focusing on the defence/military sector via a scientometric review of the academic research over the past decade.

We described a taxonomy of autonomous systems and defence research, which showed multiple links between networks, autonomous systems and society and most interestingly, the role of cyber-physical systems. The taxonomy resulted in the generation of a cluster map focusing on swarms and automation including clusters such as autonomous robotics (including swarms), the stability and control of autonomous agents, swarm algorithms, infrastructure for autonomous systems and machine learning.

The convergence of these technologies (algorithms, hardware, networks), shaped by the complex interrelationship of societal pressures (expressed or otherwise) and legal frameworks, will ultimately drive new research into swarm intelligence. To date, a few unclassified practical applications of swarms have been published in the military domain. Instead, the literature so far has largely offered algorithmic solutions based on a variety of swarm optimisations for increasingly autonomous and a complex array of networked technologies. However, research and simulations of these technologies are being conducted by the larger military superpowers. The grey literature in English is able to demonstrate examples of such swarms being funded by the US DARPA agency, including OFFSET, CARACaS, ROBORDER, Perdix and other projects [195, 398, 707]. The lack of information in the academic and public realm might make it difficult for these frameworks to be formed in a coherent manner.

The research on swarm intelligence is in its infancy. Although we conducted a thorough search of the literature, we followed the default settings of the software that include only those terms that appear in 10 or more publications. Thus, there may be interesting insights that appear in a few publications that were not included in this analysis.

Finally, we recognise that "autonomous" systems, in its literal sense, presume that there is complete freedom for an entity to make its own decisions. However, we believe that systems can be imperfect in their algorithms, which has implications such as programmed bias, nontransparency in actions and blatant errors (see, e.g. Klarin et al. [468]). Therefore, we propose that professionals who design these systems and those that operate the systems ought to bear responsibility for the conduct of the autonomous systems. Further, as the discipline matures, government legislation, standards and professional bodies will act to assign responsibility to the correct parties.

Open Access This chapter is licensed under the terms of the Creative Commons Attribution 4.0 International License (http://creativecommons.org/licenses/by/4.0/), which permits use, sharing, adaptation, distribution and reproduction in any medium or format, as long as you give appropriate credit to the original author(s) and the source, provide a link to the Creative Commons license and indicate if changes were made.

The images or other third party material in this chapter are included in the chapter's Creative Commons license, unless indicated otherwise in a credit line to the material. If material is not included in the chapter's Creative Commons license and your intended use is not permitted by statutory regulation or exceeds the permitted use, you will need to obtain permission directly from the copyright holder.

Chapter 3
Definitions, Sources and Categorisations for Thinking Swarms

Christine Boshuijzen-van Burken

The concept of "thinking swarms" gives rise to some initial reflections on the history and nature of this emerging technology. It was coined to denote artificial agents that can collaborate to achieve common goals in a self-organising manner. Several definitions, biological or sociological sources of inspiration and categorisations have been suggested over the past decades. In this chapter, I provide a brief overview of various conceptualisations and definitions of swarms found in the existing swarms literature, including sources that gave rise to these conceptualisations of thinking swarms. An important concept is autopoiesis, which is a process of self-sustainment found in living entities, and that became popular in systems thinking circles. The aim was to provide the readers of this book with a concise background to better understand the context of current swarms research and development.

3.1 Definitions

One of the earliest mentions of artificial simulations of natural swarms comes from Craig Reynolds, when he published his research on simulations of the flocking behaviour of birds [676]. A decade later, Bonabeau et al. coined the term "swarm intelligence" [135].

Beni and Wang [108] first defined an intelligent swarm as "collections of autonomous, non-synchronized, non-intelligent robots cooperating to achieve global tasks". According to them, robots will achieve this via unpredictable, non-trivial, order-forming capacities, which express themselves as intelligent behaviour.

In the *Encyclopedia of Complexity and Systems Science*, which was published a decade later, Beni stated the following about swarms: "a swarm of agents (biological

---

C. Boshuijzen-van Burken (✉)
The University of New South Wales, Canberra, ACT, Australia
e-mail: c.vanburken@unsw.edu.au

or artificial) which, without central control, collectively (and only collectively) carry out (unknowingly, and in a somewhat-random way) tasks normally requiring some form of intelligence" [106, 8896].

The origins of Beni's adoption of the word "swarm" were that it was casually suggested to him by Alex Meystal during a conference discussion, as he presented his "group" robots research: "The fact is that the group of robots we were dealing with was not just a "group". It had some special characteristics, which in fact are found in swarms of insects, i.e., decentralized control, lack of synchronicity, simple and (quasi) identical members. Important was also the size, i.e., the number of units. It was not as large as to be dealt with statistical averages, not as small as to be dealt with as a few-body problem. The number of units was thought to be realistically of the order of 10^2 to $10^{<<23}$. So "swarm" was not just a buzz word but a term quite appropriate to distinguish that type of group." [109, 2]

Keywords in the definitions of "swarm intelligence" are that "intelligence" is achieved "collectively", "randomly" and "unknowingly". For Beni, these keywords apply to biological and artificial intelligent swarming agents. This is an important observation, because there is intrinsically, for Beni, no distinction between key characteristics of a biological or an artificial swarm. What distinguishes swarms from other groups is that they are not simply performing useful tasks, but also "intelligent" tasks. "From the robotic side, swarms *self-organize* into patterns. From the biological side they *construct* ordered patterns. The *production* of ordered patterns is a characteristic of intelligence. Another is the *recognition and/or analysis* of patterns, which swarms do when they optimize a function. So, from all sides, we are led to look at the swarms as maybe doing something intelligent—"swarm intelligence"." [109, 3,emphasis added] For Beni "intelligence" in the context of swarms goes beyond pattern recognition and involves the active contribution to patterns, by entities, whether they are biological or artificial. Whether this has been an explicit choice, or whether Beni made his assumptions about biological and artificial swarms explicit through the definition, is of lesser interest, but what I will make clear in Chap. 4 is that one's inspirational source and broader view on the world, specifically, how parts relate to wholes, does matter for developing and thinking about swarms.

Bonabeau [135] added a social element to the definition of swarms, although it remains unclear how the social is distinguished from biological features: "a bottom-up approach to controlling and optimizing distributed systems using resilient, decentralized, self-organized techniques initially inspired by how social insects operate".

Zakiev et al. [860] discussed Beni's definition, as well as those suggested by others, and argued for clearer distinctions between the variety of concepts that are used, such as swarms, robotic swarms, swarm robotic systems or robotic swarms, multi-robot systems, multi-agent systems and weak robots. According to him, Beni's 2005 definition still remains popular in the swarms literature and is merely refined by others. Zakiev et al. listed the following requirements to call a group of mobile robots "a swarm" [860]:

3 Definitions, Sources and Categorisations for Thinking Swarms

1. A swarm has a scalable architecture: there is no strict requirements for a swarm to include a large number of members.
2. Inter-robot communications and sensing are limited to be only local.
3. Scalability requirement determines distributed control topology usage.
4. Members of a swarm must be simple and quasi-identical.

They further provided a comprehensive overview of additions that have been made in the literature, for example, "distributed control topology", which means that all decisions are made by swarm members independently. This requirement assumes some form of intelligence by individual members. Another key addition, made by Navarro and Matia [613], is that of autonomy, meaning that members must be capable of perceiving data from the environment and interacting with it. Şahin [694] replaced the fourth requirement of "simplicity" with "relative incapability", because it is more precise than "simple". As with many of the characteristics for swarms that are mentioned, such as "large numbers", or "local", or "quasi-identical", it is very hard to precisely define them. Further, for terms such as "intelligence" and "autonomy", there is even less agreement as to what this entails inside or outside the field. This should not pose a hindrance for advancing research on swarms, so long as the researchers are aware that some of the terminology appeals to an intuitive understanding of relevant concepts rather than universally accepted scientific vocabulary.

A recurring theme in the swarms literature is the question of optimising collective behaviour through the mechanism of ignoring some sort of environmental cues while receiving others. The origins of this direction for thinking about systems behaviour can be traced to the "autopoietic systems thinking" era, which I discuss in the section that follows.

3.2 Autopoietic Systems Thinking

Beni and Wang's definition arose in an era in which the idea of autopoiesis (cf. Von Bertalanffy [116], Luhmann [545] and, later, Mingers [584]) became popular in systems thinking circles. Autopoiesis can be understood as the basic systemic strategy to "assert itself against the overwhelming complexity of the environment" [545, 250], quoted in Valentinov [805].

Mingers summarised autopoietic systems as self-producing or self-constructing systems, a theory that was originally developed to explain the particular nature of living entities, as opposed to non-living entities [584]. In Luhmann's popular view, autopoietic systems implement a strategy for self-sustainment by developing substantial autonomy from their environments through the mechanism of growing sensitivity and simultaneously growing insensitivity to environmental complexity [545]. The theory of autopoiesis was a response to "open systems" theories, which were a response to the narrow view of "close systems" that did not consider interactions with their environment in any way when describing systems in terms of inputs and outputs. In some ways, autopoiesis could be viewed as a theory that resolved the incompatibility between open and closed systems theory.

The autopoietic view on life was first formulated by the biologist Humberto Maturana. Literally translated, autopoiesis stands for "self-creating". The classic example for explaining autopoietic systems is the living cell. It has a cell membrane and cell fluids and various biochemical components, such as proteins, organelles and a cell nucleus that all interact with one another in a highly complex way. The cell, through its internal interactions, produces the components it needs to maintain the bounded cell structure. In early systems thinking circles, this biologically inspired model became popular.

"Autopoietic systems are organised in such a way that they produce their own components – they are self-producing systems. Formally, "a composite unity whose organisation can be described as a closed network of productions of components that through their interactions constitute the network of productions that produce them, and specify its extension by constituting its boundaries in their domain of existence, is an autopoietic system" [564, 349]. Autopoietic systems are characterised by autonomy since they are not dependent on external production processes, and self-reference since their organisation closes in on itself and their structure is self-defined – it can be any so long as its supports autopoiesis." [584, 294]

Concepts such as network, autonomy and systems were being used interchangeably between the engineering and the biological realms. Against this background, systems developers took seriously the idea of replicating swarms' behaviour to navigate the complexities of the world. Swarming behaviour can be classified as an in-between systems view similar to autopoietic thinking, in which the system is autonomous internally and defines its own system's boundaries by maintaining the system's stability in a complex environment.

The idea of autopoiesis spilled over to the field of cognition, which is relevant for the idea of "thinking swarms". It inspired the "enactivist" position in cognition theory. In this view, cognition can already be found at the level of bacteria that are finding their way to higher concentrations of nutrition, or in preying behaviour. Further, perception, communication in language and thinking in humans are forms of "biological couplings" (a key term in enactivism denoting the dynamic interactions) between an organism and its environment. The enactivist position is in disagreement with the computational view on perception and knowledge acquisition, which holds that all forms of cognition are forms of representations. These differences in views on perception, knowledge, representations and environment interactions, including their inspirational sources, are relevant to our understanding of "thinking" swarms. Arguably, the incorporation of the idea of autopoiesis into recent theories of cognition, with repercussions for the field of artificial intelligence, may not have occurred had the idea been left aside in the heydays of systems thinking. Recent developments in artificial intelligence continue to draw on biologically inspired analogies for describing interrelationships between entities, such as Abbass et al. [19] who referred to Licklider's [530] extended concept of symbiosis for human machine teaming. They are, however, careful to distinguish between various assumptions about symbiogenesis, for example, cognitive and biological.

3.3 Biological and Sociological Sources for Thinking About Swarms

In general, the swarm is a metaphor from the biological realm applied to technical artefacts that relate and communicate with one another in a way that resembles the behaviour of certain animals, such as bees or bird species. Table 3.1 reveals that there are various inspirational sources for thinking about swarms and that biologically inspired metaphors are most commonly found. As mentioned before, Reynolds [677] made simulations of the flocking behaviour of birds and called them swarms. After this, many more biologically inspired models of swarms followed, each with their own acronyms and descriptors. Verbruggen [815] stated that the power of the swarm is in its natural formation. She alluded to a certain naturalness that humans observe in animal behaviour, by which swarms achieve a complex task because they "synchronise individual systems to operate as a collective with a common goal." She further set out a research program for potential defence use of intelligent swarms, but noted that any true swarming is not yet developed at a usable scale. Comprehensive swarms literature overviews are offered by Xue and Shen [845], Hou et al. [391] and Liu et al. [537], including inspirational sources; Liu et al.'s [537] overview of the swarms literature is highly detailed and informative and includes a list of metaphors, authors and numbers of citations (see Table 3.1. The table includes three additions that are not in the original table (marked with "*", added by the author of this chapter). Some authors do not refer to swarms as biologically inspired, but rather as sociologically inspired, but it remains unclear where these authors delineate between what is considered biological or sociological.

Some authors have critiqued the dominant, biological approach to intelligent swarms. Sharkey [727], for example, rejected the possible interpretations of strong swarm intelligence. Strong swarm intelligence suggests that intelligence emerges from a "group mind" from a natural, or robot, swarm. This includes that behaviours could emerge from a swarm of artificial robots in the same way as they emerge from a biological swarm. In Sharkey's view, these interpretations are dismissed as being unachievable in principle, because insects and robots remain separated by the divide between the living and the purely mechanical. There are nonbiologically inspired metaphors and classification too. Hou et al. [391] classified the methods to describe and understand intelligent swarms as follows: describing swarms in terms of task assignment-based methods (divided into market-based methods and alliance-based methods), bio-inspired methods (divided into biochemical information-inspired methods, vision-based methods and self-organisation-based methods), distributed sensor fusion and reinforcement learning-based methods. Hou et al.'s classification shows that there are multiple ways of describing and modelling swarm intelligence, each bringing it back to a core defining element, inspired by analogies borrowed from a different realm. The analogy that one is using creates different options and features, opens up different possibilities and gives rise to different mathematical problem-solving options; in short, it steers the practical outcomes of swarming research depending on the metaphor one uses and the conceptual tools at hand

Table 3.1 Comparison of swarm optimisation algorithms: the basic information and the metaphor. Updated from [537]

Algorithms	Inventor	No. of papers	Solution representation	Inspiration/solving process
Ant colony optimisation (ACO)	Dorigo [249]	18,308	Ant path	Ant foraging
Particle swarm optimisation (PSO)	Kennedy and Eberhart [457]	67,275	Particle position	Particle flocking
Marriage in honey bees optimisation (MBO)	Abbass [21]	104	Bee phenotype and Bee mating	Raising broods
Artificial fish swarm optimisation (AFSO)	Li [527]	934	Fish position	Fishes preying, swarming and following
Bacterial foraging optimisation (BFO)	Passino [647]	333	Bacterium position	Bacterial foraging
Shuffled frog leaping algorithm (SFLA)	Eusuff and Lansey [271]	994	Frog position	Frog leaping
Artificial bee colony algorithm (ABC)	Karaboga [451]	4335	Bee position	Bee foraging
Cat swarm optimisation (CTSO)	Chu et al. [199]	342	Cat position	Cat seeking and tracking
Glowworm swarm optimisation (GSO)	Krishnanand and Ghose [484]	376	Glowworm position	Glowworms' moving
Firefly algorithm (FA)	Yang [853]	3398	Firefly position	Firefly flashing
Cuckoo search (CS)	Yang and Deb [853]	3432	Cuckoo egg	Cuckoo breeding and levy flight
Bat-inspired algorithm (BA)	Yang [852]	2896	Bat position	Echolocation behaviour of bats
Fireworks algorithm (FWA)	Tan and Zhu [778]	384	Spark position	Firework sparking
Brain storm optimisation (BSO)	Shi [735, 736]	241	Human idea	Human brainstorming
Fruit fly optimisation (FOA)	Pan [639, 640]	649	Fruit fly position	Fruit flies searching

Chicken swarm optimisation (CCSO)	Meng et al. [577]	144	Chicken position	Chicken hierarchical behaviour
Grey wolf optimiser (GWO)	Mirjalili et al. [585]	2034	Wolf position	Social hierarchy and hunting mechanism
Pigeon-inspired optimisation (PIO)	Duan and Qiao [258]	130	Pigeon position	Pigeon homing
Butterfly optimiser (BO)	Kumar et al. [488]	191	Butterfly position	Male butterfly mate-locating
Spider monkey optimisation algorithm*	Bansal [96]	88*	Spider monkey position	Foraging spider monkeys
Sparrow search algorithms*	Xue and Shen [845]	453*	Sparrow position	Group wisdom, foraging and anti-predation behaviours of sparrows
Preaching-inspired swarm intelligence*	Wei et al. [828]	32*	Preachers' position	Preachers' social behaviours and their followers

given the chosen metaphor (see Chap. 5 for an elaboration on the use of metaphors). For example, approaching swarms from the reinforcement learning perspective creates different behavioural patterns than the biochemical information-inspired models. Biochemical regularities differ from economic logics and, again, from social behavioural patterns arising from sensory vision.

3.4 Thinking Swarms Versus Swarm Intelligence

Thus far, the literature cited has referred to swarm intelligence rather than thinking swarms. No literature was found on the distinction between swarm intelligence versus thinking swarms. The editors of this book invited interdisciplinary contributions to ways of thinking about swarms, as well as furthering the technical feasibility of a thinking swarms (see Chap. 1). It is clear that exhibiting intelligence should not be conflated with "thinking". Before I disentangle the notion of thinking swarms, I reflect on the various ways in which key authors speak about swarm intelligence.

Gerardo Beni and Jing Wang introduced the term swarm intelligence in [107]. For Beni [109], who admitted that the term "intelligence" is a poorly understood concept that he was not aiming to fully resolve, pattern recognition and pattern forming are key aspects of intelligence. According to him, swarm intelligence "was an emergent property which led to systems of significant power in forming patterns of matter" [109, 3].

Swarm intelligence used to be understood as one individualised intelligent member of a swarm that exposes relatively simple behaviour, which is then aggregated into an intelligent swarm, which exposes swarm intelligence, which can solve complex tasks [375]. Hinchey et al. stated the following: "Swarm intelligence techniques are population-based stochastic methods used in combinatorial optimization problems in which the collective behaviour of relatively simple individuals arises from their local interactions with their environment to produce functional global patterns." [375]

Note that Hinchey referred to intelligence techniques, suggesting that intelligence arises from the stochastic method applied to non- or less intelligent agents, rather than stating that individual agents are intelligent, as Beni and Wang suggested. In other words, it seems that in the swarms literature, there are different positions from which one suggests that swarm intelligence emerges from nonintelligent agents through their interactions with the environment (which resonates with the autopoietic systems thinking strand) or that swarm intelligence is the sum of many individual intelligent agents. These nuances in positions may be better understood after a philosophical analysis of how parts relate to wholes, which is conducted in Chap. 4. The previous discussion on autopoietic thinking and brief mention of the enacted versus computational cognition position is relevant for our understanding of thinking swarms. A computational view may consider thinking swarms as entities that collect "pictures" of the external environment and then represent these in a collective form, a "mental representation" (in this understanding, "mental" is merely a placeholder for an internal, physical structure that can store and call up signals) of an independent, and external, environment. "Thinking" in this sense is the act of creating a pattern (that constitutes the mental representation) from the many pictures that the swarm has collected. In the enacted cognition position, thinking swarms would not be collecting pictures of the external environment and merging these into a single representation. An enactivist view on thinking swarms would consider the very act of an entity going around in its environment to be thinking. In this view, thinking is not merely orientated towards a predefined final end goal of a reassembled pattern but starts by determining what the end goal should be, including finding environmental constraints that inform the possibility and creation of a proper end goal.

Understanding the inspirations underlying prior and current swarms research, including its roots in autopoietic thinking, is important. It helps engineers discern between assumptions and end goals of systems, and this may guide communication of the problem they are trying to solve and how. The authors of Chap. 2 similarly argue that this broader type of understanding swarms has a role, because creating artificial swarms is an overwhelmingly complex problem.

3.5 Discussion and Conclusion

Since the first attempts to replicate swarms behaviour via computational means, it has become increasingly clearer that simulating "emergent intelligent behaviour" similar to that in natural swarms comes at a high computational cost [537]. This is partially due to the increased complexities of the optimisation algorithms that have evolved over the past 30 years (ibid). Liu et al. [537] suggested that future swarm optimisation algorithms must be simpler in terms of their algorithmic design while maintaining high-quality problem-solving capacities. This observation leads to the general conclusion that the idea of "autopoiesis", which has been inspirational for creating algorithms in which individual agents performing relatively simple behaviours accomplish complex tasks in an autonomous way, comes at a high computational cost, which was not initially accounted for or, at least, underestimated. Thus, the idea that out of relatively simple biological or social behaviour arise complex, or intelligent, behaviours is perhaps misleading for engineers in terms of the "costs" involved. Mimicking regularities from the biological and sociological realm may not be the most efficient way for arriving at technologically highly sophisticated processes, simply because technological sophistication requires adherence to technological regularities (norms, principles, rules) that are inherently different from the norms that govern biological or social (or chemical, economical, etc.) behaviour. Indeed, in the biological realm, we may observe relatively simple behaviours and assume that complex intelligent and self-sustaining behaviours arise using little effort. We may very well underestimate the "costs" (i.e. energy, complex self-awareness) required for emergent behaviour in the realm of nature. This is strengthened by Liu et al.'s [537] observation that computation is too limited. Engineers should aim for technical simplicity while being aware that biological or sociological simplicity is inspirational rather than blueprints for solving technological challenges.

Open Access This chapter is licensed under the terms of the Creative Commons Attribution 4.0 International License (http://creativecommons.org/licenses/by/4.0/), which permits use, sharing, adaptation, distribution and reproduction in any medium or format, as long as you give appropriate credit to the original author(s) and the source, provide a link to the Creative Commons license and indicate if changes were made.

The images or other third party material in this chapter are included in the chapter's Creative Commons license, unless indicated otherwise in a credit line to the material. If material is not included in the chapter's Creative Commons license and your intended use is not permitted by statutory regulation or exceeds the permitted use, you will need to obtain permission directly from the copyright holder.

Chapter 4
A Philosophical Analysis of "Thinking Swarms"

Christine Boshuijzen-van Burken

The concept of "thinking swarms" gives rise to several philosophical questions about the nature of this emerging technology. Philosophers have long considered the relationship between parts and wholes, of which the swarm is exemplary: a swarm comprises many parts, which can function by themselves, yet when they are put together, they form a new whole, namely, that of a swarm. In the history of philosophy, the part-whole concept has been scrutinised in many contexts. In this chapter, I provide a brief interpretation of a swarm through the lens of key philosophical positions (e.g. Plato, Aristotle, Descartes). I elaborate on Dooyeweerd, who distinguished between part-wholes and enkaptic interlacements when analysing concrete reality, such as birds and nests, marble and statue, heart and human body and state and citizen. This chapter encourages developers to think through how defining relationships in terms of part-wholes, or enkaptic interlacements, influence the theoretical functioning of the swarm, its technical design and possibly how the swarm may behave in its context of use, as the context evokes new relationships of a part-whole, whole-whole or enkaptic interlaced nature.

4.1 Introduction

The concept of "thinking swarms" gives rise to several philosophical questions about the nature of this technology. It is important to seek answers to these philosophical questions, because they underpin the assumptions about the development and behaviour of artificial swarms. Assumptions about a technology and the metaphors (see Chap. 5) used to describe them, "thinking swarms" in this volume, mutually reinforce each other and often steer design choices. Expectations from

---

C. Boshuijzen-van Burken (✉)
The University of New South Wales, Canberra, ACT, Australia
e-mail: c.vanburken@unsw.edu.au

© The Author(s) 2025
S. Ng et al. (eds.), *Thinking Swarms*, https://doi.org/10.1007/978-3-031-82790-7_4

the technologies, as well as their embeddedness in society, often stem from the philosophical assumptions and normative aspirations that flow from them and that underlie design choices.

Assumptions may drive the need for internal (technical) and external (sociotechnical) harmonious functioning of swarms, for optimising trade-offs as a goal in itself, for achieving maximum efficiency or for striving for continuity, longevity and duration rather than maximising an end state as quickly as possible within a given end time.

The editors of this book invited interdisciplinary contributions to ways of thinking about swarms as well as furthering the technical feasibility of a thinking swarms (see Chap. 1). The weight of this chapter lies in the first invitation, namely, to *think about swarms*.

In this chapter, I aim to clarify possible confusions, limitations and opportunities when borrowing analogies from the biological realm to explain and develop intelligent swarms.

My contribution is a philosophical analysis of thinking swarms. A philosophical analysis typically does not require empirical data, nor does it follow strict methodologies or methods, but it is an academic endeavour concerned with conceptual clarifications, argumentation and laying bare assumptions. In this chapter, I first highlight how various philosophical assumptions about reality may lead to different conceptions of swarms by referring to diverging philosophical positions related to the nature of reality. It was not my aim to cover the entire history of philosophy. I limited myself to a rather sketchy overview of some key turns in thinking about reality since Plato. I then suggest a framework for thinking about swarms that is borrowed from Dooyeweerd's [246] theory of individuality structures. His writings about enkapsis may inspire researchers and developers of swarms in their endeavours.

4.2 Autopoietic Systems Thinking

As mentioned in Chap. 3, one of the earliest mentions of artificial simulations of natural swarms was biologically inspired. Specifically, the notion of "autopoiesis", which is a basic systemic strategy of a system to "assert itself against the overwhelming complexity of the environment" (Luhmann 1991, p. 250, quoted in Valentinov [806], became popular in systems thinking circles. "Autopoietic systems are characterised by autonomy since they are not dependent on external production processes and self-reference since their organisation closes in on itself and their structure is self-defined–it can be any so long as its supports autopoiesis." [584]

It is the idea of collective self-sustainability of individual agents amid external forces as a natural state of being of the collective whole. This chapter makes clear that one's inspirational source and broader view on the world, specifically, how parts relate to wholes, does matter for developing and thinking about swarms.

The relationship between the collected whole and the distributed parts that expose certain behaviour is a key element in swarms definitions. Philosophers have long considered the relationship between parts and wholes, of which the swarm is exemplary. A swarm comprises many parts, which can function by themselves, yet when the parts are put together, they form a new whole, namely, that of a swarm. Throughout the history of philosophy, a core concern has been to reflect on how the world fits together. "The aim of philosophy, abstractly formulated, is to understand how things in the broadest possible sense of the term hang together in the broadest possible sense of the term" [717, 35]. Within philosophy, there may be various ways of defining the aim of philosophy (some might even argue that there is no aim to philosophy). Further, outside the philosophical literature, you may find authors who try to understand how things fit together (e.g. scientists formulate abstract codes to describe causal relationships between things in reality, but this requires a philosophical assumption from the scientist, namely, that things can fit together via causal relationships and that these can be described), but I hope that Sellars' addition "of in the broadest possible sense" conveys the idea of thinking beyond disciplinary boundaries. Sellars stated that "to achieve success in philosophy would be, to use a contemporary turn of phrase, to "know one's way around" ... not in that unreflective way in which the centipede of the story knew its way around before it faced the question, "how do I walk?", but in that reflective way which means that no intellectual holds are barred" [717, 35].

This chapter should be conceived in the spirit of Sellars: thinking about swarms has an empirical and a reflective side to it and aims to inform further developments in swarms technology. Reflecting on the nature of reality and formulating how it can be that things can be experienced as distinct parts and simultaneously as coherent wholes is, thus, not new. We experience a tree as one particular whole while at the very same time we distinguish the leaves and the trunk and the branches, perhaps even the birds in it, the cool shade under it and the fruits it bears and we nevertheless experience the tree in its unity, that one tree. I am not referring to any psychological and neurological mechanism of how our eyes and brain work jointly to create a representation of a tree in our consciousness (which is a view that rests on some contested philosophical presuppositions), but I am talking about the question of how we can experience the tree as a meaningful whole while at the same time distinguish its parts. My concern is with the bigger picture, not the heuristic we use to explain particular behaviours of animals in the case of swarms (this is what happens in specific disciplines, such as biology and sociology). The question I try to answer in this chapter is whether certain philosophical frameworks (i.e. ways of thinking about the nature of reality, or how things fit together) can help us to better understand what "thinking swarms" are, what they are not and what they can and cannot do.

4.3 Biological and Sociological Sources for Thinking About Swarms

Chapter 1 showed that there are various inspirational sources for thinking about swarms, but biologically inspired metaphors are the most commonly found. I mentioned how Reynolds [677] made simulations of the flocking behaviour of birds and called them swarms. After this, many more biologically inspired models of swarms followed, each with their own acronyms and descriptors. Chapter 2 depicts numerous uses of swarms depending on the functions they perform, such as surveillance, track-and-jam unmanned aerial vehicle swarms, agricultural weeding, rescues and other functions of artificial swarms. In general, the swarm is a metaphor from the biological realm applied to technical artefacts that relate and communicate with one another in a way that resembles the behaviour of certain animals, such as bees or bird species.

Verbruggen [816] stated that the power of the swarm is in its natural formation. She alluded to a certain naturalness that humans observe in animal behaviour, by which swarms achieve complex tasks because they "synchronise individual systems to operate as a collective with a common goal." In other words, the way parts and wholes fit together in the form of an animal swarm has a "naturalness" to it, in Verbruggen's description of swarm behaviour. This observation is relevant to philosophical assumptions about the nature of reality.

4.4 Philosophical Positions Relevant to Swarms

In the history of philosophy, the part-whole concept that is important to theorising about swarms has been scrutinised in many contexts. In this section, I make very broad strokes that do not capture the nuances and details of the philosophical positions addressed, but they should give the reader an idea about the various metaphysical presuppositions that underlie explanations of "how reality hangs together" [717, 35], or, as Sellars puts it, "how things in the broadest possible sense of the term hang together in the broadest possible sense of the term".

I start with Plato, who understands reality as autozooion, meaning that the universe is a microcosmos of living beings. In the universe sits the mesocosmos of the state, which, in turn, embraces the microcosmos of man. According to Plato, all parts in the universe exist to uphold harmony in the totality of the "world-soul" universe. Reality is "animated metaphysical totality" [655]. In other words, whatever exists in our world, how that fits together is geared towards a harmonious state of affairs, because harmony is the ideal for all that exists. In Plato's world, a swarm is not a functional material accident but a reflection of the ideal harmonious state that is actively brought forwards.

Aristotle's view of how the world fits together deviates from Plato's. In Aristotle's view, the universe comprises "form" and "matter" and the two together form a "teleological whole". This means that everything that exists comes through a cause and is for a cause. Aristotle considered the relationship between state and citizens as a part-whole relationship, and he used the example of the thumb as part of the whole of the body. The thumb exists for the functioning of the body and, similarly, citizens exist for the functioning of the state. Whatever citizens do, it must be aimed at upholding the state, because without the body, the thumb has no function. In the case of the swarm, this would mean that each individual insect's or bird's behaviour is guided by the principle of upholding the swarm. The idea of a swarm is even directly found in Aristotle's work and he went so far as to call bees "divine" [515]. Lehoux argued that this remark is important, because it shows how, for Aristotle, divinity operates even at mundane levels and is an "explanation for order, proportion and rationality, even in the lowest of animals" [515]. In other words, in ancient philosophy, the idea of a "thinking swarms" can be traced, because the behaviour of bees exhibits or replicates some sort of order, proportion and rationality, which are elements found in the current swarm literature. I note that terms such as "rationality", "intelligence" and "thinking" should not be conflated. The remark about the divinity of the bee by Aristotle should be read in conjunction with what we know about his teleological view of the world. A teleological interpretation of part-whole relationships means that the whole gives meaning to the part. In the case of swarms, it means that the whole of the swarm provides meaning to the part, which is the individual agent. Thus, the individual agent exists for the functioning of the swarm.

Several philosophers have rejected the teleological Aristotelian scheme, which arose out of a view on reality that rested ultimately on the union of form and matter and which informed Aristotle's basis for understanding relationships between entities in reality. When we skip forwards from the Greeks to modernity (Descartes, Hume), we find that in the modern Enlightenment view, the universe is viewed merely as a theoretical system of mathematical physical cognitive relations. The cosmos fits together by a classical natural scientific concept of "function". It stresses the "individualistic deterministic nature of reality". In this vision, swarms of bees are exemplary of a "natural order in which everyone knows their place and fulfils their duties efficiently, thus contributing to the common good, in a kind of apian functionalism" [625]. Traces of Greek thinking can be found in social contract theory, inspired by Thomas Hobbe' s *Leviathan* and Jean-Jacques Rousseau's *The Social Contract* and which gave rise to popular versions of "political economy" theory [136]. A response to the functionalist view of reality came from the Romantics, who replaced ideas of externally observable order with appeals to emotion and an internal view when understanding how things fit together. For the Romantics, the essence of the swarm became "not an ideal polity but a mindless and malign collective, a multitude devoid of individual intelligence and free will" [625].

A lesser-known philosopher who critiqued the Aristotelian teleological account of reality, as well as the functional Cartesian system, is Herman Dooyeweerd. He distinguished between part-wholes and enkaptic interlacements in explaining how the world fits together. With enkapsis, Dooyeweerd means that "when one structure of individuality [i.e. an entity] restrictively binds a second structure …

without destroying the peculiar character of the latter" [247, p. 125 Volume 3]. For Dooyeweerd, the universe is an interwoven coherence of irreducible individuality-structures and modal spheres. Reality presents a fulness of "meaning totality" and "meaning diversity". This philosophical framework is explained in more detail in the next section.

4.5 Philosophy of Swarms

A Platonic philosophical analysis of swarms would assume that the whole of a swarm is an animated artificial structure. It is a matter of mapping the structure and characteristic of the swarm as we find it in nature onto an artificial entity and considering it animated. The swarm is thus a micro-cosmos of its own in which the various parts together form a structural whole that is a living being in itself. This view of a swarm is challenging because it remains unclear when in the development of a swarm the entity can be considered "animated", although this could be a nonissue for a Platonist because all that a systems developer needs to be concerned with in swarm development is finding the perfect form, the matter is of lesser importance.

An Aristotelian analysis is a variation of the dualistic part-whole view of the nature of reality that we find in Plato. For Aristotle, the relationship between the parts and the wholes is best explained in terms of causal relationships. The dialectic between form and matter, and which one is more important, remains unresolved. In other words, if we consider the physical properties, such as materials, sensors and hardware, used in swarms versus the algorithms, programming language and software models, it remains unclear what counts as part of the swarm, for example, whether the algorithm is as much part of the swarm as the artificial agent that together with other agents makes up a swarm remains unresolved.

If we use an Enlightenment analysis for understanding swarms, we need to start at the individual level, because, for example, for Kant, the functioning of one individual determines the function of the whole. In other words, the behaviour of individual parts determines the collective behaviour of the whole. A developer of swarms simply needs to optimise individual agent behaviour to maximise collective outcomes. If this is done well, we have successfully replicated swarm behaviour. For Dooyeweerd, reality exhibits part-whole relationships, but it cannot explain all that exists in reality. Dooyeweerd adhered to a nondualistic world view (in other words, there is not just form and matter as in Greek philosophy), but he spoke of a "meaning totality", by which things can be experienced and become meaningful in a multitude of ways. In the next section, I elaborate on the way Dooyeweerd accounted for this multitude of ways in which reality is experienced.

4.6 Multi-aspectuality of Reality

I aimed to clarify the meaning of a "thinking swarms" by analysing Dooyeweerd's theory of individuality structures. Important in the theory of individuality structures is that things and events in reality do not teleologically nor accidentally "hang together" but are experienced as meaningful wholes in a variety of ways. One of the key contributions by Dooyeweerd is his multi-aspectual view on reality [246], as well as the distinctions he made between part-whole and enkaptic relationships. The theory of multi-aspectuality forms part of a larger philosophical work, called the cosmonomic philosophy. The main ideas of this philosophy are, first, the "dynamic meaning character" of concrete things. This means that things in reality do not exist in and of or by themselves but express themselves in meaning-coherence, meaning-diversity and radical unity [758]. Second, concrete human everyday "naive" experience is important, and this pre-theoretical experience of reality expresses itself in a rich diversity of at least 15 modal aspects (see Table 4.1). These modal aspects are irreducible to one another, and their internal meaning-kernels refer backwards and forwards throughout the aspectual ladder. Finally, there exist "modal law-spheres", unique to each aspect, in which all humans, animals and objects function either as an object (passive) or as a subject (active).

The multi-aspectual analysis started as an anti-reductionist endeavour, but although the general aim of the multi-aspectual analysis is to understand the richness, diversity and unity of reality, it is not in principle against reductionist approaches, as it appreciates reductionist views in its own right, namely, because it is performed in various disciplines in scientific research. DeRoo summarised these ideas in the following way:

Table 4.1 Aspects in Dooyeweerd [437]

Aspect	Kernel meaning
Quantitative	Amount
Spatial	Continuous space
Kinematic	Movement
Physical	Mass, energy, forces, material
Biotic	Organism, life function
Sensitive	Sense, feeling, response
Analytical	Distinction, concepts, logic, pieces of data
Formative	Goals, technology, structure, processing, history, construction, techniques
Lingual	Symbolic signification
Social	Social relationships and institutions, roles
Economic	Management of scarce resources
Aesthetic	Enjoyment, harmony, humor
Juridical	Punishment, reward, due
Ethical	Self-giving love, generosity
Pistic	Belief, commitment, vision, certainty

"For Dooyeweerd, the meaningfulness of creation is not divorced from its empirical expression but is precisely tied to it—its empirical expression is, precisely, its meaningfulness and vice versa. Its being and its meaning cannot be separated—a creature is what it means and it means what it is" [233, 3]. One of the claims of the theory of multi-aspectuality of reality, is that lived reality cannot be reduced to a single aspect (see Table 4.1) of reality and that each aspect's meaning kernel can only be intuitively grasped. One arrives at a descriptive level of analysis of aspects through the method of theoretical abstraction of concrete things. Dooyeweerd spent a great deal arguing that failing to recognize the modal sphere- sovereignty between the aspectual law spheres leads to inner contradictions. He stated: "Theoretical thought is confronted with antinomies when it breaks through the boundaries between the juridical aspect of retributive justice and that of moral love and so on. In developing the special theory of the law-spheres, we shall systematically examine the antinomies arising from the theoretical violation of the modal boundaries of meaning." [248, 2:45]. Clouser [204] rephrased the aspects and highlighted a few potentially confusing aspects, because Dooyeweerd provided a slightly different meaning for how the terms may be commonly used. Against the background of the multi-aspectuality of reality, I introduce Dooyeweerd's distinction between part-whole and enkaptic relationships in the next section.

4.7 Part-Whole Versus Enkaptic Relationships

An important notion in Dooyeweerd's philosophy is that of enkapsis. In an enkaptic interlacement, the part is not absorbed by the whole, nor does the part derive its meaning from the whole, but the part exists in an interwoven relationship, in which it maintains its distinctive irreducible meaning. Dooyeweerdian distinctions have been used and applied in other disciplines [90, 91, 163, 186, 187, 323, 765, 818], but they have so far not been applied to swarms, although Zylstra's paper on "Living Things as Hierarchically Organized Structures" [872] makes a very clear contribution by using the theory of enkapsis for understanding relationships between parts and wholes in the biological realm. As we have seen, many artificial swarm optimisation algorithms are biologically inspired, and hence the relevance of Zylstra's work in this novel attempt that may inform further thinking about thinking swarms.

Part-whole relationships in Dooyeweerd's philosophical framework have the following characteristics. First, parts give up their independence for the sake of the whole. Second, the sum of the parts displays characteristics or qualities that none of the separate parts possess. An example of this is the plant (whole) and its cells (parts), a heart (part) and the body (whole) and an engine (part) in the car (whole). In all these examples, the part loses its meaning outside the whole. Unlike part-whole relationships, in cases of enkaptic interlacements, the specific character of the "parts" remains unchanged when interlaced with other parts. "Enkapsis takes place, when one structure of individuality restrictively binds a second structure of a different radical- or geno-type, without destroying the peculiar character of the latter" [247]. In other words, in the case of enkaptic interlacements, the parts do not derive their meaning from the whole; they can exist independently of the whole. The totality structure can remain intact when one part is replaced by other parts.

Examples are atoms and molecules, a bird and its nest, a house and its furniture and a sculpture and the marble of which it is made. Glas [324] rephrased Dooyeweerd by stating that atoms, nests and physical material, such as marble, continue to conform to their internal structural principles. He asserted that "the goal of nests is qualified by the fact that they serve birds as biotic object, i.e. as repository of their eggs and as shelter for their offspring. In other words, nests are qualified by their biotic object function. Apart from the bird they lose their goal and are just physically qualified compilations of physical material" [324]. I note that the notions such as biotic, physical and aesthetic here are referring to Dooyeweerd's multi-aspectual framework (see Table 4.1), which can be a source for defining the qualifying function of an entity.

In the case of the Hermes of Praxiteles statue, is the marble of which the statue is made part of the statue? According to the theory of enkapsis, the marble (the material of which the statue is made) is not part of the work of art. The marble crystals do not give up their individuality but are enkaptically bound by the statue, which is a new structure of individuality, namely, a work of art. The artist cannot undo these properties; she has to obey the qualifying functions of the marble when disclosing the aesthetic properties of the sculpture. In other words, to do justice to reality and the way we experience things and events in reality, one cannot simply claim that to one person this appears as a work of art, but that it is basically an aggregate of marble crystals. This would deny the aesthetic functioning of the statue. The two aspects (aesthetic and physical) in which the statue functions are irreducible to one another and binding of one another: the work of art is not reducible to the marble crystals.

By way of an intermezzo, I added a further distinction that can help explain what Dooyeweerd had in mind with his theory of enkaptic relationships. He distinguished between the qualifying function and the foundational function of an entity. The qualifying (or leading) function is that what gives the thing meaning, the "raison d'être", the reason for being (which is not merely its goal but a way of expressing to what an entity should contribute in the broad scheme of things and in a morally good manner). The foundational function says something about how the thing came into being and on which its existence rests. The qualifying function has a normative force, because it says something about "how the thing ought to function". If we consider the Platonic worldview, we could say that, for Plato, things exist as parts of bigger living wholes, and therefore the qualifying function of any part is always to serve the bigger whole. For Aristotle, things exist for or through a cause or purpose, meaning its qualifying function is teleological and aimed at what the thing achieves. For Dooyeweerd, things exist in and through meaningful relationships, and therefore the qualifying function of things is in the "highest" meaning sense, which is for a plant in the biotic sense, for a bird in the sensory sense, for a work of art in the aesthetic sense, for a hospital in the moral sense and for a business in the economic sense. All functioning should be geared towards the qualifying function of an entity. The foundational function often refers to the way in which an entity came into being and on which it rests. If that function fails, the entity cannot continue its existence and often could not have come into existence. For example, the nuclear family is founded in the biotic function, but its qualifying function is in the moral aspect of love. The military's qualifying function lies in the promotion of justice, but the foundational function is technical ([military] technology based). The point made by Dooyeweerd and those who elaborated on his philosophy was that an entity's qualifying functioning should guide the development of and any decisions around these entities and not its foundational functioning, because this will be detrimental. For example, in the case of a military entity, if decisions are made under the guidance of the technical (foundational) functioning, it could mean that the decision to strike a target is made simply because the ammunition (technical means) is available, rather than considerations regarding the promotion of justice and security through the use of force. I end this intermezzo with the conclusion that it is important to distinguish between the qualifying and foundational functions of swarms. I suggest that there is a genuine difference between artificial swarms and animal swarms at the level of the qualifying and foundational function. At least in their foundational functioning, artificial and nonartificial swarms differ. Artificial swarms are founded on the technical aspect. They are the result of free cultural, human forming activity, whereas nonartificial swarms are founded on the biotic (fish, birds, insects, sheep) or physical (viruses, antibodies) aspect. The latter came into being as a result of biotic or physical processes, not as a result of human formative powers.

Dooyeweerd further distinguished various types of enkapsis (list adapted from Basden [99]):

- Foundational enkapsis, in which there are necessary foundational relationships for a thing or event to exist in a meaningful way, for example, the marble is foundationally enkaptically interwoven with the Hermus statue and the online meeting platform, such as Teams or Zoom, is not part of the meeting but allows participants to have an online meeting and, together with participants, becomes a meaningful whole of an online meeting.
- Subject-object enkapsis, in which one thing can "take on" the other thing, but not the other way around, for example, the snail and shell and the nest and bird, which can exist independently of each other but have different qualifying functions.
- Symbiotic enkapsis exists, in which the relationship is mutually needed to uphold symbiotic stability, for example, a cow and the meadow (not mere grassland, because a meadow needs the cow for it to be a meadow) or parasitic relationships and corals.
- Correlative enkapsis, in which there is one-way dependency, for example, a dugong and the mangrove, a researcher and the university and a person and the society; an animal swarm may be a correlative enkaptic interlacement.
- Territorial enkapsis, which can exist in two ways: direct and indirect. The direct form, is, for example, a state and its national orchestras or national universities. A university is not part of the state but territorially encapsulated in it. National orchestras and universities are somewhat independent entities, and may function

very well without the state, but the relationship to the state is more than an accident. Further, a modern state would be impoverished as a state if it had no universities, orchestras, etc. The indirect form is exemplified by the relationship between taxpayers and schools: even if taxpayers have no children, their taxes support the schools by virtue of being in the territory of the state that takes taxes and supports the schools. The word, territorial, indicates that the relationship stops when we move out of the territory over which the state has jurisdiction.

Why is it important to consider the nature of the relationship between the swarm parts and the swarm whole? I argue that the theory of enkapsis is a theory of relationships, which necessarily implies distinct roles, responsibilities and a sense of how things "ought to relate" and therefore carries normative force. If conflations take place on the theoretical level, it may hinder progress in empirical reality. The theory of enkapsis forces the designer to think about the qualifying function of the swarm, which goes beyond mere functionality of a swarm, as would satisfy Cartesian functionalism, and beyond merely reaching a point of harmony and stability, as would be the aim under Platonic ideals. The qualifying function of an entity, specifically the swarm, carries moral force, and therefore swarms development or swarms applications need to consider cultural and societal context. What is the swarm meant to uphold? Which values should it represent? What should it strive for in the broad scheme of things? Can the individual agents perform tasks and contribute to a meaning-totality if they are no longer part of a swarm?

The list of various types of enkaptic interlacements is meant to open the different ways of thinking about swarms and where the concept of a thinking swarms might fit.

4.8 Conclusion and Discussion

Most of the literature on artificial swarms posit biologically inspired understandings of swarms, which may be because of the autopoietic influences in the systems thinking community. One of the questions I tried to answer in this chapter was whether the individual insects/birds/robots/sensor platforms are enkaptically bound by the swarm or parts of the swarm? The answer lies in whether the one structure loses its meaning when removed from the structural whole. In other words, if we remove one agent, can the swarm exist meaningfully and independently, and therefore the swarm and its agents are enkaptically bound structural whole. Some swarm developers have developed swarms in a part-whole relationship, in which one agent cannot exist meaningful independently of the swarm. To determine which is the case, a distinction between qualifying function and foundational function may be helpful.

I conclude that a robotic "thinking swarms" is a symbiotic enkaptic interlacement between one robot and the robot swarm, with its highest subject function in the logical aspect. A swarm of artificial drones may be used in an artistic manner in a drone show (perhaps even autonomously), and then the highest subject function is aesthetic. In other words, the meaning of the swarm does not lie in terms of its problem-solving capacities but in its ability to visualise harmonious patterns in the air. An animal "thinking swarms" is a correlative enkaptic interlacement between the animal and the swarm with its highest subject function in the biotic aspect. A human "thinking swarms" is a correlative enkaptic interlacement between the human and the swarm with its highest subject function in the pistic aspect. I like to propose to developers to consider what they are designing for when developing "thinking swarms". What is the model about? Is it about (replicating) behaviour, function, meaning, logical connections, physical properties, social coherence, visual harmony, enhancing ideology or something else altogether?

A Dooyeweerdian analysis would consider a swarm meaningful in the logical aspect of solving a problem, rather than:

- Imitating naturalness (Plato)
- Parts whose goal is to uphold the whole (Aristotle)
- Parts that together provide functionality to the whole (Descartes)

This does not mean that biological naturalness and artificial harmony or functionality has no place, but the swarm as we experience it cannot be explained in such terms alone.

Finally, enkaptic interlacements theory can strengthen (and critique) critiques of the reductionist view on swarm intelligence. Sharkey [728] (2006), for example, rejected strong swarms intelligence, because insects and robots remain separated by the divide between the living and the purely mechanical. Dooyeweerd's theory of part-wholes and enkaptic interlacements provides a language to express the distinction between what is living and what is not. It can mitigate the risk of ontological conflations, which may affect the proper development and use of swarms. This chapter encourages developers to consider how analogies influence the theoretical functioning of the swarm, its technical design and, possibly, how the swarm may behave in its context of use, given that the context evokes new relationships of a part-whole, whole-whole or enkaptically interlaced nature.

Open Access This chapter is licensed under the terms of the Creative Commons Attribution 4.0 International License (http://creativecommons.org/licenses/by/4.0/), which permits use, sharing, adaptation, distribution and reproduction in any medium or format, as long as you give appropriate credit to the original author(s) and the source, provide a link to the Creative Commons license and indicate if changes were made.

The images or other third party material in this chapter are included in the chapter's Creative Commons license, unless indicated otherwise in a credit line to the material. If material is not included in the chapter's Creative Commons license and your intended use is not permitted by statutory regulation or exceeds the permitted use, you will need to obtain permission directly from the copyright holder.

Chapter 5
Robots as Research and Drones as War

Unpacking Metaphors from Media Coverage of Swarm Technology

Tara Roberson and Christine Boshuijzen-van Burken

As swarm technology moves towards real-world applications and impact, technological development should be paralleled by investigation into their societal dimensions. The language used by stakeholders working in swarming technology is already fraught with semantic diffusion and confusion. Uncertainty within the field, inevitably, informs public communication. This confusion makes dialogue on the effects of this emerging technology difficult.

In this chapter, we explore the public communication of swarming technology. Specifically, we analyse media coverage to understand the metaphors that currently inform public discourse on this topic and the values that underlie this public conversation. We use automated concept analysis to review qualitative data and review key metaphors used to describe swarming technology. Our iteration of analysing the metaphors shows the lack of "neutral" metaphors in media coverage of swarming technology. Metaphors are value laden and carry normative weight. In other words, they state how the world 'ought to be'. Prudent use of metaphors is key for managing expectations and keeping watch over which avenues of research and development are prioritised.

T. Roberson (✉)
ARC Centre of Excellence for Engineered Quantum Systems, University of Queensland, St Lucia, QLD, Australia
e-mail: tara.roberson@qut.edu.au

C. Boshuijzen-van Burken
The University of New South Wales, Canberra, ACT, Australia
e-mail: c.vanburken@unsw.edu.au

© The Author(s) 2025
S. Ng et al. (eds.), *Thinking Swarms*, https://doi.org/10.1007/978-3-031-82790-7_5

5.1 Introduction

Swarming technologies are an emerging technology that have civilian and defence applications. These technologies consist of robotic and autonomous systems that can interact/act in concert with one another to achieve specific effects. As outlined at the start of this book, for something to count as a swarm, there needs to be many, not just a few, interacting systems. We currently lack a complete understanding or consensus about what a swarm is, which is at least partially a reflection of the reality that swarm development remains in its early stages. Despite this, swarming technology is of interest for a range of sectors, including agriculture, mining, space exploration and defence. This chapter focuses on the language used in media coverage to discuss swarming, and adjacent, technology with the intention of better understanding the values that inform public communication over this emerging field.

In other areas of robotics, autonomous systems and artificial intelligence, lack of agreement on key terms has hampered interdisciplinary discussion, for instance, in relation to the ethical and technical challenges of autonomous weapon systems. The Group of Governmental Experts on lethal autonomous weapons, for example, has so far failed to deliver on its task of examining issues related to these weapons in the context of the objectives and purposes of the Convention on Certain Conventional Weapons. Part of the difficulty faced by the Group of Governmental Experts was the diversity of definitions relating to autonomous weapon systems; indeed, there is a multitude across the spectrum (e.g. Taddeo and Blanchard's review of various official definitions of autonomous weapons systems [774]). Our chapter anticipated similar dilemmas for swarming technologies, for which definitions continue to be unsettled and public discourse on the topic has the potential to be highly polarised. One example of this discourse is the 2017 and 2021 release of short films on slaughterbots [216, 825] by the Future of Life Institute. These films show swarming drones that have lethal capabilities and are designed to invoke fear in public audiences.

As Chaps. 1, 3 and 4 of this book indicate, the language of swarming is fraught with semantic diffusion and confusion. The range of ideas that informs what swarms may or may not be presents early obstacles to discussion, even among the most interested and involved stakeholders of this emerging technology area. During the workshop that informed some of the research that grounds this book, swarming was defined in terms of behaviour (What will a swarm do?), embodiment (What physical properties define what a swarm is?), properties (What are the vital characteristics of a swarm?) and cognition (Does a swarm think? How would a swarm think? Could a swarm learn?). As the presence of these several questions indicates, uncertainty was a dominant theme during discussion. This was heightened by the interdisciplinary nature of the workshop, given that people who had contrasting expertise (engineering, law, philosophy, cognitive science, ethics and more) worked to find common features that could be agreed on.

The uncertainty that informed this discussion also, inevitably, informs public communication about swarming technologies. One reason for this high degree of uncertainty is the struggle to determine a common language that describes the development and use of swarming technologies (specifically, thinking swarming technologies). Layered over this is the difference between a 'thinking' swarm as discussed in the taxonomy outlined in the opening chapter of this work and the references to autonomous swarms depicted within media coverage and popular fiction and film.

In the face of this confusion around what defines and makes a thinking swarms, research informing eventual public communication around thinking swarms is necessary. The impact of swarm operations on people will be multidimensional. Workshop participants in 2022 indicated that possible negative effects (including lowered morale of workers and injury) might be partially caused by misunderstanding of swarm operations and operational contexts. Other challenges cited by participants in the workshop included barriers to community acceptance, such as invasion of privacy, damage to environment, lack of regulatory frameworks suitable for swarm technology and poor accountability and responsibility processes for swarm operations. These responses show us that the social infrastructure that surrounds this emerging technology needs to be nurtured and developed alongside engineering and research into thinking swarms. Public communication will be a key part of this. Moreover, public communication has ethical ramifications, and it is therefore important to understand the metaphors that are used in public communication about swarms.

As swarm technology moves towards real-world applications, the technological development should be paralleled by and intertwined with investigation into their societal dimensions. Work in robotics has so far considered how swarms might be engineered to be safe "from the get-go" to build public trust and confidence. As an example, see Hunt and Hauert's proposal of a 'safe swarm checklist' [402]. Trust and confidence may also require public conversations to assist us in critically reflecting on the choices ahead of us in terms of systems design. Chapter 2 further reflects on the need for technological foresight and preparedness that enable improved ethical design and deployment of autonomous systems. In addition to the need to be able to explain to the public what a novel technology, such as a swarm, is, what it is capable of and what the ethical and societal challenges are, there is a need for a more precise language among engineers that can capture nontechnical considerations. The need for a language among engineers when working on novel technologies is expressed through emerging standardisation and ontologies for systems design. One example in which the language that engineers use for the technical design of their systems is folded into the desire for ethical alignment of these systems is the IEEE 7007 Ontological Standard for Ethically Driven Robotics and Automation Systems [659]. The authors of the standard state that their work "has a significant impact worldwide in being used to teach ethical design; for both human and institutional capacity building in the domain of the ethics of AI; to create computational ethically aligned systems; to create a taxonomy to support the elaboration of public policies; and to strengthen digital cooperation across nations applied together with the other members of the IEEE P7000 family" [659]. A similar effort may be expected for the design of swarming technology.

In this chapter, we use an automated concept analysis of selected media coverage (conducted using the software tool Leximancer) to review current key metaphors used to describe swarming technology. Metaphors influence the public reputation of the field [747] and "the problems we see and the solutions we arrive at" [573] in technology development and governance. Our analysis then considers the normative questions posed by these key metaphors. This includes reflection on whether these metaphors help people think and make good decisions about these technologies and asks what else may be needed within the thinking swarms community's communicative toolbox. Metaphors have normative force. That is, they can be a force for good (supporting understanding) and bad (generating misunderstanding), which may have effects as the field develops. These effects could include changing how and where financial investments are made, creating appropriate (or not) risk perception in relation to swarms and defining who may benefit from swarm technology and who may not.

Metaphors used to advocate for or against the use of swarms in various domains, purposes or geographic areas will have concrete ramifications. An example of this occurred in 2021 when a geographic area within an Australian city (Canberra) was excluded from drone deliveries [197]. While residents of one part of Canberra in the north have access to drone deliveries via the company Wing, people in southern parts of the city are locked out of this service. This seems to be at least partially due to the advocacy of an anti-drone advocacy group that located a critical mass of residents who were against the use of the "low-flying machines" through a door-knocking campaign. The use of specific metaphors (here, "low-flying machines") around emerging technology helped nurture a specific narrative and change the direction of deployment and governance of a drone delivery service.

5.2 Method and Data

We adopted a qualitative approach to understanding how swarming technologies are depicted in media coverage. This approach drew on the metaphor analysis approach.

Metaphor analysis is a long-standing practice in science communication research. Metaphor analysis has been used to investigate how public communication between science and publics occurs [521], interrogates implicit and explicit frames in public debates [365] and research design [684] and charts the influence of language on public perception and scientific understanding of a range of topics [366].

As a rhetorical tactic in communication, metaphors can be deliberately adopted to capture the imagination of different publics. Metaphors can help popularise issues, frame particular views and even legitimise certain avenues of research over others [366]. Within media coverage, metaphors are part of a standard route to making issues concrete, engaging and newsworthy [365].

Our method was influenced by these existing studies, as well as an example of automated analysis of frames and metaphors [365], although we used a different content analysis program (Leximancer). The metaphors we examined told us something of existing public ideas of swarming technology and may inform public communication efforts in the future.

5.2.1 Data Collection

This analysis was concerned with public-facing communication on swarming technologies. In this area of emerging technologies, media coverage on this branch of robotic and autonomous system technology is generally limited, and attention is devoted to technology in immediate use (for instance, drone adoption in civilian spaces, including the example of drone deliveries in Canberra, Australia, referenced earlier). However, "swarming" concepts have emerged in public discourse as a result of the Future Life series 'slaughterbots' and because of research-focused press releases from the field of robotics.

For this chapter, we reviewed the media coverage generated by the 'slaughterbots' campaign. We wanted to investigate the dominant frames and metaphors that emerged following the release of those videos and compare this coverage with other discussions on swarming and robotics.

To collect data for this study, we focused on the periods of the slaughterbots campaign, which released two videos in 5 years: the first in November 2017 and the second in December 2021. A Google Trends search (which maps search terms commonly used in the Google search engine) for the term 'slaughterbot' over the past 5 years reflects these two periods, depicting a spike in the use of the search term in 2017 and 2021. We used this information to select periods that included the campaigns that could present a full 12 months of media coverage. Our first period was from 1 November 2017 to 1 November 2018, and our second was from 31 December 2020 to 31 December 2021. A sample of media articles written in English was then collected from a Google News search ($n = 139$; see Table 5.1 for details). We collected all media articles that appeared for the first author in a Google search and compiled information on these articles (title, publication, author, article text and publication data) into comma-separated value files. For this qualitative analysis of the coverage, the article text was of specific interest.

Table 5.1 Overview of media articles collected from Google News

Period	Search term	Number of articles
1 November 2017–1 November 2018	Slaughterbot	11
	Robotic swarms	36
31 December 2020–31 December 2021	Slaughterbot	32
	Robotic swarms	60
Total articles		139

5.2.2 Data Analysis

We used an automated content analysis software called Leximancer to review the dominant metaphors within media coverage on swarming technology. Leximancer allows a researcher to analyse content by automating the grouping of text with a high level of reliability and repeatability [148]. The software groups sentences into blocks to generate text-based statistics, which include rank-ordered concept lists to indicate the strength of relationships between tags and concepts. The software generates text-based statistics, including rank-ordered concept lists, as well as a two-dimensional map (see below).

The automated concept analysis via Leximancer generated the two-dimensional concept map represented in Fig. 5.1, which features eight themes represented by the circles at 40% theme size.

Leximancer represents concepts from the text in a two-dimensional map. Each Leximancer concept map shows the main concepts that occur within the text. These are shown by black text and grey dots. The grey links show connections between the concepts. The concepts are clustered within coloured circles or themes on the map. The colour of the theme is designated by heat mapping to convey importance. The most important theme appears in red, and the second appears in orange and so on according to the colour wheel. Each theme takes its name (the text that is the same colour as the circle) from the most frequent and connected concept within the circle. In Fig. 5.1, robots is the most common theme and then use, weapons, technology, control, fish, need and light.

The red text within the Leximancer concept map indicates the search term and period for the articles analysed through the software. The placement of the

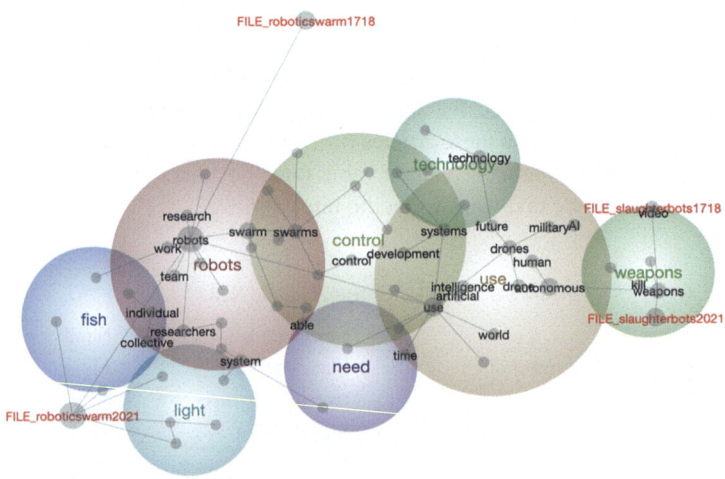

Fig. 5.1 Leximancer concept map—40% theme size

file tags indicates where the majority of articles in each group is located on the Leximancer concept map. For example, the red 'roboticswarm2021' tag on the left side of the concept map indicates that most media stories on robotics swarms during the December 2020–December 2021 period contributed terms to the analysis that focused on biological metaphors (fish, light) and robots. Meanwhile, the red 'slaughterbots2021' tag on the right side of the Leximancer concept map indicates that most media stories on slaughterbots during the December 2020–December 2021 period focused on terms related to weapons and military use.

The red file tags helped to make sense of where key themes were produced by media articles. Within the Leximancer map, we see that the articles concentrating on 'slaughterbots' in both periods focused mainly on the use of autonomous weapons, whereas articles on 'robotic swarms' were more widely concerned with the technology in terms of research developments, sensors and mimicry of biology (i.e. fish) for swarming. In addition to making a clear distinction between the use of swarming technology for research and military use, this division of terms on the concept map indicates that the 'thinking' component of swarming is more readily associated with autonomous weapons.

5.2.2.1 Key Metaphors: Robots as Research and Drones as War

Through the Leximancer analysis, we discovered five key metaphors within the media articles. These metaphors described the nature of robots, swarms and drones. They also presented values associated with the use of artificial intelligence or autonomy in relation to swarming technology and drones.

The first key metaphor in the Leximancer concept map was robots as research, informing discovery. On the left of the concept map in Fig. 5.1, robots is a separate theme concerned with research and the elements that allow them to interact with the world (i.e. sensors or light). The concept related to words including "sensors", "team", "fish", "create", "tasks" and "space". An example excerpt from this theme is included below to illustrate the meaning of the metaphor:

"Eventually, the researchers hope the robots will work in a fully autonomous swarm and be able to mine materials and construct simple structures without any instructions from Earth." ("Robot Swarms Could Mine and Build on Lunar Surface", reported in *E&T*, 2021)

The next metaphor from the analysis focused on swarming technology. In this theme, swarms were described as a collective, a team within the robots theme. It related most strongly to words including "drone", "group", "working", "program" and "team". The excerpts that describe swarms feature both engineered and biological examples. We have included an example below:

"Individual fire ants cannot swim, but they can swarm into formations of floating balls in order to cross water." ("It's Time To Start Thinking About Robot Swarms Again", reported in *CleanTechnica*, 2021)

Our third metaphor focuses on the use of drones as weapons in war. The concept "drones" sits towards the right of the concept map within the theme "use".

Drones related most strongly to words such as "drone", "military", "shows" and "target" and connected with the concepts generated by the slaughterbot articles ("kill", "weapons" and "video". The excerpts of articles related to this concept are militaristic and nationalistic in nature. For example:

"Indian security firm with drones aims to triple revenue to $3 billion." ("Drone Swarms' Are Coming, and They Are the Future of Wars in the Air", reported in *ThePrint*, 2021)

"The United States tested 103 miniature 'Perdix' drones, deployed from three F/A-18 Super Hornet fighter jets." ("Powerful Video Warns of the Danger of Autonomous 'slaughterbot' Drone Swarms", reported in *Global News*, 2017)

"These various projects all suggest that the same idea has taken root across several services: that swarming drones now represent a powerful new capability." ("Robot Motherships to Launch Drone Swarms From Sea, Underwater, Air and Near-Space", reported in *Forbes*, 2021)

From these three metaphors, we can conclude that in this domain of public communication for swarming technology, there is a distinction between 'swarming tech' as being related to research and development work by academics in the pursuit of fundamental discoveries about the world or universe and 'autonomous drones' as technology used for militaristic purposes. Although 'swarming drones' is a phrase used in some of the slaughterbot media coverage, the predominant association with the word in this corpus is more strongly linked to civil and research use cases. Autonomy, meanwhile, is very clearly linked to weaponised drones as we will see in the next section of this chapter.

5.3 Autonomy of Machines as a Threat

We have so far covered the left and right sides of the Leximancer concept map; we now turn our attention to the cross-over themes at the centre. Within the midst of the themes on robots and weapons are concepts concerned with control and use of swarming technology. As we uncover within this section, concerns about swarming technology within these metaphors are focused mainly on how human oversight can be maintained and who might use weaponised autonomous systems.

In the fourth metaphor from this analysis, artificial intelligence (AI) is positioned as a threat. In this Leximancer concept map, AI is a concept that sits inside the "use" theme. The words most strongly related to AI are "kill" and "military"; the technology is viewed as a futuristic foe or problem. Examples of the use of AI within media coverage are included below.

"Weaponisation of artificial intelligence, or military AI, is no longer the stuff of fiction." ("Wars of the Future will Have no Boots on the Ground", reported in *Times of India*, 2021):

"Imagine a future where AI weapons fall into the hands of criminals and terrorists who can use them to wreak havoc." ("A Dark View of the Future of Autonomous Weapons", reported in *Axios*, 2021)

The fifth and final metaphor focuses on autonomy as lethal and uncontrolled. The concept "autonomous" also sits within the use theme. Strongly related to "weapons", "ban", "human" and "kill", this concept is chiefly concerned with the lethal impact that uncontrolled swarming technology could have on humans. Examples from the media coverage include:

"humans at the mercy of these bots. These are autonomous micro-drones with cameras, facial recognition software and lethal explosive charges" ("Beware of drone ASSASINS: Experts Warn 'slaughterbots' that use AI to ID Their Victims could Soon be Commonplace", reported in *Daily Mail*, 2017)

"In so many words, the red line of autonomous targeting of humans has now been crossed." ('Lethal Autonomous Weapons Exist; They Must Be Banned It May not be Too Late to Put the Evil "Slaughterbots" Genie Back in the Bottle, If the World Acts Now', reported in *IEEE Spectrum*, 2021)

From this concept analysis, we see that robots were chiefly positioned as under the guidance of researchers and other technology developers. Meaningful swarms was a lesser-known idea, and "swarms" was used to describe behaviours in nature that robots might come to mimic. However, drones were generally associated with autonomous weapons and militaries. Autonomy for drones was a core concern in the media articles, and the potential for machines to make decisions without human input was positioned as a lethal threat.

From a communicative perspective, the metaphors present within this sample of media articles (drones as weapons, war; AI as a threat; autonomous as lethal, uncontrolled) mean that explicit discussions about meaningful human control and limits on autonomy of machines are necessary.

We suggest exploring the following dimensions of language associated with swarming technologies to further inform the capacity to conduct discussions or dialogue on swarming technology and robotics, autonomous systems and AI more broadly. First, the emotional appeal of language (metaphors) carries ethical weight by implying that a technology can be harmless (e.g. a swarm of fish) or dangerous (e.g. a swarm of wasps or street dogs). Second is the ability of metaphors to attribute moral agency via analogy onto artefacts that do not have moral agency, and third is that metaphors may imply assumptions around technology that may be incorrect, leading to expectations on the behaviour of swarms that may never be met. We unpack these thoughts in relation to public communication of swarming technology in the remainder of this chapter.

5.4 Ethical Dimensions of Swarming Metaphors

We propose that the narratives and metaphors used to describe swarming technology carry value assumptions about these systems, and, in this way, the choice for a metaphor forms part of the systems design endeavours. The metaphor and the technical system description cannot be fully separated. At some point, the metaphor may become settled and accepted terminology in a certain community or globally.

This has been pointed out by Branch [497], who further stated that metaphors are not simply evocative comparisons but have complex effects through entailments, which are follow-on concepts or ideas that reshape thinking, decision-making and practical outcomes [295, 321, 496, 718].

Metaphorical language may, when instrumentally deployed, have pervasive, often unintended, policy consequences. In the case of swarming technology, there is a political dimension, as this chapter makes clear through the case of the slaughterbots campaign. There are conceptual metaphors that are "systematic sets of correspondences ... across conceptual domains" that "reflect conventional patterns of thought" [496]. This is the case for words that are chosen to reflect something akin to the idea of a swarm but still carry metaphorical force. Any word that is chosen to convey the message that highly collaborative actions are taking place among inanimate individual entities carries some form of metaphorical meaning. For instance, when speaking about "multi-agent systems", there is the metaphor of agency. The term "agent" in the systems engineering community, however, has gained a settled status similar to words like "virus" or "cloud" are no longer considered metaphorical among computer scientists or the general public. Other words that allude to the idea of swarming behaviour are coordinating system, self-organised system, networked robotic system and distributed artificial intelligence, among others.

Viewed in this way, the narratives and metaphors used for describing a swarm are part of the sociotechnical complexity of systems design. Following from this, we argue in line with Friedman and Hendry, for a value-sensitive design approach to systems development (in our case, swarming systems), which includes metaphor choice. Value-sensitive design is a three-part design methodology that works to integrate values into the earliest stages of system and artefact technical design. The three stages are conceptual, empirical and technical investigation. These were designed by Friedman and Kahn in the 1990s and then further developed and applied to different domains [162, 294]. Friedman and Hendry stated that "the design process engages reciprocally with and, ... co-evolves technology and social structure. Social structures are viewed broadly and may include policy, law, regulations, organizational practices, social norms, and others" [554].

We suggest including "metaphor choice" in the make-up of a social structure because it co-evolves with technology. Metaphor choice could be considered an important step in conceptual investigations of systems design, which is one of the tri-partied steps in the value-sensitive design methodology. Metaphor choice could ultimately inform policy and governance around swarming systems, in a similar way to Branch [497] who has pointed out for the metaphors used in relation to "cyber". Branch proposed that some type of metaphors support specific ways to understand complex issues, provide discursive resources to some arguments over others and shape policy contestation and outcomes. They highlight the rhetorical effects of using cyber analogies, such as it being a "domain", and how this language has been essential to expanding the military's role in cybersecurity, and consequently for how international law is applied to cyber operations.

In the case of swarming technology, a particular interest in military application caused the slaughterbots campaign to release the video described at the beginning of this chapter. Had a campaign been launched that used an opposite scenario, for example, in which a swarm of drones was sent into an area that was geographically cut off from the world after a natural disaster (e.g. a mass earthquake similar to that in 2023 in Syria and Turkey), and the mini drone swarm finds survivors by using heat signatures and human voice recognition, or drops off essentials, that would show the power of a swarm but as part of a different narrative. The narrative around a swarm would not evoke fear, as in the case of slaughterbots, but one of hope, rescue and survivability for those situations in which traditional options are insufficient.

In this chapter, we used value-sensitive design to review how swarming technology has been described in media coverage. We considered the values that have been "conceptually front-loaded", or emphasised, by the metaphors present in media coverage of swarming technology. Specifically, we focused on how the metaphors related to the technology and the context of use while looking for visible conceptual cues related to values within the text.

The first metaphor we considered was robots as research, informing discovery. This metaphor strongly carries the value of contributing to scientific findings for the greater good of human society. As demonstrated by sample quotes from the media articles analysed, the predominant values here suggest that robotic swarms can play a positive role by supporting research.

"The team now plans to use their research to develop further underwater swarms that can perform environmental monitoring and inspect coral reefs or man-made underwater structures. They could even unearth new insights about the real fish that inspired them." ('Swarms of Robot Fish Could Soon Monitor our Oceans for Environmental Hazards', reported by *The Next Web*, 2021)

"She foresees her research aiding the development of swarm robots that are capable of performing real-world tasks such as search-and-rescue operations, transporting objects, environmental monitoring, and even space exploration." ('Researchers 3D print Multi-legged Swarm Robots Capable of Collectively Overcoming Obstacles', reported in *3D Printing Industry*, 2021)

"The team is creating two swarms: one swarm in the deep-sea and one on the ocean surface. Five or more intelligent deep-sea robot drones will be accompanied and supported by the same number of autonomous catamarans for geo-referencing, retrieval and transport." ('Drones, Robotic Swarms, Lasers and AI Enlisted in Shell's Quest to Explore Ocean Depths', reported in *JWN Energy*, 2018)

In the second metaphor, swarms are described as a collective, a team. Various use contexts for the technology, including space exploration and search and rescue functions, were described in these media articles. Overall, it seems that a swarm described as a collective or a team expresses the value of harmony and respectful collaboration, bringing about good outcomes that can only be achieved in togetherness. Again, a swarm described in this sense is likely to yield positive reception:

"Once they are sent to the moon, however, the hope is that they will be a fully autonomous swarm that can achieve appointed goals for mining and construction without instructions from operators on Earth, he said, comparing the work of his team to farmers." ('Smart Robots Target Valuable Metals With Sophisticated Lunar Mining Techniques', Reported in *DesignNews*, 2021)

"What happens if the fire suddenly changes direction and support is urgently needed elsewhere, the swarm of robots needs to be able to quickly adapt to this change and identify where the urgent support is needed. This is what our research is helping to do; our findings could be used to develop swarms of robots that are more responsive and able to make the right decisions much quicker than they currently can do." ('Less Communication Among Robots Allows Them to Make Better Decisions', Reported in *TechXplore*, 2021)

The third metaphor is drones as weapons in war. When analysing this metaphor using a value-sensitive design approach, one could use two approaches. On the one hand, if a swarm is presented in this way, it may conceal a value, namely, that a swarm could be used for good causes, such as fighting injustices via military means. However, the use of weapons in war is often perceived as a threat by society, rather than as an opportunity for doing good. We are very aware that there may be good reasons to be suspicious about weapons in war. This brings us to the second way of using the value-sensitive design lens: thinking about weapons developments and how most countries have moved away from "indiscriminate mines", which have an increased risk of civilian casualties, to technologies that can perform "high-precision lethal strikes", hence mitigating risks of civilian casualties.

In some conversations, the metaphor of weapons in war might carry the value of ethical improvement of military technology. In the media articles we considered for this chapter, this value was present but not prevalent. Instead, the prominent focus was on the threat of autonomous swarms:

"As terrible as the power to kill individuals or groups remotely from thousands of miles away seems, more terrible is the evolution of military drones toward autonomous attacks without any contemporaneous human supervision." ("Technology that 'empowers the individual' can threaten all of us", reported in *Resilience*, 2021)

"Major governments are pouring billions into developing powerful AI weapons that can hunt and strike targets without the need for human intervention." ("This MIT Professor Says That Drug Cartels May Soon Have Access to 'Slaughterbots'". Reported in *Wonderful Engineering*, 2021)

"The confluence of commercially available technologies, like AI and drones, is progressing warfare into its next predicted evolution, where any force can now deploy a swarm into combat." ("An Air Force 'Way of Swarm': Using Wargaming and Artificial Intelligence to Train Drones", reported in *War on the Rocks*, 2018)

In the fourth metaphor, artificial intelligence is positioned as a threat in which the use of AI will inevitably lead to problems when it (inevitably?) falls into the hands of the wrong people. The value that is expressed here is concern for safety and loss of human autonomy:

"There are growing concerns that drones could get into the hands of rogue actors as technology and AI systems continue to develop, but Frantzman said there is 'no evidence' yet of killer drone warfare." ("War of the Drones: US at Risk of 'Losing

Drone Superpower Title' to China as Fears Mount ' slaughterbots' May Get in Hands of Rogue Powers", Reported in *The Sun*, 2021)

"The FLI[1] and groups like the Red Cross and autonomousweapons.org argue weapons that kill by an algorithm rather than human judgment are immoral and a threat to global security." ("Future of Life Institute Continues Call to Ban AI-powered Drone Swarms", Reported by *Drone DJ*, 2021)

The fifth metaphor is autonomy as lethal and uncontrolled. The value expressed here relates to the fourth and third metaphors (i.e. concern for loss of life and concern for human autonomy). The scenario imagined through this metaphor is future-focused; it is concerned with highlighting and preventing the eventuality that swarms will one day apply lethal force:

"Experts say leading militaries are kidding themselves if they believe they can control the spread of these advanced new weapons." ("Terrifying Rise of AI 'slaughterbots' programmed to kill", Reported by *NYPost*, 2021)

"One area of work where robotics may be useful is security, but this has to be under control and well regulated, without risks of serious escalation." ("10 Challenges for the Future of Robotics", Reported by *Imperial College London*, 2021)

"The film is the researchers' latest attempt to build support for a global ban on autonomous weapon systems, which kill without meaningful human control." ("'Slaughterbots' Film Shows Potential Horrors of Killer Drones", reported by *CNN*, 2017)

This iteration of analysing the metaphors shows the lack of "neutral" metaphors in media coverage of swarming technology. Metaphors are value laden and carry normative weight. In other words, they state how the world "ought to be" [718]. In Chap. 4, the relationship between swarms and views on the world is explored in more depth. In the metaphors we explored, we see the use of emotional appeals in media coverage used to imply that the technology can be both harmless and dangerous. In addition, the applications of swarming depicted via these metaphors suggest assumptions around the design and use of technology that may be incorrect. This could potentially lead to expectations in terms of the behaviour or use of swarms that cannot be met in the future.

5.5 Conclusion

In this chapter, we explored aspects of swarming technology's public profile by drawing on media coverage. The qualitative data we analysed within this work provided an illustrative example of concepts (metaphors) that currently inform public communication about this topic and the values that underlie this conversation.

---

[1] Future of Life Institute.

Ultimately, the swarming research community will need to be able to answer the question of 'Why should we want swarms?' and address public concerns prompted by the use of this technology in public spaces. Responding to public concerns in an early phase of development is important. Responses could include engaging with and taking on values expressed through public concerns during the design of swarming technology.

As swarming technology continues to mature, developers, users and regulators in the sector will need to be aware of the metaphors that inform public understanding of the technology. This is vital because metaphors can have long-lasting effects on investment, risk perception and the building of trust in relationships, potentially influencing the regulation and adoption of swarming technology. Prudent use of metaphors is key for managing expectations and monitoring which avenues of research and development are prioritised.

Extensions of this area of research might consider how public dialogue could expand research on swarming technology by identifying additional possibilities for the use of swarming technology while improving understanding of public sentiment. This dialogue might devote time to considering swarm functions that relate to governance, for instance, by considering details of legal and ethical frameworks and technical compliance, and to analysing terms surrounding the use and testing of swarming technology.

Open Access This chapter is licensed under the terms of the Creative Commons Attribution 4.0 International License (http://creativecommons.org/licenses/by/4.0/), which permits use, sharing, adaptation, distribution and reproduction in any medium or format, as long as you give appropriate credit to the original author(s) and the source, provide a link to the Creative Commons license and indicate if changes were made.

The images or other third party material in this chapter are included in the chapter's Creative Commons license, unless indicated otherwise in a credit line to the material. If material is not included in the chapter's Creative Commons license and your intended use is not permitted by statutory regulation or exceeds the permitted use, you will need to obtain permission directly from the copyright holder.

Part II
Thinking Swarms: Behaviour

Chapter 6
Swarm Hermeneutics

A Preliminary Review

Gavin Mount and David Beesley

Using hermeneutics as a tool to situate swarm activity within social contexts, this chapter presents a framework and methodology for understanding human interpretation of and interaction with thinking swarms. It reviews contemporary scholarship on the social ontology of technology and emerging research on applying hermeneutics to swarm activity. Hermeneutics allows for evaluating decision-makers' ability to interpret swarm behaviour in complex security contexts. Case studies of drone and swarm activities illustrate the interface between autonomous systems and society, highlighting interpretive challenges. The chapter proposes using scenario-based simulations to immerse stakeholders in hermeneutics and decision-making processes. Organised into four sections, the chapter introduces hermeneutics, applies it to drone studies, reviews scenarios posing interpretation dilemmas and discusses developing scenario-based simulations to explore hermeneutic frameworks in future swarm scenarios. This approach aims to enhance understanding of how human decision-makers observe, contextualise and interpret drone swarm behaviour, offering insights for security operators, analysts and governance decision-makers.

6.1 Introduction

This chapter casts a human lens on swarm assemblages by offering a framework and methodology to explore how human beings interpret and interact with thinking

---

G. Mount (✉)
School of Humanities, University of New South Wales Canberra, Canberra, ACT, Australia
e-mail: g.mount@unsw.edu.au

D. Beesley
School of Media and Communication, RMIT University, Melbourne, VIC, Australia
e-mail: david.beesley@rmit.edu.au

swarms. We introduce hermeneutics (the theory and methodology of interpretation) as a methodology for situating thinking swarms into a broader social context. This chapter reviews the contemporary scholarship on the social ontology of technology and provides a preliminary review of the emerging scholarship on hermeneutics that has been applied to swarm activity. Hermeneutics provides a sophisticated framework for evaluating the capacity of decision-makers to "interpret" swarm activity within the context of complex, remote and hybrid security contexts. Case study analysis of drone and swarm activities can be examined to reveal the critical interfacing between autonomous systems and the social context and how these interactions generate complex interpretive challenges. Such episodes need to be examined in their sociopolitical context for the purposes of extrapolating hypothetical scenario-based simulations to provide a range of stakeholders with the opportunity to learn from immersive experiences in hermeneutics and decision-making. Combining our hermeneutic framework with an evidence-based simulation design will further understandings of how human decision-makers observe, contextualise and interpret drone swarm behaviour.

The chapter is organised into four sections. Section 6.1 introduces hermeneutics as a way of thinking about the interpretive challenges that arise within human-machine interactions more generally. Section 6.2 situates hermeneutics as a plausible methodology within current drone studies to reveal four hermeneutic frameworks, from trust to suspicion, that are applicable to these domains. Our goal in reviewing these studies was to understand interpretive frameworks used by security operators, analysts and governance decision-makers. Section 6.3 surveys recent scenarios in which drone swarm activity has generated acute interpretation dilemmas. Reviewing these scenarios informed a thematic analysis that was used to categorise scenario types. Section 6.4 provides an overall discussion and presents a proposal for developing scenario-based simulations that investigate hermeneutic frameworks across a range of plausible future swarm scenarios and envisions how hermeneutic frameworks could be applied.

6.2 Ontology and Hermeneutics

There are no facts, only interpretations and this is also an interpretation. [622]

The emphasis of the present discussion is firmly orientated around the human experience as it relates to thinking swarms. Exploring the human relationship with automata raises phenomenological questions about how autonomous artefacts are perceived by, used for, and have an influence on human communities. It also raises an ontological tension between an engineering ontology of technology concerned with what technology is (and how it operates) and a humanities ontology concerned with the social consequences of technology [587]. Numerous studies on the ontology of technology have observed the ascendancy of science over humanities [720]. The emerging field of social ontology seeks to provide a corrective to this tendency in an effort to "clarify how, in what ways and to what extent, technology is a social phenomenon" [505, 71].

Once viewed through the lens of social and human interaction with technological artefacts and assemblages [228], we can discern a range of ontological positions. In general terms, the social ontologies of technology have documented the widespread cultural effects of technological change, how the "intrinsic powers of material artefacts" have been harnessed to extend human capabilities and how inscription is a mechanism for transferring scientific ideas into "a material operation of creating order" [502, 207]. The social ontology approach ought to inform automata design and development principles. Once technologies become enacted in the world, we must attend to the social and political character of the artefacts and cannot "ignore the contexts in which those objects are situated" [838, 135]. Social ontology is concerned with not only the creation of material artefacts but also how they "come into being through a process of social negotiation, conflict resolution … and which ideas, values and social relations become concretized in particular artefacts" [505, 59–60].

A social ontological approach also permits a wider understanding of how the social manifestation of scientific innovations are always a "revolutionary" [485] and an "essentially contested" [305] process. New technologies do not transform into social facts in a passive manner but are highly mediated through socio-economic, political and cultural dynamics. The philosophical understanding of this process ranges from tracing the impact of these changes [412]) to more sceptical perspectives advising caution against the "illusions of machine fetishism" and its tendencies to constrain our thinking [383] or to imbue the technological assemblage with unwarranted metaphysical properties that are not deserved [630].

The relationship between ontology and hermeneutics employed here is informed by Gadamer's notion of an "always already" situatedness of all forms of understanding [303]. Gadamer asserted that to understand or interpret anything, we must already be in the world as a necessary precondition for interpretation. To illustrate this idea, consider how our respective interpretations of "thinking swarms" are already informed by certain presumptions and preconditions of "thinking-ness" and "swarm-ness" and our pre-imagined relationship with these assemblages.

The hermeneutic process is often represented as cyclical in which the cognitive states of presumption, precondition or prejudice inform the plausible understandings, explanations and interpretations. This cyclical process has been outlined as an interpretive experience in which a new understanding is achieved, not on the basis of already securely founded beliefs. Instead, a new understanding is achieved through renewed interpretive attention to further possible meanings of those presuppositions which, sometimes tacitly, inform the understanding that we already have [316].

Although we may desire a single or universal understanding [650] of what a "thinking swarm" is, hermeneutics is premised on the assertion that it is impossible to rely on a "unitary, organized and systematic image of the world because of the multiplication of special sciences, which does not allow for any sort of unified vision" [812, 87]. The idea of multiple ontologies was characterised by Heidegger as a "battle of world views" [361, 71] and developed by Gadamer [303] through his notion of the "distorted mirror". Building on this image, Rorty also embraced the "potential fruitfulness of alterity" [658, 150] to assert that coherent meaning could only be achieved through acts of conversation across divergent vocabularies [686]. Mutual understanding might be achieved through the process of adopting a different vocabulary than our own. This notion of sharing vocabularies is a useful mechanism to engage in a genuinely interdisciplinary practice of interpretation that connects the ontology of material artefacts with their complex enmeshment with society.

The relevance of hermeneutics in an operational context is that it encourages greater emphasis on the precognitive and metacognitive processes that influence decision-making. As we have argued elsewhere [102], operational environments are becoming increasingly more decision-centric. Further, as assistive and autonomous technologies become more integrated into operations across a range of domains, the expectations on human decision-making becomes more nuanced and sophisticated. The domain of warfare is underpinned by high-speed and complex technologies that are "intended to enable faster and more effective decisions ... while also degrading the quality and speed of adversary decision-making" [201, iv].

As more systems become automated or autonomous, a number of commentators have warned that "the rapid diffusion of and growing dependency on [artificial intelligence] technology at all levels of warfare" will "counterintuitively increase the importance of human involvement in these tasks" [435, 1]. Decision-making frameworks such as the OODA loop (Fig. 6.1) highlight the complexity of human observation and orientation, which precede decisions and actions.

To summarise, our ways of thinking about thinking swarms are necessarily divergent. Engineering and social ontologies each place different expectations on the concept and phenomenon of what thinking swarms are. Our efforts to interpret these phenomena are also informed by prior states of cognition, namely, presumptions, preconditions and prejudices. Ricœur [682] explained interpretation as a process

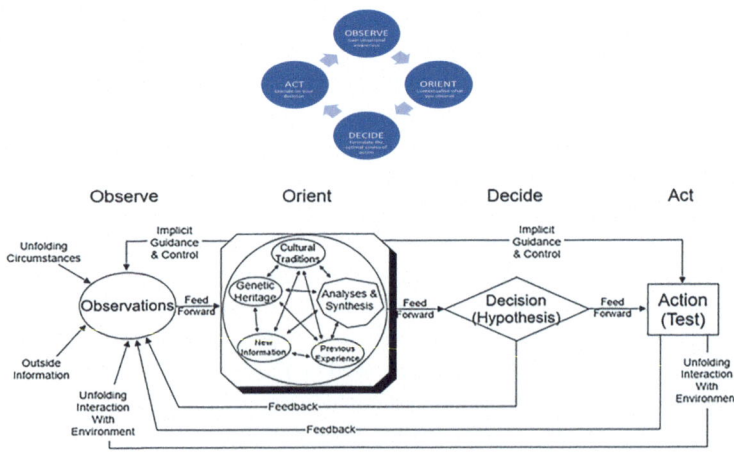

Fig. 6.1 Boyd's OODA loop, simplified and complex [435]

through which we make what was initially alien one's own by enacting it with meaning. Acknowledging the "battle of world views", one way hermeneutics seeks a pathway through this impasse is through conversation across divergent vocabularies.

6.3 Hermeneutics of Swarm Activity

> The focus of subjectivity is a distorting mirror. [303]

Hermeneutics has direct practical relevance for decision-makers, especially in the context of their interactions with systems of automata. The term "hermeneutics" is gaining traction across a range of contemporary swarm studies. Although many of these applications have emphasised the importance of embedding and enhancing trusted digital architecture and autonomy systems (see [159, 231], others have broadened and deepened the analysis to incorporate "cultural and political dimensions" [29]. The growing importance and application of hermeneutic theory is evident in studies seeking to evaluate the capacity of human operators and analysts to interpret individual incidents and to assist them in the task of determining whether they are being influenced or orchestrated by adversary measures [318]. Hermeneutics is viewed by some as a critical methodology required to meet the challenges of operating in a "post-truth" era [63].

Drawing on both the previous discussion and a preliminary review of this contemporary research, we propose four hermeneutic frameworks, from trust to suspicion (Fig. 6.2). These frameworks also inform a more extensive methodology which we have used to evaluate how security operators, analysts and governance decision-makers interpret swarm activity [102].

A hermeneutics of trust affirms the relationship between truth and method [303]. In the current context, it is concerned with the assurances of robust design, regulation and monitoring of the digital architecture, algorithms and event plans; however, this is only part of the interpretation challenge [236]. Vigilant decision-making processes must be maintained and robust systems may still be disrupted, hijacked or weaponised by adversaries. Risk management and countermeasures may be required to test and respond to anomalous behaviour. Increased trust can also lead to complacency and vulnerabilities to adversarial influence.

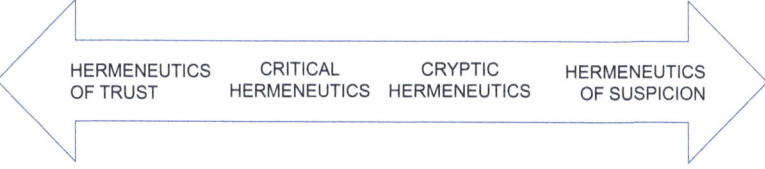

Fig. 6.2 Hermeneutic framework

References to the hermeneutics of trust are constantly employed in contemporary swarm scholarship. Karlsud and Rósen [452] reviewed the use of surveillance drones in UN peacekeeping operations. They found that "drones can dramatically improve information-gathering capacities in proximity to populations at risk", support surveillance of smugglers and spoilers and increase peacekeepers precautionary obligations under international humanitarian law in targeting situations [65, 452]. Nevertheless, the perceptions of human rights groups have been largely sceptical and influenced by debates on the negative effects of drones. Building trust among a range of peacekeeping stakeholders is crucial to the effective implementation of swarms in peacekeeping operations.

Another wide-ranging study examining the "use of large-scale swarms of [unmanned aerial vehicles] as a combat and reconnaissance platform" [43] sought to identify and address a range of challenges associated with the scalability of these semi-autonomous systems. Increasing flock size creates additional strain on human operators and makes it more difficult to maintain trust and an efficient and robust infrastructure. The study advocated for more careful design of patrolling routes to improve the trust and reliability of these systems. Research centres such as the Stanford Internet Observatory have sought to address a "major challenge in building trustworthy technology" and demonstrate how systematic patterns of "coordinated inauthentic behaviour" [334] are managed in response to social and political justifications.

Critical hermeneutics situates these digital artefacts and processes within a broader sociopolitical and cultural context. The complex interfacing of human communities and drone swarm activities within the framework of digital life is contentious [130]. There have been a number of appeals for a hermeneutics of public sociology using a "distinct interpretative tradition that participates in contemporary discussions seeking answers to the social-related questions posed by the public" [389, 309]. Such an approach requires a situated understanding of swarm activity that is able to monitor the implications of social reaction to these artefacts, ranging from more optimistic to more pessimistic views (see Chandler and Fuchs [184]; Bigo et al. [123]). Another application of critical hermeneutics is evident in the scholarship that seeks to examine the potential mass use of highly normalised surveillance technologies already embedded within everyday digital life (see [550, 868]). The critical hermeneutic perspective is also useful to reflect on how drones and swarms have emerged as a medium of representing the social and political context [141, 759].

Cryptic hermeneutics is a disposition that actively seeks to reveal and unmask systemic bias or nefarious elements within phenomenon being examined. A significant focus of this scholarship has been on algorithms. Observing that "people encounter, employ and socialize with technologies and systems based on algorithms on a very ordinary basis", Anderson [47, 1480] sought to understand how we live with them in our everyday lives and daily routine encounters with algorithmically mediated and constituted systems. Drawing on Gadamer's philosophy in *Truth and Method*, hermeneutics has grappled with the elusiveness of a single "truth" and recognised that interpreters are not free to interpret cultural artefacts. Anderson presupposed that "we do not have direct access to the meaning of algorithms in the same way as we do not have direct access to the meaning of other cultural artifacts" [47, 1483]. It follows that trust is not forged through the discovery of a universal or inner "truth", but through cultivating an understanding of how we make sense of algorithms in our everyday lives. Delageniere [227] employed the term "cryptic hermeneutics" to reveal the limitations of positivist epistemology conventionally used to understand intelligence failures made by the CIA during the Cold War and advocated a critical discourse analysis approach. Gerz et al. [318] proposed strategies to interpret when an adversary "attacks in both physical as well as the cyber and information world to disturb an operation".

A hermeneutics of suspicion incorporates interpretive theory and method to grapple with what one writer has described as the "poetics of conspiracy" [254] and allows for sophisticated forms of deciphering, decoding and unmasking multiple sources of knowledge and layers of meaning. It challenges the "foundational interpretative schemes ... [and] justification of scientific-technological and political projects in the modern world" [831, 5]. Radical hermeneutic approaches have emerged in contemporary swarm-related studies to examine interpretive behaviour and societal responses in relation to "post-truth" political contexts [54, 63, 124]. This disposition has been described as one in which "there is no escape from false consciousness. Its goal is to deconstruct – decipher, decode, or unmask – the reality or truth of the meaning of all notions or ideas that we take for granted and show these meanings to be entirely contingent and relative" [710, 138].

In Table 6.1, we compiled the application of hermeneutic frameworks to studies related to drone swarm activity.

Table 6.1 Theory and application of hermeneutic frameworks

Hermeneutic context	Intellectual influences	Concept	Applications to swarm behaviour
Trust	Gadamer	Affirmation of the relationship between truth and method	• Surveillance drones used for peace keeping [452]
			• Patrol and reconnaissance swarms [43]
			• Trust of everyday algorithms [47]
Critical	Habermas	Situating interpretation within the public sphere	• Hermeneutics of public sociology [389]
			• Monitoring pessimistic and optimistic social attitudes [123]
Cryptic	Heidegger	Interpretation, deciphering and decoding nefarious digital activity	• Interpretation of algorithms [47]
			• Discourse analysis of intelligence failures [227]
Suspicion	Ricœur	Disposition of incredulity towards grand narratives	• Understanding and interpretation in the context of post-truth society [710]

6.4 Swarm Scenarios

> The smoke alarm went off in the hallway upstairs, either to let us know the battery had just died or because the house was on fire. [229]

Interpreting thinking swarms can be challenging. When placing swarms in the context of the social domain, there is a persistent tendency not to see the world from different perspectives, which allows interpretive mistakes to happen. The intrinsic value of the following case studies is that they require a hermeneutic analysis—or interpretation—through a social context. Swarms can be considered assemblages comprising aggregates of socially and culturally embedded systems in an evolving confluence of previous technologies and cultural forms. Swarms combine software and hardware with an entanglement of platform capabilities, human skills and environmental and geophysical factors including atmospheric and geographical relations [285, 373] enmeshed with ethical and regulatory aspects as integral elements of the assemblage.

These assemblages produce vast amounts of real-time data. Interpretation of these data-rich hyper-speed assemblages is a common dilemma for command, control and analysis. In this context, hermeneutics, or the theory and methodology of interpretation, is the critical challenge. Scenario-based simulations, regarded as a mode of authentic learning [304], can provide a scaffolded foundation for professional expertise. By using realistic narratives in line with existing problems, as well as plausible future scenarios, educational goals can be achieved. We suggest the use of scenario-based simulations around four hermeneutic modalities to enhance the capability of decision-makers to anticipate and respond more effectively to complex interpretation challenges.

The following scenario considers the interpretation challenges created from a single consumer-drone encountering a swarm of human stakeholders and invites a hermeneutic lens to be cast on the subsequent response. The scenario is based on an incident in which an alleged drone sighting over Gatwick airport (Fig. 6.3) shut down one of Britain's busiest transport hubs. Further, it touches on second-order effects and the ongoing questions about maintaining drone-free zones and the integration of drones and other autonomous technologies into controlled airspace.

6.4.1 Scenario 1: Single Swarm Element

In December 2018, pilots reported sighting a drone flying near the perimeter of Gatwick airport in the UK [721]. This caused a sustained disruption to air traffic control operations and forced the airport to close, leading to the subsequent cancellation of flights. The incident affected an estimated 160,000 passengers, and potential costs were estimated between US$25 million and US$60 million.

The Gatwick incident, caused by an alleged sighting of a single consumer drone, prompted a surge of second-order effects, or excessive countermeasures, of

Fig. 6.3 Gatwick airport, UK (adobestock.com)

increasing socially disruptive potential. Proposed counter-drone measures ranged from the use of trained eagles through to powerful jammers to block the mobile phone/command and control link spectrum. Electromagnetic pulse weapons, snipers and high-powered lasers were also suggested. In hindsight, these seem like a disproportionate response to take down a single $2000 consumer drone and overstate the actual risks to modern, multi-engine commercial aircraft posed by a consumer drone weighing less than 2 kg, which, from a kinetic and risk assessment perspective, would be more akin to a cricket ball hit or bird strike [203]. If the goal of the adversary in this scenario was to provoke excessive countermeasures, the interpretive challenges and second-order effects would be vastly more complex if the single drone element was replaced by a swarm. As Brian Garrett-Glaser [310] put it, this sort of scenario is recognised as a "wicked problem" by the Federal Aviation Administration and is an acute interpretation challenge.

When conceptualising thinking swarms, we often reach immediately for a conventional viewpoint involving dark visions of a murmuration (Fig. 6.4) and swarms of weaponised drones and other robotic military hardware as used by nation-state actors in an international context [29, 110, 182, 292]. However, at a national level, swarms are increasingly being used in the civilian realm for performance and entertainment purposes involving the use of hundreds of nonweaponised drone platforms.

After the release of the Parrot AR.Drone at the Las Vegas Consumer Electronics Show in 2010, the innovation trigger for contemporary consumer drone cultures, one of the immediate applications of use was to perform mass choreographed "dance" displays controlled by a single human operator. The Parrot AR.Drone

Fig. 6.4 Murmuration of starlings (adobestock.com)

was awarded Toy of the Year at the 2010 CES and sales literally took off, but perhaps for unexpected reasons. There was little consumer interest in the augmented reality gaming aspects of the Parrot AR.Drone; instead the emergent properties of the drone platform attracted most interest. Third-party apps, such as Digitalsirup's DroneDance, were quickly available on the Internet and app stores, which gave users the ability to fly a number of AR.Drones simultaneously and perform choreographed sequences, a precursor to the vastly larger drone swarms now being used by Intel [211] and Ars Electronica's Spaxels [267]. In 2012, the Ars Electronica Futures Lab premiered the world's first drone swarm lightshow (Fig. 6.5) in Linz, Austria [268]. By the 2020s, sophisticated mass drone swarm lightshows using hundreds of contemporary consumer-grade drones were flourishing; however since 2020, there has been a spate of drone lightshow failures.

The following lightshow scenarios highlight some of the unique interpretative challenges posed by this public drone swarm phenomenon.

6.4.2 Scenario 2: Disrupted Swarms

There is a persistent belief that drone platforms are infallible, yet techno-fail is a real issue. Even though the small consumer-grade drones used in lightshows are not weaponised, a falling 2 kg drone, let alone tens or even hundreds of falling drones, could cause significant injury or physical, psychological, environmental and economic damage.

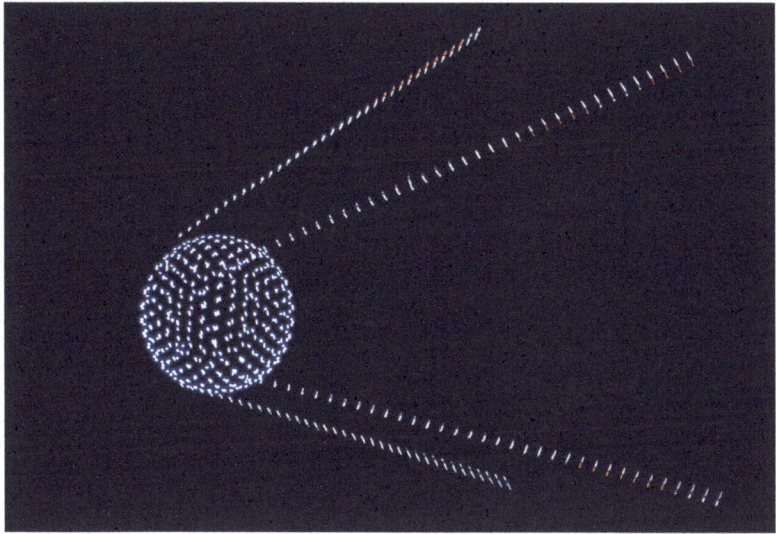

Fig. 6.5 Drone lightshow (adobestock.com)

On Sunday 20 November 2023, thousands of people turned out to witness the City of Light drone lightshow spectacular involving a choreographed display of 500 drones above Elizabeth Quay in the Western Australian capital city of Perth. Perth drone company Drone Sky Shows conceded the City of Light event was "an expensive show" after 50 of its drones fell from the sky in front of the assembled onlookers [575]. One in every 10 drones in the 500-strong lightshow malfunctioned. An internal investigation was launched to determine what occurred. Each drone cost about A$2000 equating to approximately A$100,000 of drone hardware ending up at the bottom of the Swan River. Divers attempted to retrieve the drones to determine the cause of the mass failure. A certain amount of attrition is anticipated in drone lightshows because of wind interference causing drones to hit each other, as well as the occasional snapped propeller. The loss of two to three drones per show is the norm. In this instance, the loss of 10% of the swarm was abnormal, and GPS and wind interference were considered possible causes. At the time of writing, investigations were still ongoing. The incident was reported to the Civil Aviation Safety Authority, and, luckily, there were no injuries to any spectators because of a 120 m exclusion zone maintained between the drones and onlookers. The City of Perth is awaiting the findings of the review before proceeding with any future planned lightshows.

In Germany on 29 April 2022, the Amsterdam-based artist collective, formerly known as Studio Drift, compromising Lonneke Gordijn and Ralph Nauta, had to cancel its commissioned drone lightshow performance *Breaking Waves* because of "aggressive disruption of the airspace" according to their Instagram account (Studio.Drift, 2022). The kinetic installation of *Breaking Waves* consisted of 300 illuminated drones programmed to move together in a wave-like pattern around the facade of the Elbphilharmonie concert hall in Hamburg to mark its fifth anniversary. According to the artist, "The work is a conversation between the building and the moving light drones ... an attempt to build a moment of connection between humans, machine and the environment." [440].

In coordination with the aviation security authorities of Hamburg and Lower Saxony, the performance was cancelled at short notice. According to the Studio Drift Instagram post, during dress rehearsal and the premiere, the art event was massively disrupted by "foreign high-speed drones" [767]. There were several collisions and, as a result, numerous drones crashed. The incidents were unique disruptions to air traffic that required a reassessment of the security situation. Given that it could not be ruled out that such incidents would again be committed by anonymous drone pilots, the remaining performances were cancelled in the interest of the safety of viewers and employees. In art circles, this incident was deemed an act of sabotage [100], but to date there has been no official response from regulatory authorities.

In Taiwan in 2020, 800 drones were designated to perform a lightshow at the Taiwan Lantern Festival in Taichung City. Of the 800 drones used during the performance, 48 crashed during the lightshow performance. A YouTube clip of the incident shows the damage that commenced a little over 2 minutes into the performance, in which numerous drones drop onto the increasingly panicked crowd below [255]. The failure was allegedly caused by terrestrial jamming equipment [723]. Official reports state that electromagnetic interference was detected, yet the origin of the interference was not discovered.

It is unclear whether this disruption to the lightshow was deemed a deliberate hacking attempt; however, it has been postulated that the "interference" could have been a deliberate act of protest by drone operators who opposed Taiwanese government regulations that were yet to take effect that would require mandatory drone registration. More sinister and speculative motivations for the lightshow hijacking were the potential for anti-propaganda messages to be displayed in the sky through to deliberately sending the drones diving into the crowd to cause mass panic and introduce a sense of fear, uncertainty and doubt towards the technologies involved [256].

A plausible extension of these lightshow scenarios involves a performance by foreign manufactured ("red-listed" or "grey-listed") drones: a state actor using a hardware exploit or embedded "back door", or a motivated hacker, taps into facial recognition software and "active track" modes of flight for nefarious purposes and havoc ensues [590]. Again, does (or should) this register on our threat-perception threshold? Further, what can we learn from how humans interpret and interact with aggregates of systems, especially when these trusted systems go wrong?

6.4.3 Scenario 3: The "Slow" Swarm

Increasingly and as evidenced by the 2021 Ukraine-Russia war, the dual-use properties of consumer-level technologies are becoming evident, specifically commercial off-the-shelf drones. These developments, combined with the briefly mentioned concerns regarding "foreign high-speed drones", necessitate reconsidering the potential disruption of dual-use technologies and challenge our preconceptions and interpretation of swarm interactions.

Consumer drones, sold in their thousands, are now sophisticated platforms capable of autonomous flight and equipped with advanced vision systems, collision avoidance technologies and satellite positioning. Industry estimates provided to the domain regulatory authority, the Civil Aviation Safety Authority, suggest that there are well in excess of 150,000 consumer drones currently in Australia (Civil Aviation Safety Authority, 2018). The Chinese manufacturer Shenzhen DJI Sciences and Technologies Ltd, better known by its trade name DJI, has undisputed market dominance in the consumer drone sector [633, 769], which has led to concerns about data security and integrity. This scenario poses the notion of tens of thousands of trusted consumer drones opportunistically and asynchronously capturing flight data including aerial footage and sending the data stream back to servers located on foreign soil [603].

Large data sets are required to train artificial intelligence, and DJI potentially has a massive database of imagery, GPS data and flight telemetry acquired globally from their consumer aerial platforms. By default, the end user is connected to DJI's cloud servers, which keep flight records and telemetry data of every flight unless one deliberately "opts out". China's capabilities in handling massive volumes of data are well-documented, including Beijing's facial recognition programs that rely on an ubiquitous network of surveillance cameras and its proprietary GPS system enabling real-time tracking of the Muslim minority in Xinjiang [536].

These data security concerns were taken seriously enough that in 2020 the US Department of Interior officially "grounded" all DJI drones [574] and the US Military has lists of "blue", or trusted and deemed approved for military use, and "red", not trusted, consumer drones, which focus on country of manufacture [802]. Unsurprisingly, DJI drones are not on the blue list.

These examples echo similar concerns reported from China in which "slow" swarms of Tesla electric vehicles were barred from the city of Chengdu on 8 June 2022 ahead of a visit from President Xi Jinping [415]. Some Chinese military sites have similarly forbidden Elon Musk's flagship product. The 8 June ban seems to be out of concern that the vehicles' impressive array of sensors and cameras may offer insights into meetings of Beijing's senior leadership. China published rules in 2021 essentially prohibiting automotive companies from transmitting data to servers outside China's borders, specifically acquired video and geolocation data [536].

Is the asynchronous mass surveillance potential of tens of thousands of consumer drones interpreted as a swarm? Given that our interactions with the "slow" swarm are so heavily mediated through intermediate channels of information, it is not the swarm we see but the individual and synthesised version of the swarm. Are we even cognisant of this as swarm activity? Should we be concerned about the emergent properties of large-scale systems and the potential of "grey" swarms comprising dual-use consumer technologies?

The aforementioned scenarios challenge conceptions of swarms and pose authentic problems of interpretation. How do human beings interpret and interact with thinking swarms? How can interpretation be made more robust and decision-makers, the human swarm, more resilient?

6.5 Methodological Implications

> O divine art of subtlety and secrecy! Through you we learn to be invisible, through you inaudible and hence we can hold the enemy's fate in our hands. [800]

Building on this preliminary review, we propose the development of scenario-based simulations to research, test and train hermeneutic frameworks and their applicability to complex swarm scenarios. Using simulation technologies and immersive platforms combined with digital ethnographic methodologies, problems of human interpretation were explored to gather further information and contribute to the creation of training and testing tools in an iterative process.

Simulations are a mechanism to explore and hypothesise and systems to perturb, interact and inform. Simulation scenarios must be meaningful and it is important not to confuse simulations and visualisations. Visualisations are graphical depictions often with high levels of photorealism; however, simulations simulate an actual set of behaviours that then have implications for methodology, practice and design. Even though swarm systems are deterministic in that they initially obey deterministic rules, emergent properties can, and most probably will, arise. Therefore, scenarios need to evolve and require further simulations to interpret. Simulations can prepare humans to recognise and discern threats arising from disruptive smart technologies and innovations occurring in the consumer realm, as well as contributing to a growing taxonomy of threat scenarios.

We propose to build scenario-based simulations to model hermeneutic (interpretation) challenges arising from the mosaics [201] of human decision-making and machine interactions. Informed by substantial pedagogical foundations, simulations provide a powerful learning tool to simultaneously access internal cognitive processes and immerse participants in real-world scenarios. To enhance and maintain authenticity, our process of scenario selection used subject matter experts to review the interpretive challenges and decision points of real events. These case studies were then codified and "storyboarded". These cases were then used to extrapolate plausible future scenarios. The immersive nature of simulations provides participant and learners with the opportunity to build their own "internal model of the phenomenon" [681] and to "recreate complex processes" [839], Fig. 6.6.

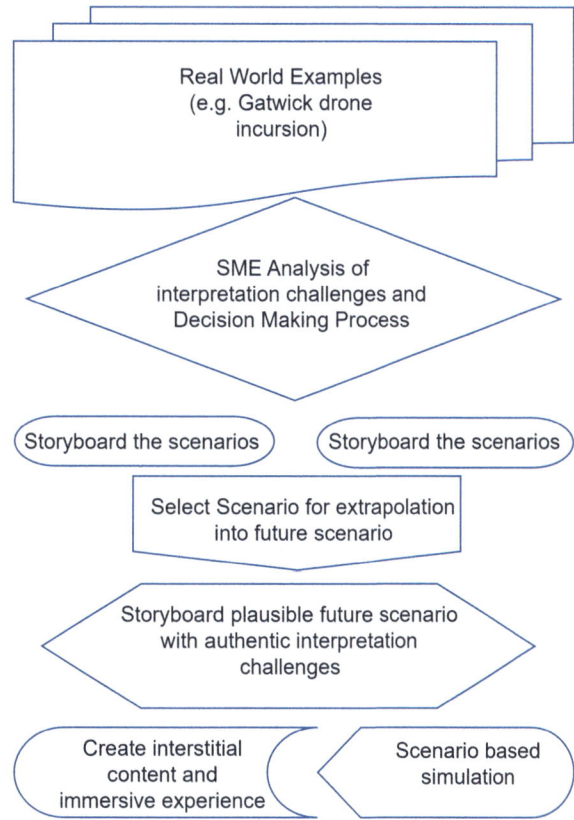

Fig. 6.6 Building an authentic scenario-based simulation

The authentic scenario-based simulation procedure (see Fig. 6.6) identifies distinct interpretative challenges and discrete decision points within a plausible scenario. Participants must make sense of the simulated phenomenon in terms of hermeneutic contexts (trust, critical, cryptic, suspicion). The scenarios can also simulate individual perception and group dynamics to explore various types of orientation challenges within a social and organisational context. As our understanding and application of thinking swarms develops, further scenario-based simulations can be designed to explore interpretative challenges.

The application of our hermeneutic model will vary in its emphasis depending on the roles and identity of actors and their relationship to thinking swarms systems. For example, a designer of a thinking swarms system should be required early in the process to interpret the legal and regulatory requirements (see Chaps. 11 and 12). The emphasis will be on establishing a hermeneutics of trust but also considering social ontologies (critical hermeneutics), addressing inherent risks within the system (cryptic hermeneutics) and, ultimately, testing the system in its design phase under conditions of strategic uncertainty (hermeneutics of suspicion). Likewise, a commander connected to a lead drone should be reassured that the relationship is one of trust. They must also be sufficiently informed and vigilant to consider potential risks of algorithmic bias and adversarial disruption. Observers of thinking swarms are not passive actors but potentially targets of nefarious operations because thinking swarms are engaging in their own forms of "observation". Conversely, counter-swarm observers recording swarm behaviours with a view to locating exploits are observing the swarm through a cryptic hermeneutic lens. The two sets of observers interpret the same set of events through completely different hermeneutic models, and it is the relationship between the identity and the role of the agent that, to a degree, determines the hermeneutic context in which swarms and swarm behaviours will be interpreted.

Although it may be intuitive to assume that certain roles are more aligned with particular forms of interpretation, we argue that this relationship should be understood more in terms of a range of interpretative challenges. In the humanities and social sciences/social studies field, the notion of using a range, or "circuit", as an analytical framework, for example, the "Circuit of Culture" [257], is an established method. Our process is informed by contemporary pedagogic studies that have evaluated the value of simulations as educational and training tools. Simulations offer an opportunity to develop problem-based learning and an immersive and persuasive experience [132] with the purpose of developing an insight into the complexity of the world.

One of the challenges associated with the rapid integration of intelligent assistive technologies such as thinking swarms into human society is that the tools can create decision-making complacency. Recent studies examining the integration of intelligent autonomous systems into the defence sector have recognised the accelerated uptake of systems [445] while cautioning that "[artificial intelligence] is not a simple substitute for human decision-making" [327]. A social ontology approach to these issues highlights how the interface between technology and society varies from one context to another in terms of how they create space in society for these digital agents [320]. One of the risks with human-machine teams could be that they generate cultures of acceptance or enthusiasm. Within defence organisations, there is also a discernible risk of a "warrior culture" being created among small teams and a lack of an "index of suspicion" among command leadership [14].

6.6 Conclusion

Human and civic problems are becoming technical and our technical problems are becoming human and civic. [196]

As thinking swarms become an increasingly plausible phenomenon, we will need to develop our capacity to understand and interpret their integration into society. Hermeneutics presumes that there will be multiple and divergent interpretations of these artefacts and assemblages. Drawing on contemporary swarm scholarship, our hermeneutic framework and practical scenarios of swarm activity, it is evident that these swarm events require layers of understanding and interpretation. In some scenarios, questions of trustworthiness were identified and addressed through mechanisms of accreditation and regulation. In other cases, the events needed to be interpreted within a broader social context. The Gatwick incursion scenario and the drone lightshow scenarios from Germany, Taiwan and Australia highlighted how quickly these matters can become entangled within contemporary politics (or poetics) of conspiracy. The slow swarm scenario challenged our notion of what constitutes a swarm and highlighted how multiple layers of meaning and nefarious surveillance mechanisms operate simultaneously within a swarm event.

Understanding and interpreting thinking swarms behaviour requires an interdisciplinary conversation between ontologies concerned with both what swarms are and how they interact in the world. This chapter focused on the challenges for humans when interpreting thinking swarm activity. Although increased interdependency with assistive, autonomic, autonomous and thinking swarms may reduce (human) cognitive load in some areas, the demands on human hermeneutics for difficult (and wicked) problems will become more intense. The future challenge will be to consider how thinking swarms may assist with these more complex problems and in doing so how they may engage in their own interpretations of human communities.

Open Access This chapter is licensed under the terms of the Creative Commons Attribution 4.0 International License (http://creativecommons.org/licenses/by/4.0/), which permits use, sharing, adaptation, distribution and reproduction in any medium or format, as long as you give appropriate credit to the original author(s) and the source, provide a link to the Creative Commons license and indicate if changes were made.

The images or other third party material in this chapter are included in the chapter's Creative Commons license, unless indicated otherwise in a credit line to the material. If material is not included in the chapter's Creative Commons license and your intended use is not permitted by statutory regulation or exceeds the permitted use, you will need to obtain permission directly from the copyright holder.

Chapter 7
Identifying and Predicting Hidden Coordinated Behaviour Using Synthetic Language Narrative Models

Andrew D. Back and Beth Cardier

Swarms are often considered within the context of a large group of coordinated agents. Here, we consider a different social problem: Is it possible to identify and predict the behaviour of a coordinated swarm of agents within a larger swarm of unknown agents? Recent neurocognitive research indicates that properties of social agency may be captured within mechanisms for language. Accordingly, we considered new information-theoretic dynamical system models with which the behaviour is described using small language models. In this new modelling paradigm, systems are characterised in terms of short-term probabilistic micro-events forming self-emergent, synthetic languages. This approach creates the possibility for novel language-based signal interpretation, capturing social activities. In this frontiers style piece, we propose a model that incorporates this framework using independent component analysis as an approach to separate coordinated agent behaviour. Some experiments are presented, which indicate the potential feasibility of this methodology.

A. D. Back (✉)
School of Electrical Engineering and Computer Science, University of Queensland, St Lucia, QLD, Australia
e-mail: a.back@uq.edu.au

B. Cardier
Institute for Integrated and Intelligent Systems, School of Information Communication and Technology, Griffith University, Brisbane, QLD, Australia
e-mail: b.cardier@griffith.edu.au

7.1 Introduction

Artificial intelligence (AI) models are being developed to predict future outcomes by analysing social agency's temporal connectedness. The challenge lies in understanding the complexity of behaviours, events and probable outcomes, which requires models that comprehend not only features but also contextual relationships, which many systems lack.

Swarms have been considered as a model for a multitude of agents that somehow cooperate to achieve a task of higher-level complexity than is possible to be solved by one agent alone. Swarms are typically understood as synchronised movements of multiple agents who have low levels of individual intelligence, leading to emergent swarm intelligence [135, 455]. Such physical swarms are well-known and have been considered a way to solve complex technical tasks and potentially build sophisticated AI systems.

We propose a new approach to understanding the world, focusing on information rather than physical swarms. We suggest that the world can be observed through signals, movements, behaviours and language that constantly swarm. We raise the question of whether a better model for understanding the complexities of the world and its various systems and agents can be derived from observing simple messages constantly being passed around.

The classical view of science over the past 100 years suggests collecting as much data as possible to create the most accurate model for a system, ignoring potential noise and errors in measuring data. However, our view suggests that the concept of noise may need re-evaluation.

This chapter discusses a novel model approach that suggests noise can be viewed as swarming elements transmitting information, potentially enabling the identification of intelligent agents beyond the reach of conventional methods.

Biological organisms often mimic host behaviour to stay undetected and perform nefarious purposes, similar to humans hiding in large crowds. Current methods for addressing this problem rely on specific domain knowledge and exhaustive methods to discover individual agents, which have significant problems. If domain knowledge is unavailable or connections are not apparent, it becomes a difficult task.

We explore the possibility of hidden social agents within larger groups who have a common intent. We consider a novel method to discover hidden agent behaviour within swarms using an information-theoretic method in line with the concept of swarms of information elements.

Camouflage is a method of evading predators by appearing the same as the surroundings [426]. It involves adopting the same visual appearance and making organisms almost invisible. A more sophisticated form of disguise involves hiding not only appearances but also behaviours, aiming to perform a particular behaviour while appearing the same as every other agent in a larger group.

Steganography involves concealing significant information in plain sight [576, 605, 692, 841] and is often found in viral infections, such as autoimmune diseases and carcinogenic tumours, such as Epstein-Barr virus [514]. Viral determinants mimic host antigens, causing reactive T cells to destroy host tissues [557, 864]. The task of detecting coordinated activity within a swarm holds significant potential for AI and defence applications because rapid assessment is needed to identify hidden agents within a larger domain of agents [28, 225].

Hidden coordinated groups may exist within a larger swarm and seek covert activity for social control and influence [219]. These agents may be physical or cyber [336, 641], or have a subtle linguistic form found in various media and communication [824]. Detecting hidden influences without knowing what to look for ahead of time is crucial for national security and intelligence. Understanding these hidden influences provide significant capability for preventing potential threats.

Predicting social phenomena and behaviours is a further step that can be considered. Recent developments in neurobiological mechanisms appear to be critical to the development and rethinking of mathematical AI models that might represent social agency [127, 660].

A significant challenge in neurobiology is the question of how semantic memory encodes knowledge about people, objects, events, relationships, beliefs and intentions in distributed networks of neural systems [126, 222, 450]. Distributed brain regions of neural activity have been demonstrated to represent semantic information; however, the mechanisms of content and grammatical aspects are weakly understood [48]. Distributional semantic models in which meaning is inferred by probabilistic characteristics such as word co-occurrence have become of significant interest [344]. These models have demonstrated effectiveness in learning relationships between concepts [433] and word meaning [493]. This is achieved through the association strength between concepts using probability distributions within topic models and cosine similarity between semantic vectors [491].

Open problems include how a model of semantic memory that can successfully explain behaviour in one task would be able to explain behaviour in an entirely different task [490] and how methods for social and communicative aspects of language processing can be established [392, 785]. One possible avenue for a solution is the concept of amodal hubs, which contain no semantic features but provide a bridging mechanism to represent and retrieve semantic similarity among concepts even when they may be different in modality-specific attribute [648].

Recent neuroimaging studies have shown evidence for both modality-specific and supramodal representations of semantic memory in regions of the temporal and inferior parietal association cortex [125]. The concepts of amodal hubs and supramodal representations provide support for the concept of prediction according to nonspecific learning. It has been demonstrated that the social world contains important statistical information for language learning and processing [432].

Semantic models of social understanding and prediction may need to incorporate concepts in which meaning acquisition is based on a distributed system, enabling social relationships to form. These systems should enable independent multi-agency meaning acquisition from smaller pieces of information, connecting various concepts that have not been previously observed.

This chapter presents a frontiers-style study that is not a definitive methodological treatment of building models for solving problems. However, according to early results and successful advances in information theory, which underpins much of the observed world, it is a possible starting point.

7.2 A New Form of Model Using Synthetic Language

A significant concept within intelligence is to form models that provide a degree of understanding, and this understanding is experienced through the ability to solve problems. At the heart of problem-solving is the ability to predict future behaviour [46, 462]. This chapter presents a novel approach to predicting outcomes from a narrative source. The work presented here is a conceptual framework that is not complete, but in an on going state of development. It is proposed to stimulate the direction of this challenging area in a way that differs considerably from any known techniques.

Prediction using numerical data has its own challenges, although typically there are certain assumptions made about the data that makes the task of finding a model more achievable. For example, it may be assumed that the data are ergodic and stationary [302]. However, this chapter considers prediction in complex scenarios of swarming information. Thus, how do we characterise and understand what is meant by prediction in this context?

Humans are able to comprehend the flux of the open world in part because of their ability to make narratives. The world is a swarm of cues, both external (spatial, physical, social) and internal (memory, sensory). In response, the human mind similarly swarms, creating unfolding structures that enable us to make decisions when navigating that world. Internally, this comprehension does not manifest as a single logical prescription, but instead is more akin to a quantum field, in which threads of inference and speculation vie for connection and dominance. Out of this flickering array, some ideas or plans may become validated depending on their salience and supported histories. Eventually, in a moment of decision, the field yields a leading pattern, in which some elements are determined to be trusted enough to enable us to commit to an action. This trusted history must be flexible enough to change when new information is received, but only when it is confirmed to be reliable enough for inclusion.

A goal for natural language processing is the automatic comprehension of text and stories [339, 376]. The idea dates to the work of Charniak who proposed a model for answering questions about children's stories [188]. More recently, attention has been given to models that will be capable of understanding social norms, human behaviour and commonsense [680]. Our interest is more than stories and encompasses the idea of predicting behaviours. Unlike the particular considerations of story comprehension based on written language, we consider the basis of language itself within the context of information swarms. Specifically, by information swarms, the idea is that there is some form of rich, complex narrative that is occurring, being conveyed by a form of language and going beyond the usual stationary, limited task models. How should we proceed in this case?

The concept of determining collective understanding from a swarm of smaller pieces of information arises within many fields. Solving complex optimisation problems through the emergent behaviour of colonies is well-known [129] and raises intriguing questions for model development [181]. A classical view of scientific models is to form a unified model that acts like a "central brain" and an overarching algorithm to extract meaning from the observed data. However, the computational capabilities apparent in swarm-like models present the idea that novel, bio-inspired models deserve strong attention [450].

Moreover, there may be many non-stationary subsystems [58] governing behaviour rather than a single stationary model [691]. This concept calls for a rethink of how to approach modelling [120]. Another viewpoint is that instead of having a swarm model solving a single complex task, there may be a swarm of smaller messages all leading to a more considered understanding of the situation [38]. This approach seems more akin to human-level intelligence. The problem is not measuring the data to form a model but recognising that the many smaller nuanced pieces of information are not mere data points but are rich within themselves and lead to a rich, complex model [849].

Humans learn to survive in the world through an understanding of behaviours and their relationships to objects. AI needs to be founded on actions, relationships and behaviours, which builds on treating language as encompassing any dynamical systems behaviour, not only human language. Our key idea is that social actions and behaviour are not just described by language; they are forms of language, not human language but the types of language that describe almost anything from any source—physical movements, sounds, changes in behaviour, the relative movement of agents and even intent, awareness, beliefs or capacity. The way we address this uses a new paradigm for AI called synthetic language and information topology (SLAIT) [87].

This novel information-theoretic approach provides a method of learning social agency and relationships by extending Amari's information geometric model. By using changes in the shape of relative distributions, we show how this forms a mathematical proxy-style of model that captures meaning, and we propose a method to predict narrative outcomes.

The SLAIT information topological approach stems from a reconsideration of Shannon entropy [725, 726]. Entropy-based measures have been applied to a range of tasks, including face recognition [731], human gene mapping [528], emotion detection [465], covert communications [319], language description [371] and drug discovery [300]. However, although entropy is useful for characterising the probabilistic nature of language, a limitation is the dependence on large amounts of data [80, 552, 597]. For real-time language applications, we developed an algorithm for estimating accurate distributions using limited data and that improves estimation speed significantly [85].

A convenient model for determining the distance between distributions is relative entropy [487], defined as

$$H_R(X; Y) = \frac{1}{2} \sum_{i=1}^{M} p(x_i) \log_2 \left(\frac{p(x_i)}{p(y_i)} \right) \quad (7.1)$$

However, another way to consider probabilistic divergences is through information geometry, in which each distribution can be considered to exist as a point in a statistical manifold. Such manifolds are not necessarily flat, as in a usual Euclidean space, but may be curved. Curvature of the Riemannian probability space is measured using the relative entropy and transport distance between distributions. This is made explicit in the Ollivier-Ricci curvature [636]. The Ricci curvature has found application in numerous areas to characterise high-dimensional complex probabilistic data, including Internet topology [620], cancer studies [697] and phylogenetics [830].

However, in terms of recognition of the flow of dialogue in synthetic language, our interest is in forming a global view of the probabilistic nature of language with local features. Can this be extended further to estimate synthetic language structure? We extended the normalised Ollivier-Ricci curvature to an information topology space, and the probabilistic structure of synthetic language sequences was obtained by contrasting distribution sequence trajectories traversing a statistical Riemannian manifold. A more detailed explanation of the concept of information topology is found elsewhere in this volume and in [86].

We applied this new information topology model to autonomous segmentation of unknown synthetic language without any semantic or other linguistic knowledge. In this chapter, we consider the application of this approach to discovering hidden agents within swarms. The idea is to derive a model that (1) separates the behaviours of agents and (2) provides a mechanism to analyse the otherwise hidden synthetic language output using information topology. Here, we demonstrate a method of achieving the first stage of this process.

7.3 A View of Language for Narrative Prediction

7.3.1 Learning to Predict Narratives

The development of AI for complex real-world problems requires a model that incorporates contextual narrative information from multiple sources and estimates their trustworthiness for optimal future outcomes [28, 42, 225, 601].

Machine comprehension and narrative prediction face challenges because of the infinite richness of possible meanings and outcomes in human language. Although supervised learning methods using labelled data are common [53], they rely on the expectation of future narrative trajectory. Although determining topics and key aspects of a narrative is possible, predicting a future narrative that has never been observed remains challenging. Therefore, it is important to consider methods that can provide predictions without prior learning or contextual data.

Current neural network methods may use various methods to store semantic knowledge [346], including word embedding [326, 518], character embedding [466], sub-word embedding [719] or various combinations of embeddings [192]. A common aspect of these methods is that they acquire a large semantic database to form predictions. Suppose, however, that an entirely new situation is encountered for which there are no stored representations.

An example of this idea is where a person recognises horses but has never seen a zebra. If semantic knowledge is provided about the characteristics of a zebra in relation to a horse, then recognition can take place. Some challenges exist with this approach, including problems of bias because of projection domain shift; the presence of universal neighbours in space, known as the "hubness problem"; open set recognition, assumptions regarding training; and testing data classifications [299, 657, 843].

Machine reading comprehension is a complex field that differs significantly from unsupervised classification and zero shot learning methods. It faces challenges because of language being non-stationary in various ways, including statistical time series properties and meaning variation according to symbolic components. Therefore, classification methods such as recurrent networks or hidden Markov models are needed using an appropriate language model [662, 663].

A zero shot learning method for modelling non-stationary systems based on using a deep convolutional neural network and causal representation model to jointly extract invariant causal signal features was proposed in [395, 501]. A method that detects changes in the class dictionary and incorporates new classes on the fly was derived in [26] as a means of handling nonstationarity. Advances in machine comprehension of language aim to develop a probabilistic model capable of predicting narrative outcomes without semantic input or previous observation. The challenge lies in finding a model that incorporates the dynamic connectedness of language, moving beyond static feature classification or state-based models. There is a gap in language modelling representational architecture that could offer a better approach for machine comprehension and narrative prediction.

Unlike human language with its infinite richness [653], synthetic language is based on the idea that the behaviour of many systems may be treated within a simpler framework. In this case, probabilistically framed behavioural events derived from dynamical systems may be viewed as words within a synthetic language. We describe this method below and develop an approach for using it to perform narrative prediction.

7.3.2 Probabilistic Prediction of Language

Language is generally understood to be formed through an ordered set of words, conforming to some grammatical rules [698]. Evidence suggests that consistent rules of grammar develop rapidly even with new languages. Moreover, this is found to occur with languages other than spoken or written forms, for example, sign languages [698]. Hence, unlike narrative prediction methods that rely on semantic knowledge, there are predictable aspects of language because of the various structural levels present, from probabilistic [558] to semantics and syntax [755].

Behavioural and neural evidence indicates that prediction in language comprehension is primarily explained by probabilistic models and at multiple levels and grains of representation [492] and evidenced by observations such as the garden-pathing effect [282]. Contextual predictability in language and its influence on lexico-semantic processing is evidenced in a range of situations [62, 559].

This multirepresentational hierarchical architecture is becoming evident as an important consideration in machine learning comprehension [733, 823]. It suggests a trade-off between the potential accuracy of semantic detail and yet being easily fooled and the lower accuracy, rapid predictability of using a multilevel approach with probabilistic methods but more general reliability [117, 244, 844].

In recent work, to determine sentence-like structure in synthetic languages, we proposed an information topological model for autonomous segmentation without any semantic or other linguistic knowledge and compared it with an unsupervised semantic disambiguation model. By not using semantic information, the probabilistic model demonstrated potential advantages by avoiding adversarial problems, thus supporting the findings of behavioural and neural evidence [492, 519] and indicating the potential for employing multilevel models with probabilistic inference and semantic-level processing for greater accuracy [476], as we noted in [87].

In the following development, we outline a novel approach for narrative prediction that avoids the direct semantic approach but follows a method founded in the concept of information and probability theory. One way of thinking about this model approach is that it provides a mathematical formulation of the concept of "if it seems roughly as if we have seen this before, even though it might be nothing like what we have observed, then we can predict the future".

The combinatorial richness of syntactic human language occurs because of a small set of probabilistic elements [629]. These primitives combine to form words, sentences and phrases, extending to longer narratives that can be understood in terms of probabilistic principles such as Zipfian laws [481, 558, 865, 866].

A range of narrative prediction methods has been developed that incorporates simple probabilistic principles. A method of narrative event prediction based on a state event model with the assumption of narrative coherence was proposed in [183]. A neural network, which was proposed in [333], learns word-based event vector-space representations and their cross-coherence. Learning narrative events by a scaled graph neural network model in the form of a knowledge graph was presented in [529]. Skip-grams (n-grams allowing for skipping symbols) [347] were proposed for a model to acquire script knowledge of event prediction [427]. A semi-supervised model using generative pre-training based on Transformer networks [811] was proposed for language understanding tasks in [664]. A long-short-term memory model generating story endings was proposed in [341] and encodes context clues spanning story context. A model using hidden variables to weight contextual storyline semantics was proposed in [189].

7.3.3 Entropy-Based Analysis of Language

In the context of language processing, statistical models of symbol sequences are of interest and can be defined by the probability $p(\Omega) = p(s_1, \ldots, s_N)$ where Ω is a sequence of N symbols $\{s_i\}$. Shannon formulated the idea of novelty contained in a message according to the information content, rather than the features [725, 726]. The way in which this concept is developed uses the notion that information (and hence a form of meaning) is conveyed by the level of "surprise". This is measured using the concept of entropy, in which the single symbol Shannon entropy is defined as

$$H_1(X) = -\sum_{i=1}^{M} p(x_i) \log_2(p(x_i)) \qquad (7.2)$$

Prediction of the next symbol in a sequence can be computed using the Markov property by considering the previous block of symbols in terms of the conditional probabilities as

$$p(s) = \prod_{i=1\ldots N} p(s_i | s_1 \ldots, s_{i-1}) \qquad (7.3)$$

Hence, provided that the symbol probabilities can be estimated, it is possible to determine the n-gram probabilities and n-gram entropy of language sequences.

Entropy has been applied to a wide range of applications, including statistical keyword detection [371], phylogenetic facial recognition [461] and drug discovery [300]. Entropy rate has been used to measure the complexity of language [45], heartbeat variability [656], visual salience attention [822], animal vocal complexity [459] and patterns of behaviour [813].

To compute n-gram entropy [725, 726] requires a large amount of data, and the reliability is questionable for $N > 5$ [709]. Entropy provides a way of characterising language behaviour and hence is of interest to potentially form predictive dynamic models using short-term probabilistic structures. However, classical plug-in entropy-based algorithms require a large number of samples to adequately model the probabilistic structure [80].

A number of entropy estimation algorithms using limited data have been derived. An estimation method for short symbolic sequences of dynamical logistic map time series models was proposed in [516]. A computationally efficient method for calculating entropy based on a James-Stein-type shrinkage estimator was proposed in [358]. The Nemenman-Shafee-Bialek estimator uses Dirichlet distributions, and a Bayesian prior construction method for power-law probabilities was proposed in [618]. An efficient algorithm based on a coincidence counting for estimating entropy of natural sequences using limited data was proposed in [85].

Ma introduced the concept of coincidence detection models that have been applied to efficiently estimate entropy [552, 597]. In accordance with this earlier work, we extended this approach to derive an entropy estimation method that incorporates linguistic constraints into a new nonequiprobable coincidence counting algorithm that has been shown to be effective for tasks such as entropy rate estimation with limited data [87].

The model we propose for computing entropy with the nonequiprobable symbols is in line with the idea of forming a model such that for any given probabilistic entropy order M, each symbol of a specified rank r can be treated as being equiprobable. For natural sequences, a model can be defined for each symbol, indexed by rank using the Zipf-Mandelbrot-Li law described in [80].

Adopting a probabilistic model according to the symbolic rank, we define

$$F(n; r, M) = 1 - \prod_{h=1}^{n} (1 - P_h(r, M)) \tag{7.4}$$

where $P_h(r, M) = h\gamma'/(r+\beta)^\alpha$ and for iid samples the constants can be computed defined by the procedure in [525]. Hence, defining

$$f(n; r, M) = F(n; r, M) - F(n-1; r, M) \tag{7.5}$$

We have $E_r[n] = J'(n; r, M)$ and

$$D_r(M) = \sum_{n=0}^{M} nf(n; r, M) \tag{7.6}$$

Hence, a per symbolic rank model for estimating M can be obtained by prescribing $J'(n; r, M)$ in (7.6) and then inverting this to determine $\widehat{M_r}(D)$. Having estimated $\widehat{P}_h(r, M)$, the entropy can then be estimated as

$$\widehat{H}_1(r, X) = - \sum_{h=1}^{\widehat{M}} \widehat{P}_h(r, M) \log_2 \left(\widehat{P}_h(r, M) \right) \tag{7.7}$$

which defines the rank r Shannon entropy estimate. The advantage of this model is that a very limited amount of data can be used to estimate entropy. This means that it is possible to apply entropy methods to characterising the short-term behaviour of dynamic sequences.

7.4 Synthetic Language and Information Topology

Dynamic systems can be considered in terms of hierarchical, co-probabilistic output sequences in which the relationships between output states are described by emergent synthetic languages [81]. Although we use the term language, it may or may not necessarily meet the same understanding of language used in the field of human or even animal languages [24, 654].

Our attention is directed towards the usefulness of treating observed dynamical systems outputs as a form of language from a mathematical perspective, rather than seeking to consider the various aspects of such languages, including sociolinguistics, morphologies, lexical-functional syntax, grammars, functional composition or categorical distinctives [150, 215, 272, 273]. This approach has demonstrated effectiveness by the potential capability of diagnosing neurological conditions from listening to conversations; translating biological behaviours into human language equivalents [81]; forming synthetic languages that characterise human, machine and human-machine teams; and predicting sentence boundaries without semantic knowledge [87].

7.4.1 Symbolisation

A first step in synthetic language is to extract the primitives that form words. We derived symbolisation algorithms that extend the concept of scalar and vector quantisation [565]. The symbolic elements were constrained using an expectation-maximisation (EM) method [598] to have a Zipf-Mandelbrot-Li distribution that approximates the behaviour of language elements [85]. A brief overview of the algorithm is given here.

A symbolisation algorithm partitions the input space, for example,

$$x(t) = X_i : U_i \leq u(t) \leq U_{i+1} \quad \forall i, i = 0, \ldots, M \tag{7.8}$$

where $U_0 = \inf(u(t)), U_M = \sup(u(t))$, where $\{U_i\}$ are chosen according to any desired strategy, for example, randomly, such that $U_i < U_{i+1} \ \forall i$, and $U_i \sim \Phi(\mu, \sigma)$. An adaptive symbolisation algorithm based on a constrained EM algorithm in which the learnt probabilistic regions correspond to synthetic primitives of a synthetic language alphabet was proposed in [87]. For a Gaussian mixture model, adaptive support regions can be placed on each mixture to provide a set of probabilities, and we introduce a parametrised model as $\widetilde{p}_M(x|\varphi) = \sum_{k=1}^{K} \alpha_k \widetilde{p}_k(x_i|z_i = k, \varphi_k)$ with usual Gaussian parameters $\varphi_i = [\mu_i, \widehat{\mu}_i, \Sigma_i, \widehat{\Sigma}_i]$, and a probabilistic bound associated with each component is defined by the hyper-volume with edges $\{\boldsymbol{\alpha}\}, \{\boldsymbol{\mu} \pm \widehat{\boldsymbol{\mu}}\}, \{\boldsymbol{\Sigma} \pm \widehat{\boldsymbol{\Sigma}}\}$, resulting in probabilities

$$p(x_k|\varphi_k) = \int_{\mu_j-\widehat{\mu}_jk}^{\mu_j+\widehat{\mu}_j} \cdots \int \widetilde{p}_k(x_k|z_k=1,\varphi_k) \quad j=1,\ldots,d \qquad (7.9)$$

which can be computed using a Genz-Bretz quasi-Monte Carlo integration [315]. A linguistically constrained EM algorithm can be derived to estimate model parameters. We determined the probabilities associated with each cluster, sorted them by probabilities and aligned them to a theoretical probability distribution, where $e_H(x|M,\varphi) = H_z(M) - H_0(x|\varphi)$ and

$$H_0(x|\varphi) = -\sum_{i=1}^{M} p_s(x_i|\varphi_i) \log_2 p_s(x_i|\varphi_i) \qquad (7.10)$$

where $p_s(x_i|\varphi_i)$ are the rank-ordered probabilities, and the linguistically constrained EM update equations follow readily [87]. The entropic error was minimised progressively as a constraint by adapting the mean and variance of each cluster. This continued until the cluster probabilities converged as indicated by the entropy.

7.4.2 Information Topology

A key to machine comprehension is measuring the properties of particular language elements. Our approach was to consider how an information-theoretic model could be used to model synthetic language sequences and provide a form of meaning. Without absolute grounding, it is difficult to assign meaning to sequences in the usual sense. However, we suggest that there are other ways to consider meaning.

One approach to consider meaning without having any teaching inputs is to extend the usual concept of unsupervised learning. In this case, the aim was simply to measure the difference between multiple classes and then label those classes. In the case of synthetic language sequences, this becomes more challenging. We utilised the probabilistic information for the symbolic sequences. Although it is also possible to include other forms of information such as visual features, our particular interest was in deriving meaning from the behavioural sequences.

A convenient starting point to measure the distance between two probability distributions is to use relative entropy. Relative entropy (also known as the Kullback-Leibler divergence [200, 487]) is defined as

$$H_R(X;Y) = \frac{1}{2} \sum_{i=1}^{M} p(x_i) \log_2 \left(\frac{p(x_i)}{p(y_i)} \right) \qquad (7.11)$$

The idea was to measure the probabilistic distance between synthetic language elements and use this to characterise the sequences. Although relative entropy is useful for contrasting pairs of distributions, this raised the question of how to contrast a sequence of distribution pairs. Probabilistic divergences have been considered through the concept of information geometry [44], in which each distribution can be considered to exist as a point in a statistical manifold. The full treatment of information topology was omitted; however, we provide a brief outline of the ideas below.

There are a number of ways in which information topology can be used to infer meaning. The key idea is that it provides a mechanism to connect the probabilistic language elements together such that the topology provides a way of deriving meaning. Put another way, the shape, the holes and the connectedness of the language elements can form emergent patterns that may then be used as the basis for a higher level of recognition. Unlike directly using the data and having a classifier learn to recognise shapes and patterns from raw data, here the classifiers operate on patterns of language representing behaviour.

One starting point to derive a form of meaning from synthetic language is to consider the short-term statistical curvature over the language sequence [86]. The idea is that it is not only the distance between probability distributions that may be useful, but also the curvature and potentially the shape of the manifold can act as a mathematical proxy for changes in information and hence provide a method of understanding meaning. We provide a brief outline of the approach below. The sectional curvature $K(v, w)$ at a point $x \in X$, between distributions on a Riemannian manifold (X, d), (X is a metric measure space equipped with distance d), with two tangent vectors $\{v, w_x\}$ is defined over all directions w where [636]

$$d = \delta \left(1 - \frac{\varepsilon^2}{2} K(v, w) + O(\varepsilon^3 + \varepsilon^2 \delta)\right) \quad (7.12)$$

A simplified formulation is the Ricci curvature that averages $K(v, w)$ over all directions w. The Ricci curvature has found application in numerous areas to characterise high-dimensional complex probabilistic data, including Internet topology [620], cancer studies [697] and phylogenetics [830]. The Ollivier-Ricci curvature is a coarse approximation of the Ricci curvature given by

$$\kappa(x, y) = 1 - \frac{W_1(u_x, u_y)}{d(x, y)} \quad (7.13)$$

where $\{u_x : x \in X\}$ is a family of probability measures on the manifold and $W_1(u_x, u_y)$ is the Wasserstein-1 transportation distance given by

$$W_1(u_x, u_y) = \left(\inf_{\xi \in \Pi(u_x, u_y)} \iint d(x, y)^p du(x, y)\right)^{1/p} \quad (7.14)$$

where $\Pi(u_x, u_y)$ is a set of all couplings between measures u_x and u_y. The transportation distance from u_x to u_y represents the shape of the curve or the effective distance between the spheres u_x and u_y and $d(x, y)$ is the minimum path distance between vertices on a graph between the centres of u_x and u_y. The relative entropy, or Kullback-Leibler divergence, provides a measure of the distance between the elements of ranked-order probability distributions. Hence, as a means of applying a probabilistic curvature model to synthetic language, we introduced a normalised Ollivier-Ricci curvature measure defined as

$$\widetilde{\kappa}(x, y) = 1 - \frac{\widetilde{W}_1(u_x, u_y)}{\widetilde{H}_R(u_x, u_y)} \tag{7.15}$$

where $\widetilde{H}_R(u_x, u_y; \mathbf{x})$ is the normalised relative entropy across $\mathbf{x} = \{x, y\}$. Similarly, the normalised Wasserstein-1 transportation distance $\widetilde{W}_1(u_x, u_y)$ was given by

$$\widetilde{W}_1(u_x, u_y) = \frac{1}{(1 - \xi_L)} \left(\frac{W_1(u_x, u_y)}{\xi_H} - \xi_L \right) \tag{7.16}$$

with scaling factors $\{\xi_L, \xi_H\}$ given by

$$\xi_L = \inf_n \{W_1(n), n \in [1, N_a]\} \tag{7.17}$$

$$\xi_H = \sup_n \{W_1(n), n \in [1, N_a]\} \tag{7.18}$$

A normalised matched relative difference entropy using interpolated spline functions was introduced to address issues of nonequal cardinality of $\{u_x, u_y\}$. Similarly, the matched normalised Wasserstein-1 transportation distance $\widetilde{W}_{1m}(u_{\widehat{v}}, u_{\widehat{z}})$ was given in the same way, using scaling according to (7.16)–(7.17) as before. When applied to synthetic language sequences of symbolic data, this provides the starting point for extracting the information topology.

A next step to obtain meaning from the distribution properties of synthetic language sequences is to shift our view from the Ricci curvature distance to a multidimensional perspective. Hence, the manifold properties form the basis for characterising language sequences within a topological framework. Instead of treating data as the key basis for our models, this approach uses symbolic elements of information. This simple concept is non-trivial to implement; however, the advantage of this is that it provides the basis for reformulating some well-known model approaches.

In the next section, we provide an outline of how we propose to explore narrative prediction using a synthetic language model.

7.5 Information-Theoretic Narrative Prediction

7.5.1 A Narrative Prediction Model

How is it possible to build AI that understands concepts of narrative? It is useful to consider this from both an information-theoretic perspective and models of narrative. We may expect that there is a middle ground in which the concepts of narrative can be implemented and observed within an information-theoretic framework.

As a basis for this approach, we first considered how syntax can be understood. Syntax can be considered as an array of cues that vary in importance or reliability. Some of these cues will emerge as critical to the story's outcome and others will not. There are multiple ways a story weights those cues to direct the reader's "snap" to the desired interpretation. Structures such as zooming, repetition and retroactive reinterpretation all signal to a reader that a piece of information has agency. Numerous connections exist between such narrative states. Humans cross-reference multiple connections between fragments of information to disambiguate [870] or interpret [788]. In a narrative, this increases recall [312] enabling the reader to prioritise the pieces of information in their overall interpretation. Similarly, the proposed approach aims to assign a measurement of trustworthiness to the data fragments to determine their reliability or importance of communication.

In conventional numerical prediction models, importance is according to variance measurements. High variance signals indicate low levels of confidence and vice versa. This enables an optimal prediction to be formed. In our proposed approach, the idea was to incorporate a similar confidence measurement for inputs. We then required a method of estimating some form of proxy for the variance or confidence in a linguistic-narrative sense. Several narrative-based methods may be adopted here to structure those results, though these may be considerably more complex than a simple numerical variance measure. The following are some structures that were explored in terms of weighting cues at the syntax level.

In narratives, repetition occurs when a cue returns at a later time in a new context or attached to new details so that it is "same-but-different" [169]. This allows a reader to follow an entity through a series of states to understand its progression or its dimensionality. In a story, the persistence of that cue weights it as significant, as causal importance can be indicated by connectedness and detail [176]. Repetition can also be an aspect of zooming, in which the same entity is encountered in states of increasing or decreasing scale. In both cases, the emphasis indicates that the cue is important to interpretation [175, 176].

In narrative zooming [171], the granularity of the state changes, so the reader can gather more details about the way in which a situation is deviating from a prior state. This extra information is gained by comparison with the prior state, the frequency of occurrence and the details within those occurrences. More states may also be examined within a given period to increase this zoomed detail further. Zooming likewise weights a cue as causally significant, signalling that it will influence the story's outcome [171].

Retroactive reinterpretation occurs when a new detail connects streams of information in a new way, changing the weighting of their cues and altering interpretation [173]. This is a complex extension of repetition and requires an example to explain. Consider the film *Memento*, in which the protagonist, Lenny, behaves like a detective, assembling clues to find his wife's killer. This is particularly difficult for Lenny because he cannot make new memories because of brain damage, and he does not carry a human-style "history" of events in his mind; instead, he writes notes on polaroid photos as a substitute for recollection. There is a turning point when the audience realises that Lenny's interpretation of his wife's death might not be reliable because a prior scene is replayed with a critical difference. In that scene, Lenny is shown pinching his wife's leg. In the replay, the pinch is replaced with a needle and he injects insulin into her leg instead.

This single point of information enables the audience to superimpose Lenny onto a character in one of Lenny's stories, who accidentally killed his wife by injecting her with too much insulin because of his lack of memory. This new piece of information changes the relationships already built by the story and, with this change, the interpretation of them. Retroactive reinterpretation will be a needed capability in a swarm of data sources whose collective readings may snap to one interpretation and later detect new information that changes the weighting and connection among cues, requiring an understanding. These concepts indicate that predicting narratives by systematically incorporating various linguistic measures into a mathematical model may be possible.

The aforementioned narrative tools operate at the level of syntax. At a higher level of discourse, we propose emulating the global interpretive structure of a narrative, which works at the level of a theme ("the central or dominant idea"), which we refer to as a "concept trajectory". A concept trajectory provides a context for that system, which further constrains the attribution of weight to cues by informing decisions to include or disregard signals about the overall story so far. To emulate this, we propose a method based on measuring the "shape" of the information topology, which defines the behaviours of the narrative using synthetic language.

This method can include history, intentions, personality, identity and other features as described. An observed state consists of some function of the actual state. The elements of these sets are defined using qualitative and quantitative models as coarse or fine-grained as required, equipped with corresponding metrics.

7.5.2 A Linguistic Kalman Filter

One of the most successful methods that has been used for prediction in terms of numerical dynamic models is the Kalman filter [447]. This method provides a recursive algorithm for handling multiple sources of data to provide an accurate prediction. Kalman filters are a well-known model for estimating the state of a system using input data, a system state model, an observed output and a process of updating the model according to errors between the observation and the predicted model. It has had an extraordinary impact in many fields over the past 60 years, from putting people on the moon, guidance and control of aircraft, economics, computer vision and machine learning to environmental applications.

This raised the question of whether it is possible to derive a linguistic Kalman filter-inspired model approach for predicting the outcome of complex agent narratives. Specifically, we considered the synthetic language approach as a basis for deriving linguistic Kalman filter predictive models.

There are a few possible approaches to developing Kalman filter-inspired models for narrative prediction. Normally, the Kalman filter is applied to value-based signal processing models. The aim was to consider the Kalman filter in language-based systems to determine trust, specifically narrative structures. Hence, one approach is to investigate how to use the features of narratives at various levels to achieve this, namely, weighting cues (syntax), concept trajectory (global history) and content (entities and relationships within the detected situation). In this way, developing a linguistic Kalman filter requires a very different approach than the numerical system using the Cartesian coordinate system, and instead a probabilistic approach is required. The challenge of this is that simple mathematical constructs—such as forming a state-based model, estimating variance of observations and determining the differences between states—is considerably more complex than in the traditional approach. Specifically, if we consider states in terms of probabilistic entities, then one way to consider these concepts is from an information-theoretic perspective [80, 81]. This is a highly appealing approach and could be adopted to the SLAIT framework by deriving models that correspond to the equations for a Kalman filter. However, instead of doing this within a signals framework in the traditional sense, it would be done in a linguistic sense using synthetic language [85–87].

However, for synthetic language systems, a method for estimating the linguistic or semantic states is required. Specifically, given the expected difficulty of providing a teacher for such unknown languages, we were interested in unsupervised methods for extracting linguistic states. We subsequently required a model of the difference between these states.

If the states are numeric, then the difference is trivially computed. However, how do we define the difference between linguistic objects? Semantic cognitive models provide a mechanism of storing long-term context that may provide the basis for modelling conceptual differences [122, 345, 377, 378, 429, 666, 856, 857]. However, the problem with this approach is that it requires long-term direct learning experience within the specific domains of interest. The likelihood of extensive sparsity is a particular problem with this approach [217, 311, 560, 599, 737, 760], and although this can be overcome within linguistic n-gram data using smoothing models [191], this presents a challenge for more extensive lexicons of semantic objects.

Another approach is to recognise that probability distributions of words or phrases can in some way act as a proxy for lexicality, which is the basis for unsupervised probabilistic algorithms, used, for example, for parsing and sentence comprehension [208, 520, 683]. With this approach in mind, it indicates that a measure of semantic relative entropy could be introduced as a method of linguistic state difference estimation, independent of which particular semantic aspects are defined as states.

In another approach, a series of Kalman filters is applied to incorporate a stream of information and provide an assessment of the input at various steps to provide an optimal prediction of the narrative trajectory. In the next level, multiple Kalman filters can be used to construct multiple competing models, which are then used to create a "swarm" of new predictions. Using these predicted models, it is possible to assign a weighting to each of the inputs such that a weighting given to each input provides the optimal state estimate.

While this approach was of strong interest to us, in the next section, we consider another approach to using Kalman filter prediction for narratives in which the narrative is transformed to a different domain before the Kalman filter is applied.

7.6 An Information-Theoretic Model of Linguistic Prediction

Observing many inputs and numerous signals presents a challenging problem of interpretation: how to determine the current state of the world. Within an array of information, some cues will be more reliable or important than others in terms of accurately interpreting a situation and those that follow. Appropriate decision-making thus depends on the correct selection and weighting of these cues during interpretation.

To preface the context of our approach, we considered the well-known task of automatic speech recognition [637]. Speech recognition generally requires a process of unsupervised word segmentation in which an observed corpus Υ consists of a number of strings $X = [x_1, \ldots, x_m]$, where it is assumed that each has been generated by a single, statistically consistent language model G. A series of words $W = [w_1, \ldots, w_a]$ given a set of acoustic features $S = [s_1, \ldots, s_b]$ of a speech signal is found by creating a statistical model for the posterior probability of the words given the acoustic features and searching for the word sequence $\widehat{W}$ that maximises this probability [790]. This is known as the maximum a posteriori probability, or the Bayes inference principle, where

$$\widehat{W} = \arg\max_{W} P(W|S) \tag{7.19}$$

$$= \arg\max_{W} \frac{P(W)P(S|W)}{P(S)} \tag{7.20}$$

$$= \arg\max_{W} P(W)P(S|W) \tag{7.21}$$

The component $P(W)$ is referred to as the language model and $P(S|W)$ as the acoustic model. Although numerous variations exist, this typically results in three steps: (1) an acoustic model to estimate $P(S|W)$ using a hidden Markov model or recurrent neural network, for example, (2) a language model represented by the conditional probability of the next word given all the previous ones and (3) a decoder to perform a hypothesis search to determine a word sequence with the maximum a posterior probability by searching through all possible word sequences in the language.

The inherent difficulty of automatic speech recognition is the sparseness of observations in comparison to the vast richness of language [104]. Even with extremely large corpus sizes, many possible combinations of words may never be observed. Therefore, Step 2 is especially difficult computationally and a reason why such large databases are required for training and smoothing techniques are required to overcome the sparsity problem [191].

Prior work has established that a synthetic language model is able to characterise conversation between people as a series of states in which the deviations between them indicate whether that discourse is diverging from its prior pattern [87]. Narrative prediction models applied to social behaviours aim to predict the likely behaviour. Models applied to social discourse aim to predict the likely narrative outcome (the "story ending").

A common neurobiological viewpoint is that for most languages, the semantic properties of words are related to their phonological and orthographic surface properties only through the arbitrary conventions of a particular vocabulary [125]. However, recent evidence has shown that across many languages, more semantically similar word pairs are also more phonologically similar [223]. Information-theoretic analysis describes variation in linguistic form and meaning, in which word meanings represent efficient partitions of semantic space and word length enables efficient communication [322, 593]. These results, combined with the observations discussed earlier regarding amodal hubs and semantic memory consisting of both modality-specific and supramodal representations, provide support to the development of the model proposed here.

Narrative prediction models are challenged with exponential richness of language expressions in contrast to the inevitable sparseness of observations. Hence, the model needs to be capable of generating unseen predictive outcomes that can learn from a limited vocabulary size of synthetic language and then generate novel potential predictions that are close in some sense.

The representational structure required for a model that has supramodal properties raised the question of exactly which properties should be used. We propose that topological information trajectories may form the basis for such a model. Conceptually, the idea is that many storylines are processed to extract the topologies. Our approach used an information topology method using reverse proxy mapping.

In brief, the idea of the proposed approach can be summarised as follows.

1. Learning an information topology language model from symbolisation of sensory narrative inputs from corpora or observational data acquisition (see Fig. 7.1). The model retains the association between symbolic sequences and the topological trajectories.
2. In the recognition stage, the information topology trajectory is determined from an input narrative and matched to the closest learnt contextual information topology trajectory in the language model (see Fig. 7.2).

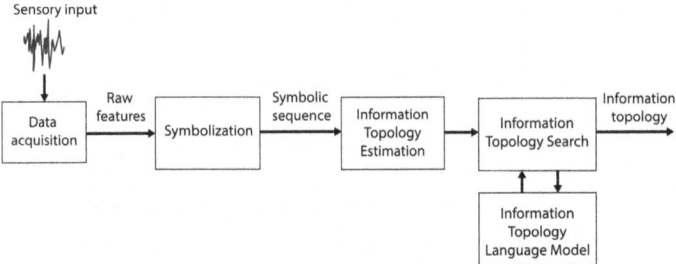

Fig. 7.1 The first stage of the narrative prediction model is to obtain an information topology language model by symbolisation of sensory narrative inputs. The input data could be from corpora or observational data acquisition. The data acquisition front end uses a transducer to convert some input signals to raw features, such as transformed physical data, which are to be symbolised

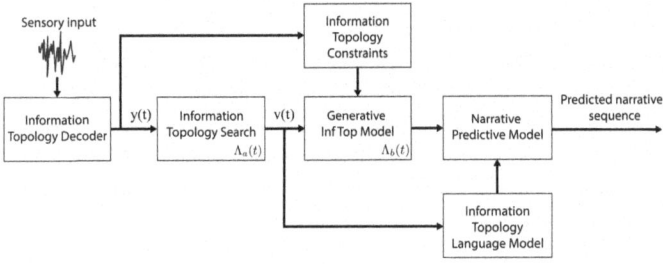

Fig. 7.2 In the proposed model, narrative prediction is achieved by transforming the input narrative to an information space. Information topology is used as a proxy to mathematically capture the characteristics of the narrative. This permits a model to be formed and then future predictions made when the information shape and trajectory are similar. A reverse mapping is used to constrain the information to the original domain to produce the final predicted narrative output

3. It is postulated that information topologies capture the essential characteristics of language sequences and the contextually constrained trajectories of nearby and previously observed information topologies that have known outcomes can be reverse mapped to the novel input sequence to provide a narrative prediction (see Fig. 7.3).
4. In the information prediction stage, the highest ranked subsequent topological trajectories according to the topological language model are reverse-mapped to the input information topology trajectory.
5. In the narrative prediction stage, a set of novel information topologies are generated using the reverse-mapped topology and constraints from the input symbolic sequence. The most probable topology is associated with the symbolic sequence to determine the predicted narrative outcome.

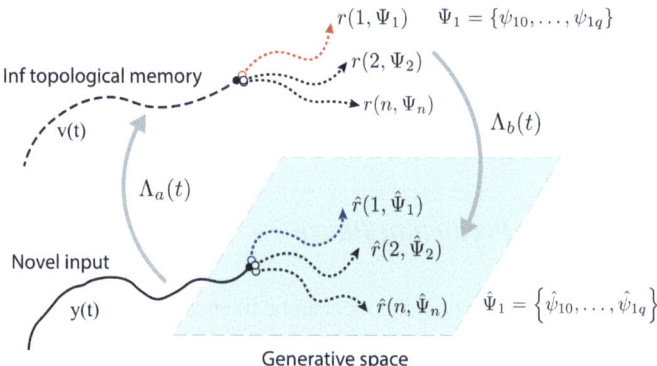

Fig. 7.3 A conceptual view of the information topology search process. The symbolic sequence from the synthetic language input is supplied to the model and transformed into an information topology. At any given point, the current input $y(t; n)$ is matched to the closest information topology $v(t; n)$ where the language model captures the manifold of subsequent outcomes. Each trajectory $v(t; n)$ is associated with a set $\{r(i, \Psi_i)\}, i = 1, \ldots, n_b$, which corresponds to the potential narrative outcomes Ψ_i of each trajectory. The most probable predicted outcomes are mapped back to the current input to provide predicted outcomes $\hat{\Psi}_i$

7.6.1 Information Topology Language Model

In this stage, for any given narrative input $\Omega(n)$, there exists an information topology trajectory $v(t; n)$ that corresponds to a symbolic sequences at particular time point t. Segments of the input narrative are used to form information topology trajectories; however, given that the topology is formed across a symbolic level, word segmentation is not required. In this sense, the meaning is not associated with words but with the information shape of the sequence. For each $v(t; n)$, a distance metric is used to measure the similarity of new trajectories, that is, $v(t; m)$, and then trajectories are clustered. This can be achieved using any number of methods, for example, an entropy-based k-means clustering algorithm, an EM algorithm using the assigned distance metric or other approaches more directly in line with the semantic, phonological or orthographic input. Note that unlike language models based on word segmentation and probabilities as is the case for speech recognition, here the model is formed by determining information topologies and associations with their original symbolic sequences. The clustering of topologies is a considerably easier task than building word segmentation and identification.

7.6.2 Recognition Phase

In this stage, recognition occurs in multiple phases. In the first phase, the input narrative is mapped to an information topology trajectory $y(t; n)$ and matched to the closest contextual information topology trajectory $v(t; n)$ in the language model through a search process $\Lambda_a(t)$. The information topology trajectory matching can be performed in various ways, for example, using a normalised relative difference entropy method [86].

7.6.3 Information Prediction Phase

The information topology language model can be likened to a swarm, in which there is a sea of data, corresponding to a multidimensional array of information topology manifolds. Each dimension corresponds to a narrative dimension, indexed by the symbolic sequence (e.g. using synthetic topics), while each manifold corresponds to the set of known topological trajectories. In an initial sense, each trajectory $v(t; n)$ is associated with a set $\{r(i, \Psi_i)\}, i = 1, \ldots, n_b$, which corresponds to the potential outcomes of each trajectory. However, by treating each of these trajectories across the multidimensional array of manifolds with their associated observed potential outcomes over time as a swarm of informational data, then a Kalman filter prediction model [447] can be readily derived. Hence, the most probable trajectory can be determined. Allowing for there to be an array of potential outcomes according to the multidimensional narratives, a set of possible trajectory set $\{r_o(i, \Psi_i)\}, i = 1, \ldots, n_c$ is produced from this stage. The idea is that the consistent information topology outcome in one dimension might correspond to a potential outcome in another dimension (though it may not be observed in the language model).

7.6.4 Narrative Prediction Phase

The novel aspect of the proposed approach is that the information topologies represent a lower dimensionality space than the raw language space that provides a basis of predictability. Through a process of forward search in the information topology language model, reverse topology mapping, constrained topology and symbolic generation, a narrative prediction can be determined without any direct semantic knowledge.

In this stage, the reverse topology mapping set of subsequent information topologies $\{r_o(i, \Psi_i)\}, i = 1, \ldots, n_c$ is contrasted with the symbolic input sequence to generate a new set of topologies through the process $\Lambda_b(t)$. The most probable information topological output is determined from the generated set by using a further Kalman filter prediction process from the input information topology $y(t; n)$. The narrative prediction is then formed by association with the most probable symbolic sequence.

In the learning stage, a surface of narrative information topologies is learnt from the presented data. The reason a surface is required is that for any given sequence in a narrative, there exists a multitude of possible outcomes. Hence, even though a similar sequence may have been observed previously, there are many potential directions in which a narrative may unfold. Classical machine learning models based on neural network architectures that have a typically fixed structure, and the process of learning is to form statistical models, which reduce the many possible variations to a smaller representative set embedded within the neural network model. In contrast, the approach proposed here is one that captures the many possible variations; however, it is not that every narrative is learnt, but the information topologies. This avoids the problem of memorising every single possible narrative but, rather, builds a form of ontology, in which the information topology is learnt and then indexed by the narrative.

Up to this point, we have developed a proposal for predicting social narratives using synthetic language. These narratives may be derived from almost any dynamic agency—human behaviour, movements, biological entities, complex systems and cyber systems, to name a few. However, suppose the agents are enmeshed within a swarm of noise. This is not just the regular noise that we are accustomed to thinking about in classical signal processing models. Rather, we considered the idea that in the real world, especially in terms of defence and intelligence applications, there may be swarms of agents and behaviours. The task we proposed is to discover hidden coordinated agents seeking to remain hidden. In the following section, we consider this problem. The methods described to date can be applied to this next stage.

7.7 A Model for Detecting Hidden Agents in Swarms

7.7.1 Aims

The aim of this proposed method can be understood as follows. Suppose there is a large swarm of observable agents, of which some are cooperating in some manner but with actions that are hidden or disguised in some manner. By this, we mean that there is a hidden group intent, but this is not immediately or easily observable. This might be a result of intentionally performing various actions that hide the true, underlying movements and are possibly different for each of the hidden agents.

Such a swarm may be physical or virtual, including potential hybrid, multidimensional systems. The "movement" we refer to could be in a physical sense of a traditional swarm, for example, agents within a large crowd, or it may be some form of digital or cyber activities. Another possibility is that this may be viewed in a communications sense. For example, we might receive a multitude of messages over time, some of which overlap in their information. However, the question is, which are cooperating together with a common, shared intent versus the remainder that, although they may share some common local activity, are not cooperating? Thus, despite the medium or specific form of cooperation, there appears to be a common underlying problem.

For example, suppose an agent is intending to move from "a" to "b" as the first step in a coordinated sequence of movements but hides this by moving from "a" to "c" to "e" to "b". Another cooperating agent might have the same intention, but instead move from "a" to "d" to "f" to "b". At the same time, we might observe other unrelated agents who move across these pathways, even going between the same points. Yet, over time, it is assumed that there is a group of coordinated agents who share a common goal that is hidden among another larger swarm that sometimes has similar movements but are unrelated.

7.7.2 Conceptual Design

The idea is that simply observing a swarm of agents might be misleading and not reveal what we are really interested in, which is the smaller coordinated group of agents who have a common task to perform. Hence, how can we discover this underlying true coordinated movement? Moreover, is it possible to discover which of the swarms are coordinating their movements?

Some possible approaches to this might include looking for certain characteristics of the agents that might highlight their motive or mission. However, such approaches are problematic, because visual disguises mean that they do not necessarily work. Other approaches might be to set particular tests or look for some identifying behavioural features. These approaches tend to rely on the agent's behaviour matching the test that we have decided a priori to look for. However, this may not work and suffers from the disadvantage of requiring human expertise and time to implement. Moreover, the agent only has to make some minor changes to evade detection.

7.7.3 Separating Behavioural Streams

Independent component analysis is a mathematical approach that extends the concept of correlation analysis [83, 84, 408]. Hidden coordinated agents have missions that are correlated to each other but less correlated over time with other agents. In practice, the actual movements or behaviours of agents are complex, with significant common aspects and small deviations between hidden cooperating agents and the rest of the group.

From a mathematical perspective, it may not be sufficient to use an algorithm that takes long-term average behaviour. Instead, a better approach is to use an information-theoretic viewpoint. This approach allows for large amounts of common behaviour to be discounted because of its lack of useful information while allowing for very small amounts of data to be weighted more heavily.

This approach is closer to human experience than a simplistic averaging, correlation-based approach. Human behaviour often discounts useless data based on the idea that it is not telling us anything new. Instead, learning and focus of attention rely on a model of relative difference, in which information is conveying something novel that we did not know previously.

Although a range of algorithms has been proposed to enable mixed signals to be separated according to information-theoretic principles [409], there are some key areas in which this approach could be used, but the theory has not been developed. Some early work has been conducted to derive feature vectors from linguistic data using independent component analysis [142, 382]. We propose a method of extending independent component analysis to linguistic streams to separate behavioural language.

In independent component analysis, a set of signals is assumed to exist with the provision of statistical independence between the original sources. The signals are then mixed together in some way that effectively masks the original signals. The aim of independent component analysis is to obtain an estimate of the unobserved true sources. Here, behaviours are transformed into a synthetic language in a form suitable for independent component analysis. The goal is to discover an underlying hidden mission by a pseudo agent, demixing observed behaviours to remove extraneous activities and uncover the true agenda. However, this approach is only suitable for numerical value-based systems. Here we introduce a linguistic form of independent component analysis to separate language streams.

7.8 Independent Information Analysis

7.8.1 *Symbolisation of Input Space*

Symbolisation extends the concept of scalar and vector quantisation [317, 335] to map a sequence of continuous or discrete values to a symbolic sequence. The goal of symbolisation can be differentiated from quantisation in that the properties of determining language primitives may be very different from simply efficient data compression or even fidelity of reconstruction. For example, these properties may include metrics of robustness, intelligibility, identifiability and learnability. In other words, the properties, goals and functional aspects of language elements are considerably more complex in nature than those used in vector quantisation methods.

Consider a hidden behaviour that can be described initially in terms of a signal, as shown in Fig. 7.4. This might represent a physical behaviour or potentially a goal-orientated social behaviour in some multidimensional space. Here, it is not necessarily the precise signal value that represents the behaviour but the approximate position or dynamic movement encoded in some manner such that the usual independence assumptions are not in line with the actual signals but the system behaviours. Hence, the way we approached the investigation was to derive a synthetic language from this behaviour using a symbolisation algorithm as indicated above. The symbolisation process was applied to all measured input data sources.

7.8.2 Symbolic Quantisation

The model aims to identify hidden behavioural groups by separating mixed behavioural streams and processing synthetic language streams through independent component analysis. Quantisation can be performed according to a specific performance metric to process these streams. As a means of providing a simple demonstration, we quantise the symbols according to their ranked probabilities. The resulting quantised behavioural stream for this case is shown in Fig. 7.5 for a synthetic language of 10 symbols. Note that there is no particular requirement that the appearance of the curves in Figs. 7.4 and 7.5 should be similar because this depends on the characteristics of the symbolisation and quantisation algorithms.

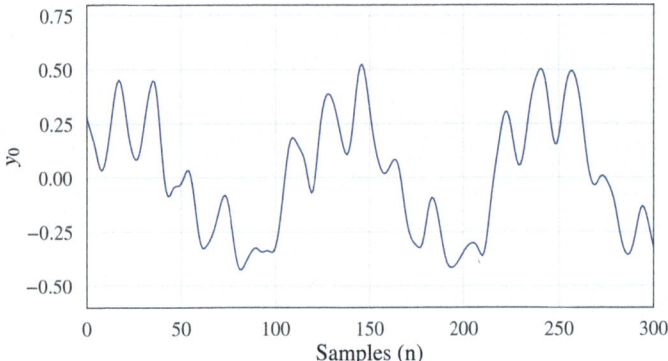

Fig. 7.4 A representation of some unobserved behavioural properties for a coordinated group of agents within a physical or virtual swarm

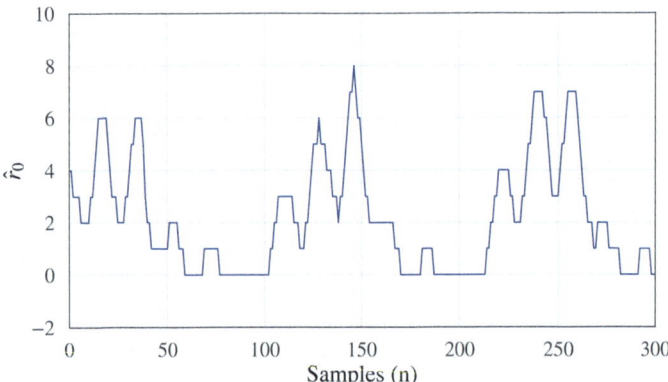

Fig. 7.5 The symbolic ranked probabilities of the behaviour of some coordinated group of agents. In this case, the synthetic language has 10 symbols, hence the quantisation of the behaviour

7.8.3 Quantised Independent Information Analysis

The aim in this stage was to obtain streams of hidden language sequences to be obtained from an observed continuous behavioural space. The resulting output sequences could then be processed in the subsequent stage using synthetic language information topology models. The proposed independent component analysis model separates streams of synthetic language behavioural sequences from numerical data streams, focusing on information streams instead of data. Key approaches include deriving symbolisation and quantisation algorithms, considering factors such as maximum intelligibility, quantisation factors, adaptive algorithms, algorithmic constraints, quantisation error properties, parametrisation, properties and limitations, experimental results and performance analysis.

Continuing the example above, we further assumed that there was a range of behaviours within this swarm that were performed by other agents unrelated to the ones of particular interest. The aim was to separate the behaviour of the coordinated agents of interest. We can consider a representation of this in Fig. 7.6.

The idea is that a hidden agent will not expose its motives or desired behavioural aims overtly but will mix behaviours so that none are immediately obvious. The observed behaviours of coordinated agents in a crowd may possess a common goal, but the actual agent behaviour is affected by others immediately nearby.

A mixture of behaviours can be considered in a number of ways. For example, it is possible that the behaviours are a probabilistic function of the various agent behaviours. Agent "a" might wish to go to "'b", but is influenced to go to "h" with some probability. Similarly, another related agent has a similar task but is influenced to first go to "g" with a different probability. A possible representation of these mixed behaviours is shown in Fig. 7.7. It can be noted that the observed behaviours are very confusing, with no easily identified group of agents.

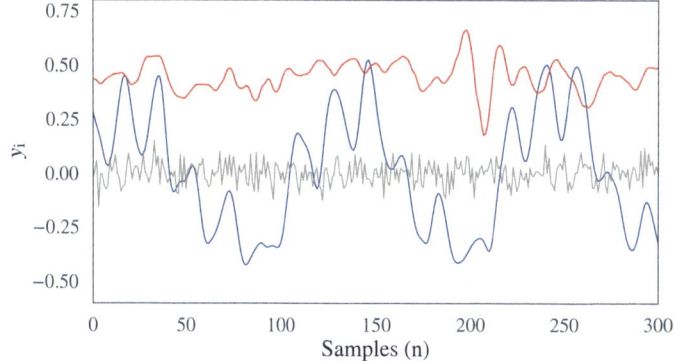

Fig. 7.6 A representation of some coordinated behavioural signals (blue) and other unrelated behaviours, all of which are not directly observable, but occur within a swarm of agents

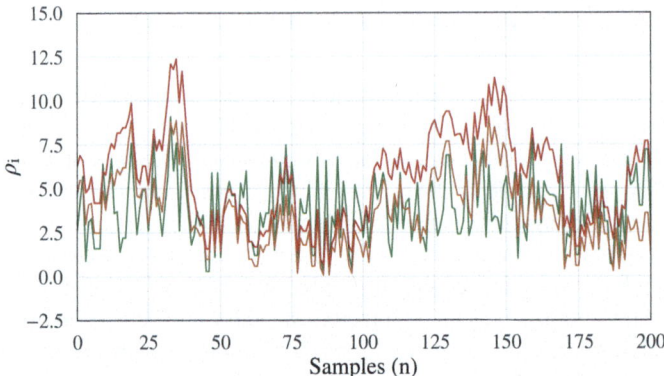

Fig. 7.7 The observed swarm behaviour using a simple synthetic language representation. The problem is that there is an underlying coordinated behaviour that we would like to discover, yet it appears that all of the agents are behaving in an approximately similar manner as expected within a swarm. Clearly, this is a challenging problem because it appears that all agents are performing similar tasks, and the hidden group is not readily observable but occurring within a swarm of agents

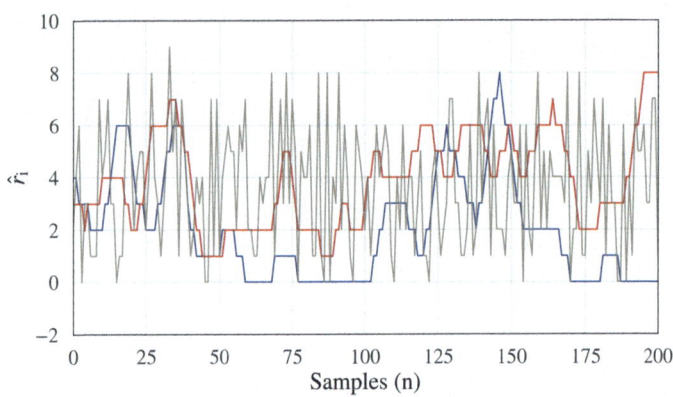

Fig. 7.8 A simple synthetic language representation of some coordinated behaviour (blue) and other unrelated behaviours, all of which are not directly observable but occur within a swarm of agents

The quantised symbolic representation of the mixed behaviours, following the same approach as before, is shown in Fig. 7.8.

The results of applying quantised independent information analysis to the behavioural synthetic language symbolic sequence representations are shown in Fig. 7.9. It can be readily observed that the hidden behaviour is extracted from the swarm behaviour with reasonable accuracy (see Fig. 7.10). By comparing the results with Fig. 7.5, it can be readily observed that the extracted behaviour shows the hidden behaviour of the coordinated agents.

7 Identifying and Predicting Hidden Coordinated Behaviour Using Synthetic...   123

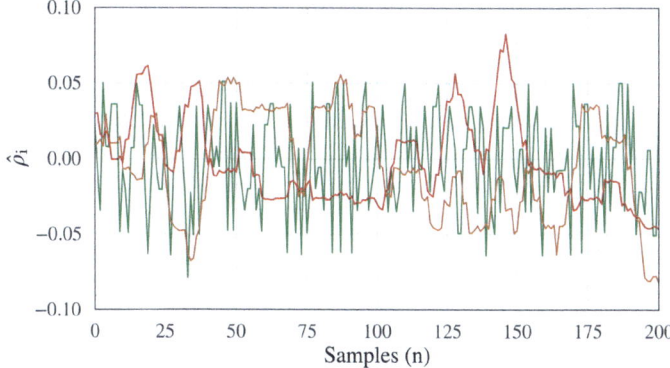

Fig. 7.9 Applying the synthetic language ICA (SLICA) approach enables the hidden agent behaviour to be extracted from the observed swarm behaviour

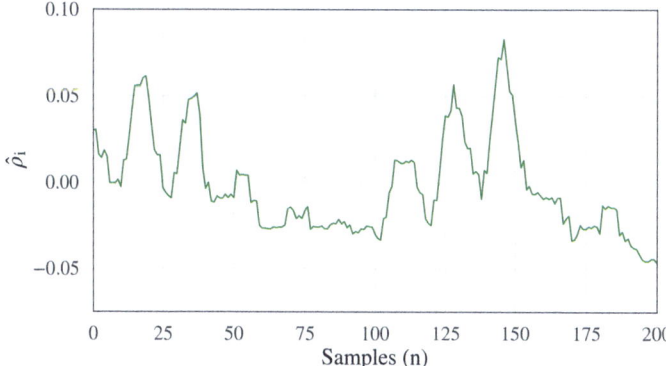

Fig. 7.10 The proposed synthetic language ICA (SLICA) approach is demonstrated to extract the hidden agent behaviour from within the observed global swarm behaviour (compare with Fig. 7.5)

In contrast to this, we can see other behaviours that are of no particular interest. Hence, it is a simple matter to determine the agents that are coordinated within the swarm.

This approach of determining coordinated hidden agent behaviour is shown to be effective in this example. Our purpose was to demonstrate the idea, and more complex strategies may be considered both in terms of the ways in which agents' behaviour is modelled and in the algorithm used to extract the behaviours.

7.9 Conclusion

The problem of narrative prediction by AI models is a particularly challenging problem. We suggest that this will be a crucial element in future AI systems. We propose that current models of representations are inadequate to deal with the complexities encountered. Hence, the concept of synthetic language and information topology provides a useful basis for addressing social agency.

Instead of treating all data points as low-level swarm elements of equal merit, this work suggests that there are ways to develop models that learn from all scenarios, and by treating them as narratives, it is possible to use the learnt properties in the future. It is exactly these characteristics that are required for future intelligence models. Using recent neurobiological evidence, a model for narrative prediction was proposed for behavioural analysis and prediction. We proposed a highly novel approach for narrative prediction model based on transforming narratives into a representation in the information domain utilising the concept of synthetic language. The idea was to predict in one information-theoretic domain and then transform the prediction back to the narrative domain. Constraints were employed to realign the predictions.

The solution utilised methods from the field of engineering. Specifically, the concept of transformation was used, and the problem of trajectory estimation was based on a well-proven methodology known as the Kalman filter. Hence, framing the problem of narrative prediction in an information space and applying a well-understood framework of estimation enable an entirely new approach to this task.

Hidden coordinated agents operating within a swarm of activity present a significant problem for areas of security and intelligence. In this chapter, we explored a novel approach for discovering these hidden agents and their behaviours. Methods that require extensive background information are disadvantaged by not only the need to know what to look for in advance but also the large amount of human expertise needed to build models. However, the problem remains that despite large amounts of data, it is incredibly difficult to detect new agents and new behaviours.

Currently, apart from using identifiable connections, such as family or known relationships, there are few, if any, algorithms or techniques to autonomously discover cooperating agents or influences based purely on observing behaviours. Therefore, the proposed model is intended to provide a method of discovering these hidden agents and automatically revealing their connectedness, including related agents or influences that have never revealed themselves before. A key benefit of the methods being proposed is that they require little data or prior training, as with typical machine learning, and can give a response in real time.

Although the proposed model described here demonstrates the potential to discover hidden behaviours using independent information analysis, the next stage is to analyse these streams. The way we achieve this is to analyse the recovered synthetic language behavioural streams using information topology, which is beyond the scope of this chapter. However, it is evident that the work described here indicates the potential for discovery of hidden agent behaviour using synthetic language combined with an independent component analysis approach. There remains various questions to address and so we propose this as a starting point.

7.10 Further Work

Our next steps will be to build on this initial foundation and formalise the particular algorithms required for the proposed framework. We have developed symbolisation, tokenisation and phrase boundary detection algorithms. Our future work will focus on developing various models to perform the tasks indicated in this chapter. This includes models for autonomously determining the alphabet size of synthetic languages and deriving models of linguistic state and state variation using semantic relative entropy for linguistic state difference estimation as a basis for a linguistic Kalman filter. The next stage of development will be to derive specific algorithms for determining appropriate information topologies to capture characteristic representations for narrative prediction in line with the concept of amodal hubs, an algorithm to match to the closest learnt contextual information topology trajectory in the language model (recognition stage), and a model to reverse-map information topology sequences to generate synthetic language narrative predictions (see Fig. 7.3). A synthetic language independent component analysis model will then be considered as a basis for identifying the cooperative swarms.

Acknowledgments The authors gratefully acknowledge funding support from the University of Queensland and the Australian Government through the Defence Cooperative Research Centre for Trusted Autonomous Systems. The TAS-DCRC receives funding support from the Queensland Government.

Open Access This chapter is licensed under the terms of the Creative Commons Attribution 4.0 International License (http://creativecommons.org/licenses/by/4.0/), which permits use, sharing, adaptation, distribution and reproduction in any medium or format, as long as you give appropriate credit to the original author(s) and the source, provide a link to the Creative Commons license and indicate if changes were made.

The images or other third party material in this chapter are included in the chapter's Creative Commons license, unless indicated otherwise in a credit line to the material. If material is not included in the chapter's Creative Commons license and your intended use is not permitted by statutory regulation or exceeds the permitted use, you will need to obtain permission directly from the copyright holder.

Chapter 8
Intelligent Swarming Narratives

Situating 'Noise' Within Interacting Information Fields

Beth Cardier, Andrew D. Back, Jessica Korte, and Pauline Pounds

Swarm intelligence requires an ability to process information at the level of systems, rather than as individual data points alone. This chapter considers an approach to processing swarms of information in which 'noise' is identified by situating it within system-level structures that are drawn from narrative. Narrative is an information-structuring process that characterises anomalies by forming and reforming system-level structures around it. It uses a range of perspectives and ontologies as stepping stones, linking them directly and indirectly to identify the behaviour and influence of agents that are unexpected. We considered how this identification process operates in two examples: the fluid flows around a whisker drone and a convoy of troops suddenly attacked by a farmer. In the second example, we explore how Back's synthetic language technique could be amplified using narrative-like structures to produce emergent synthetic narratives. The goal is to conceptualise an underlying structure for the organisation of information by which swarm intelligence could construct a common operating picture. A desired outcome is for a swarm of agents to be more responsive, be able to react appropriately to on-

---

B. Cardier (✉)
Institute for Integrated and Intelligent Systems, School of Information Communication and Technology, Griffith University, Brisbane, QLD, Australia
e-mail: b.cardier@griffith.edu.au

A. D. Back
School of Electrical Engineering and Computer Science, University of Queensland, St Lucia, QLD, Australia
e-mail: a.back@uq.edu.au

J. Korte
School of Computer Science, Queensland University of Technology, Brisbane, QLD, Australia
e-mail: jessica.korte@qut.edu.au

P. Pounds
Robotics Design Lab, EECS, University of Queensland, St Lucia, QLD, Australia
e-mail: pauline.pounds@uq.edu.au

the-fly situations in the real world and be more collectively intelligent. We consider how this might be made possible by combining approaches that connect information at various levels of a system, by a meta process of connecting the heterogeneous research ontologies in our group.

8.1 Introduction

Swarm intelligence requires calculations by and about a changing multitude. That swarm must also assess the environment through which it moves, interpreting a surrounding "swarm" of incoming information about the open world. Harnessing combined data in this manner requires a critical ability: interpreting systems of signals by systems of agents, rather than individual signals by individual agents, that is: swarms of swarms. A multitude of agents has the power to surround and capture a situation from many perspectives and formats, and even to extend an initial situation model. However, multiple information formats can also be difficult to integrate, generating noise. How can noise be managed so that intelligence can gather across numerous systems?

This chapter aims to reposition some key assumptions about the role of noise when modelling information in swarms. Noise has been defined as the "irrelevant or meaningless data or output occurring along with desired information" [579]. We propose that noise is not actually meaningless or irrelevant, a perspective based on experiences in our four projects at TAS-DCRC. Many discussions are consolidated here. Pounds' drone navigation must distinguish between motion disturbances that indicate an unexpected change of physical situation and those produced by other kinds of interference. Korte's automatic interpretation of human sign language depends on a machine's ability to assess a blur of natural human movements and identify a spine of meaning within it. Back's approach to machine learning reads an entire field of information and characterises perturbations among successive states, eliminating traditional notions of "noise" altogether. Cardier analyses narrative to understand how "noisy" signals can be aligned across heterogeneous information fields, to indicate an expended situation model. In all these projects, noise has the potential to indicate information about real situated environments, even though it is currently out of reach for most methods.

How can that off-grid signal be harnessed? Together, we examined a possible beneficial role of "noise" in situated intelligence. This chapter does not propose an algorithmic solution; instead, it aims to expand the conceptual space in which to design intelligence for the next generation (or beyond) of uncrewed aerial vehicles (UAVs) or uncrewed autonomous agents (UAAs). A more specific goal was to understand how "noise" could be situated using relationships among multiple information fields, so it can accurately inform a common operating picture. This notion was primarily explored in relation to narrative structures, in the example of signals hidden within the data stream of a whisker drone, and also in the more challenging example of identifying a new human threat in an open world.

Noise is not usually considered to be a positive contributor to swarm intelligence (or any intelligence). Current approaches to swarm intelligence [97, 851] tend to focus on scalable behaviours inspired by animals and natural systems, such as flocking, bee colony optimisation, and flower pollination. The algorithms driving these models replicate those repeatable behaviours and adjust according to obstacles, always driving towards a particular set of operations. The notion of intelligence in artificial intelligence takes a different stance, in most cases aiming to replicate or enhance particular aspects of human cognition and reasoning. Our focus concerned intelligence under open world conditions in relation to understanding what kind of situation is unfolding and what kind of agent is driving it. Adaptive, combinatorial, aggregative, and unexpected properties are key. This chapter examines the role of noise to better understand how artificial intelligence in a swarm could support the registration of these properties into an expanded situation model.

8.2 The Problem with Representing Systems

Noise provides a signal from an otherwise dead zone. No single representational system can account for everything, and so there is always a limit, past which phenomena do not register. Beyond this threshold, activity might be invisible, or perhaps appear as partial or blotchy, noise. This beyond-edge quality can potentially supply additional information about an open-world situation in a network of multiple agents.

For the past 80 years (and arguably longer), scientific models have operated by removing everything that is seemingly irrelevant to the key parameters of an experiment [296]. The aim has been to formulate an "ideal" model that follows the Newtonian maxim of obtaining the pure movement, uncluttered by our own observational distortions. Among other things, this has produced a focus on discreteness and reductionism. "Noise" is omitted from the resulting data because it is believed to be irrelevant or meaningless [579]. However, as context and real-world effects become increasingly important to fields such as autonomous systems, and these systems do not yet have a reliable means of reacting to unexpected situations, it becomes clear there is something missing from reductionist approaches.

For a human who is negotiating a path through the open world, small variations in routine can have significance. In fact, "noise" can sometimes be the very thing we use to make rapid life-saving assessments. Our example in Sect. 8.5, regarding a farmer striking a soldier with a digging tool, explores the importance of incorporating seemingly minor fluctuations into the risk assessment of a situation. When measuring complex systems, we are thus in a bind. The complexity of the world compels us to discard extraneous information so we can quantitatively measure and assess it. However, it is also these very pieces of information that might be needed to make critical decisions.

How does "noise" inform intelligent decision-making? A starting point is to set aside the connotations of uselessness that are typically associated with "noise".

We see information differently; it is not just data. Instead, we framed it within information theory and entropy, in which information characterises the level of surprise [276, 724]. This "surprise" can be understood further by considering it in terms of narrative, which has been described as being a means of modelling events that are a surprise [155, 631, 786]. Ochs and Capps observed that narrative reconciles "what was expected with what was experienced" [631], drawing attention to the relationship between an anomaly and an initial frame of reference against which something unexpected is registered. At the foundation of this activity is causal reasoning, in which cause and effect are attributed to particular agents by registering divergence against a frame of reference [265, 606]. Not only is a story able to characterise an anomaly, but the tale then tracks how its interactions unfold after that, as they "deviate from normalcy" [332]. To achieve this, narrative builds a new frame of reference, the process of which is referenced here.

Let us relate this to swarms by bringing some of those properties into an information theoretic frame. The initial state of "expectedness" can be understood in terms of an information field. An information field is "defined by the common ontology in the form of a normative structure where the human and artificial agents, involved in the decision process, abide by the norms" [36]. An information field can also be thought of as the overarching rationale, affordances, and syntax of a communication system, in which the "norms" are the common understanding about how those carry meaning and the shared use of them. A common ontology is the structure by which information can be expressed, and also by which it is limited.

An autonomous swarming entity has a specific set of information fields that it is designed to process. When the critical action falls outside that scope, or interaction occurs in ways that were not anticipated, noise emerges rather than information. In a whisker drone, the fields being monitored include states that can be physical, representational, or temporal. This might also be expressed as the form of information received or transmitted, whether from sensor measurements of the surrounding world, received commands or communications, progresses through the mission tree, or the internal states of the aircraft.

A narrative likewise relies on fields of information, and these serve a similar purpose of providing "norms" against which developments can be interpreted. Also similar to the ontologies in information science, those referenced by a narrative can take the form of explicit information, such as semantics, syntax, and linguistics. They can also emerge from systems of structures, such as an ontological perspective ("character") through which the story is told over an entire tale, or the ongoing history of its storyworld. These are both explicit and implicit, because they comprise of text and a reader's inferences about how these sections can be bridged [417]. Relevant ontologies also include information that is external to the story, such as general cultural knowledge and well-known stories in similar genres, as well as genre itself [298]. Together, these multiple ontologies identify and track agents through narrative developments [170].

The first point of divergence between knowledge structures in narrative and default information science concerns how "surprise" is understood. Current knowledge representation systems would model the "surprise" entity as an explicit part of the system (and thus, at another level, it is not really a surprise). This kind of surprise is within the world of the situation model. Narrative surprise is closer to the disruption described by Kuhn [486] and Thagard [782] in relation to scientific revolutions, in which information emerges that is partly incommensurate with the existing reference framework. This incompatibility makes accommodating it difficult initially.

To incorporate an anomaly, narrative uses multiple structures to adjust the initial frame of reference. This transformation is similar to that described by Kuhn and Thagard, and even Boden [131], as a transformation of conceptual structure. A story provokes this process throughout the tale, to continually adjust its frame of reference until its conclusion, when identification of both the anomaly and the new situation model is complete. This is one of the ways a tale facilitates the interpretation of unexpected events.

How is an anomaly actually identified if the initial reference framework cannot accommodate it? The answer lies in collaboration among a "swarm" of multiple ontologies.

8.3 Characterising Unexpected Agency Using a Swarm

A narrative depends on some sort of surprise, a divergence from expectation. However, if the information is unexpected, how can it be accommodated into the story?

In a narrative, an anomaly is accounted for via the use of multiple reference frameworks, which each define different aspects of it. For example, the phrase "once upon a time" activates the fairytale genre as a reference, because an identical and well-known phrase occurs in that ontology [170]. However, if this phrase is followed by the words "there was a science fiction movie alien", the affordances of a fairytale ontology are not enough to interpret them. Instead, these two ontologies would need to be related to each other for accurate interpretation of the entire phrase to occur (A similar example has been analysed in [172]).

Narrative performs these connections. In the above example, the story would indicate how these two very different networks, "fairytale" and "science fiction movie alien", relate to each other as the situation unfolds. Those connections are built across many types of information fields. For instance, linguistic devices direct how explicit details from each ontology relate. The ontologies provide affordances that can overlap or be adjusted towards each other. Strategic gaps, sequencing, and proximity indicate how nonexplicit structures relate. At the same time, the action in the unfolding story indicates how these new details inform the evolving situation model. Each process enhances and disambiguates the others [869]. As a whole, this activity creates a meta-ontology that is derived from parts of other ontologies. It is the primary point of reference for all prior information, as well as the next fragment of incoming information in a story. An illustration can be seen in Fig. 8.1. In this

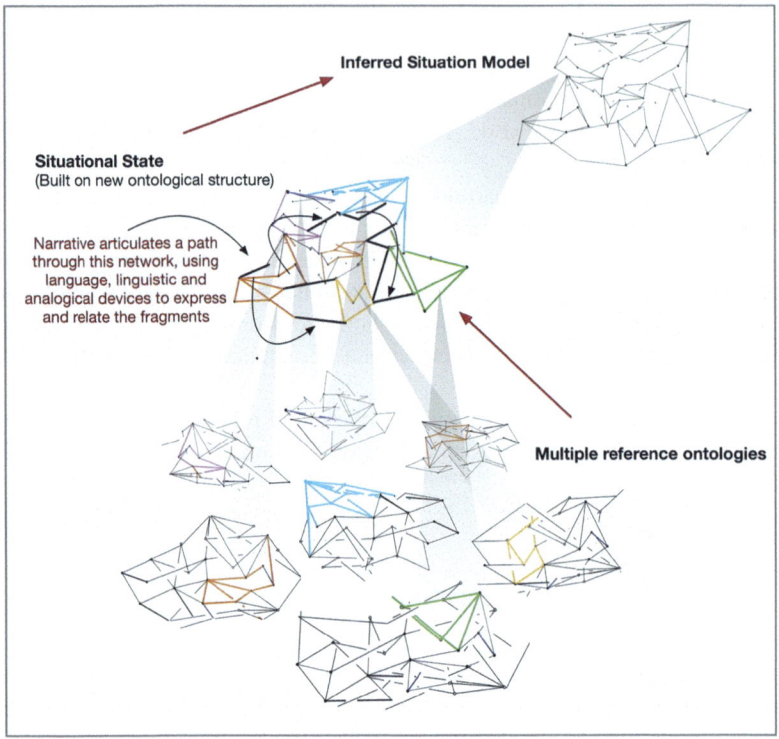

Fig. 8.1 Narrative characterises an unexpected piece of information using a "swarm" of reference frameworks

manner, a narrative derives its own reference framework for the interpretation of a continuous stream of non-normal conditions.

At the bottom of Fig. 8.1 are the many frames of reference that can bear on interpretation, each with its own ontological structure and "norms". Matches for different parts of the anomaly can be found in aspects of numerous ontologies. The relevant fragments of those ontologies are connected with each other according to the relationship made possible by linguistics and language devices in the story. At the same time, these are calibrated against the action unfolding in the recounted situation. All these tiers operate together to generate a situation model. New "norms" emerge from this combined entity.

It must seem strange to a traditional information scientist that multiple and differing ontologies can simultaneously provide a frame of reference for a single piece of information. This is possible because only a small and relevant portion of each ontology is used in the collaborative framework built by the "swarm". We refer to this limited area as an ontological fragment.

A fragment is the area of a reference framework that is directly relevant to interpreting information. Devlin described how information activates a corresponding frame of reference via its salience [238], 'salience' being the elements and their ontological fragments that influence interpretive effect at that specific state. A fragment is the part of an ontology that is directly salient to the incoming text at hand. A narrative makes a kind of "copy" of that fragment and then uses it in a new network, bringing with it some limited meaning from the system-level framework as well. The implicit ontological information beyond the area relevant to the salient fragment cannot be used. For example, after the phrase "...there was a science fiction movie alien", the reader would not be able to guess which alien or movie without further information. They also will not know how much influence each ontology will have over the unfolding situation until the story tells them. The reader thus waits for guidance from the story to understand how the rest of the meta-ontology is structured.

There are parallels in information science, in terms of leveraging multiple information fields. For example, an elemental datum is a number or value that represents some measured attribute of a system (e.g. temperature of a component or number of items in stock). However, data are rarely atomic and uncorrelated with any others in isolation. The relationships between data across time (a signal) and across a space (e.g. vector fields, graphs) are what allow for system models' predictive power, such as extrapolation of a curve. These relationships may be extended to more complex systems, consisting of a plurality of measurements, in which identifying the underlying causal factors may be challenging, for example, particle velocities in a hurricane, opinions posted to cyberspace, or transmission of a virus through a population. To identify the significance and role of an element of noise, we must view it within the context of the multiple systems it inhabits. Given all the information fields available in a drone swarm, we can imagine correlating an unexpected signal through these various fields to physically situate it or characterise the kind of influence it has, by noting the affected domains.

To understand better how these fields might be correlated, we return to the question of identifying noise, but in a narrative context. Barthes and Duisit stated that narrative "does not acknowledge the existence of noise (in the informational sense of the word). It is a pure system: there are no wasted units, and there can never be any, however long, loose and tenuous the threads that link them to one of the levels of the story" [98]. The reason noise does not appear in a story is not because it cannot appear. It is because once it appears, it is incorporated into the situation model through storytelling structures. "Noise" is a piece of information that is not connected to a story *yet*. When unexpected information appears, it comes with an expectation that the anomalous piece of information will be made relevant using the integrative structures of narrative, perhaps at great effort.

A story characterises unexpected conditions [170] by linking information across systems. Usually, many systems are linked simultaneously. Here are some of the system-level structures narrative uses to achieve this.

8.4 System-Level Narrative Structures: Context

In scientific reductionism, there is no noise because it has been left out to establish a clean frame of reference. In narrative, there is no noise because storytelling structures eventually connect it to the rest of the network. An overview of the means by which narrative forms these connections is explored, with swarms in mind.

Narrative makes connections within and across levels of abstraction. Defining exactly what those levels are is a matter of debate among narrative theorists. The levels have been characterised in a few ways, but the underlying principle is a spectrum from (1) syntactic representations (the actual text and its meaning) to (2) systemic relations (the meaning produced from numerous structures as a whole). Barthes refered to increasingly abstract levels that depend on each other—"phonetic, phonological, grammatical, contextual" [98]. Structuralists refer to the abstract level as fabula, which concerns thematic elements, versus syuzhet, which refers to the chronological structure of events [218]. Todorov distinguished between the story, which concerns a "syntax" of characters and discourse, which embodies "tenses, aspects, and modes" [98]. All these models of narrative are valid and can be found in its ecosystem, depending on the question being posed. In this chapter, distinguishing between syntactic and systemic relations is sufficient.

Meaning operates differently in the levels of syntax and system. A fragment of syntax has an explicit interpretation, but it also derives meaning from its role in the whole ontology. For example, the word "wolf" refers to a woodland canine. In the story "Red Riding Hood as a Dictator Would Tell It", the word "wolf" explicitly refers to a woodland canine, but in the context of the story, that canine is also an anthropomorphic fairytale character, which is actually a foil for a dictator, who is telling the story as a form of propaganda [172]. In the second context, the word "wolf" refers to a very different creature. Anticipating the future actions of a "wolf" (syntax) will depend on knowing which context it is operating in (system). The system-level identity adds additional information to the syntactic meaning. The system-level also establishes the "norms".

An important feature thus emerges from the establishment of a fragment within an ontology: the edge of its knowledge structure, its limit. This is also a limit of its norms. (Interestingly, the ontology could also be seen as a "fragment", albeit larger, in the sense that it is a limited reflection of the larger open world.) When a fragment is lifted out of its original ontology, it loses the "norms" that went with it. It loses most of the details of that system-level structure that contributed to its original meaning. It does carry some of that original meaning into its new situation, but this must interact with and be subsumed to the emerging meta-ontology.

This feature enables a fragment of syntax to be resituated in a new ontological network by a narrative. It is a form of nesting (see Fig. 8.2), in which syntactic meaning is preserved while the original norms bend or blend towards the norms of a new situation, using devices such as adaptation, analogy [278], and inference [417]. These devices operate together to create a best fit of multiple states, such as

8 Intelligent Swarming Narratives

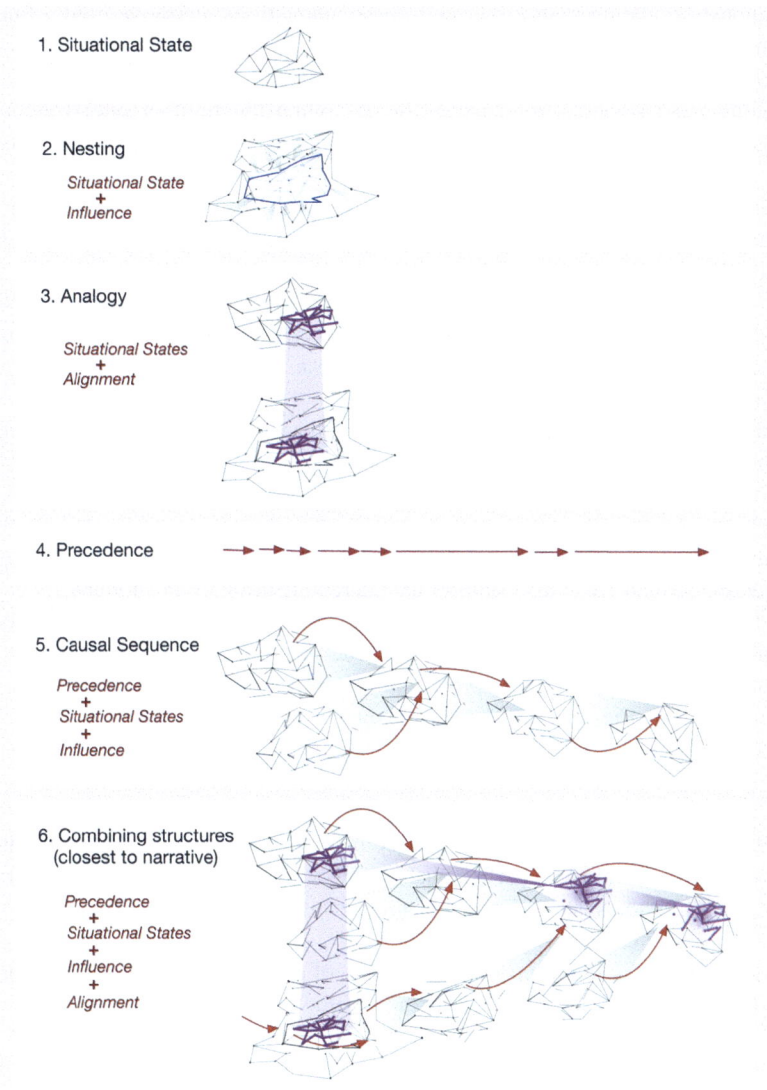

Fig. 8.2 Increasingly complex system-level narrative structures

semantics, relationship structure, and situational similarity, in determining how this structure becomes nested.

In this sense, narrative is engineered to work with the limits of representation. A story is a machine that composes and recomposes fragments from numerous ontologies to represent the unexpected conditions that arise within it to reflect a continuously updating reality. Through these adaptations, narrative guides human reasoning about unexpected entities.

This is how narrative achieves what current formal methods cannot in terms of linking heterogeneous information fields. When the limit of one structure is reached, a mirror entity that has been nested in another situation can take over as the point of reference, supporting interpretation using its own numerous connections. Connections can occur across and among levels of information fields: they are "distributional (if the relations belong on the same level) and integrative (if they straddle levels)" [98]. Each fragment is supported by numerous networks that operate over multiple levels. In this manner, fragments that were once limited by their original ontology can be directly or indirectly linked, similarly to stepping stones, to play a part in carrying the reader across a situation that is too complex for any single framework.

Narrative has numerous system-level connective structures to achieve this. Some common ones are listed below. Each structure type has its own "norms", that is, structural regularities that can act as an additional frame of reference for interpretation. New information can be arranged within, and triangulated by, these systemic structures and their established relationships with each other. In this manner, the structure can also support the characterisation of an anomaly.

Below are five common system-level structures. These operate in different dimensions: three are states, building the structures that give situations form, and two are sequential, tracking the progression from one state to the next. These can be combined (see Fig. 8.3). Key aspects of their structure are shown in Fig. 8.2. Two properties, influence and salience, are also listed.

States

Situational State [332] Multiple points of information are assembled into a network of relationships, which establishes how each entity relates to each other and how they fit into the system as a whole. Given this, it also indicates the affordances for action in a given state.

In a story, this is indicated by descriptions of situations that show what can be named, and how it is named, and how those things are related. Here is an example from Arundhati Roy's novel, '*The God of Small Things*' "May in Ayemenem is a hot, brooding month. The days are long and humid. The river shrinks and black crows gorge on bright mangoes in still, dustgreen trees" [688].

Nesting One situational state is embedded in another. The norms of the embedded situation are influenced by those of the situation in which it is housed, substituting it for its own if necessary.

In a story, nesting can be indicated in a range of ways. A common method is to place a story within a story, such as in Mary Shelley's *Frankenstein*: "He then told me, that he would commence his narrative the next day when I should be at leisure" [732, 39].

Nesting can also occur within imagery, as in this passage from Virginia Woolf's *To The Lighthouse*: "There was the old grey cloak she wore gardening (Mrs McNab

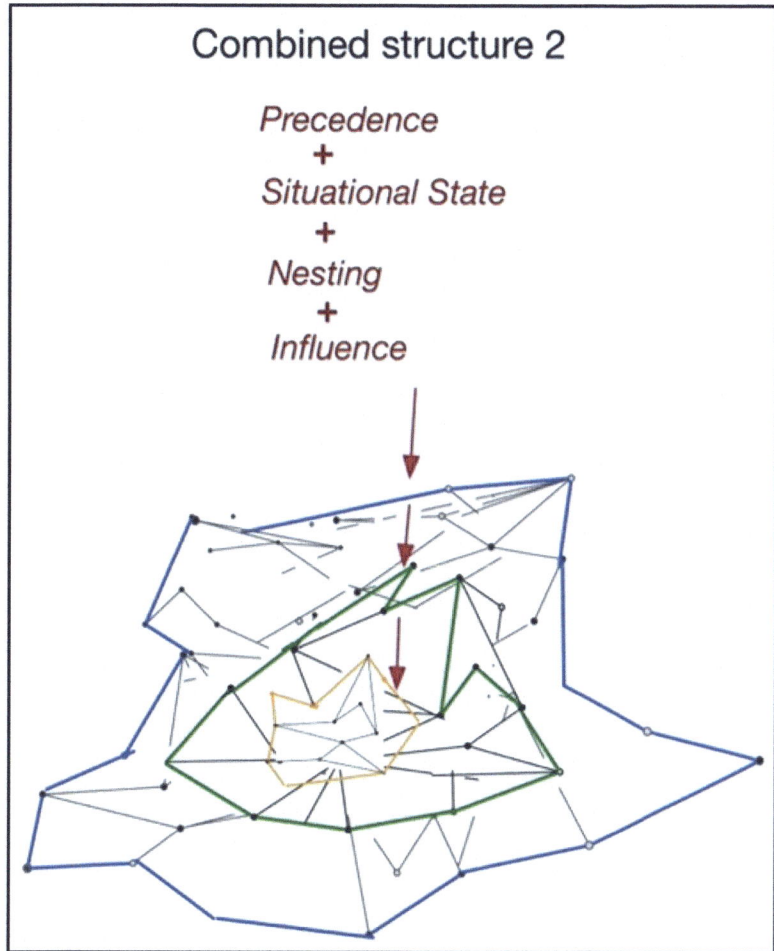

Fig. 8.3 A combination of system-level structures, precedence, and nesting

fingered it). She could see her, as she came up the drive with the washing, stooping over her flowers (the garden was a pitiful sight now, all run to riot, and rabbits scuttling at you out of the beds)— she could see her with one of the children by her in that grey cloak" [840].

Analogy Analogy aligns similar structures between a focus situation and others in the network according to matching relationships [314]. Alignments do not depend on matches of syntax and can be formed in spite of some asymmetries in the relationship structures. Analogy identifies similar situations across heterogeneous information fields.

Most people are familiar with how analogy operates in narrative, both at a syntactic and thematic level. From Virginia Woolf's *To The Lighthouse*: "As she lurched (for she rolled like a ship at sea) and leered (for her eyes fell on nothing directly, but with a sidelong glance that deprecated the scorn and anger of the world—she was witless, she knew it), as she clutched the banisters and hauled herself upstairs and rolled from room to room, she sang" [840].

Sequences

Precedence Events are sequenced according to precedence. This order is often easiest to detect as temporal [787], but it can also occur in the order of integration, for example, incremental information in a story can gradually reveal how an unknown agent has been driving a situation, as in the case of detective fiction.

In a story, this concerns the way one event follows another: "The king died and then the queen died" [287].

Causal Sequence Both events and precedence are combined, along with a determination of digression from "normality", and the additional factor of "influence" to demonstrate how affordance permits this digression to occur [265].

In a story, this concerns the reason one event follows another: "The king died and then the queen died of grief" [287].

Properties

Influence The ability of one structure to adjust the structure of another towards itself. A narrative will designate one structure as dominant at any given time. This dominant entity will adjust all entities in its network towards its own norms.

In a story, this can be seen in the example "Red Riding Hood as a Dictator Would Tell It", when the word "wolf" is used and is interpreted as a dictator spreading propaganda. The influence of implicit reference frameworks endows the term "wolf" with the additional meaning of being a foil for a dictator [172].

Salience The match between a chunk of information and a corresponding chunk of a supporting ontology, informed by a best fit of syntax, semantics, relationship structure, or situational similarity.

In a story, this would be an operation of inference between text such as "wolf" and the network of meaning used to identify it. Whether that meaning is "woodland creature" or "dictator spreading propaganda" depends on whether any other information is part of that interpretation.

See Fig. 8.2 for an illustration of these structures and how they can fit together.

All of the above system-level structures use relationships among multiple fragments of information. Each structure can also be combined with the others, for example, nesting and precedence operate together.

The structure in Fig. 8.3 would be found in a movie such as *Memento* [627], in which time is not the primary means of organising information. Instead, the driver is discovering the driving agencies of the story by delving deeper into its situations as the tale unfolds. These system-level structures can all be commonly found in narrative and enable it to adjust and build rich and unexpected conceptual structures.

Similar structures could be used together to identify and characterise noise across multiple information fields.

8.5 Identifying Useful Noise

What are the parameters by which to identify noise? In any applied system, great benefit can be gained if we can differentiate between kinds of noise that are useful to the goal at hand and those that are not. A key factor in the research of Pounds and Korte was identifying interference that has no agency (such as a glitch) versus interference that indicates the emergence of a new condition (as part of a larger structure or pattern). The "system structures" of narrative could assist with this, in both straightforward and complex cases.

The identification of useful noise again concerns a determination of norms and divergence. For example, in the subtle signals around a drone platform, a goal would be to distinguish between noise that is a blip and noise that recurs in different manifestations across numerous information fields. Recurrence could indicate that the condition creating the noise is persistent, and perhaps has agency that can affect the situation.

Identifying normal behaviour versus abnormal fluctuations could be achieved by linking numerous signals or information fields using narrative's "system-level" structures. For example, if multiple fields are linked in temporal order, the noisy signal might be revealed as occurring or changing consistently over time. If different kinds of sensors capture it, linking their signals at a particular time or place might reveal the anomaly from numerous dimensions, allowing a fuller characterisation to be developed by "situating" it. When information fields are connected, sequences of "noise" may also identify the nature of that unexpected agent's influence throughout the system or perhaps many fields are correlated in a way that reveals a new overall structure of "norms", so a larger situation or entity can be revealed.

Below are two images that provide an example of how these correlations could work in relation to real spaces. On the left is an image in which "noise" low confidence identifications by a neural-net object classifier (in this case, one of the authors) are culled and only high-confidence images remain. On the right is the same image in which low-confidence identifications are retained. Notice that the barrel, treads, and turret of the tank are all individually identified, but the disruptive camouflage prevents the overall vehicle from being assessed holistically. Recognising the logical proximity of these low-confidence detections is the key. Although the confidence of each discrete part is low, clustering the tank identifications should improve an overall score for the correct identification (Fig. 8.4).

Information fields can be linked into systems, combining to provide supplementary information that can assist with its identification. Switching between system and local levels provides a contextualisation of the "noise" that, in this case, can add even more relevant information (that the tank is camouflaged).

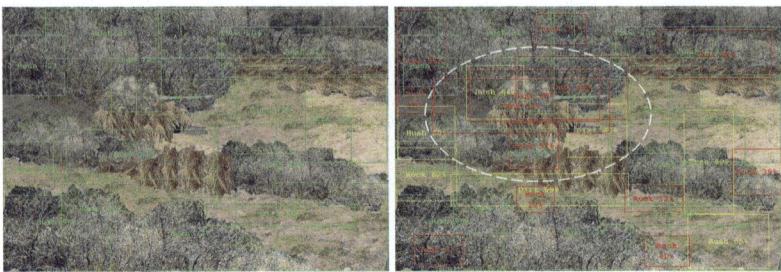

Fig. 8.4 Simple thresholding of scene identifications (left) misses a cluster of low-confidence identifications of a threat that could have been identified as meaningful through mutual proximity (right)

Correlations can be based on straightforward criteria such as location, time, and its role in a situation. They can also be based on systemic criteria, such as the kind of situation or alignments of structure. Finally, they could be based on extra system criteria, such as a similarity with structures in other, similar conditions (in narrative this would operate as analogy [278] or genre [298]). By linking multiple situations, it becomes possible to characterise noise and identify the kind of agency it might have.

A tougher example can be found in the human domain. Imagine a convoy of military vehicles that is driving through farmland. As farmers dig ditches along the roadside, one stops, striking the dirt with his hoe. He lifts the hoe above his head and strikes a soldier with it instead, killing him. Legally, the farmer was not part of a combat situation until he used his farming tool as a weapon; civilians are protected. The sudden change of legal identity of this person, from farmer to attacker, is similar to other circumstances in which ambiguity of classification causes a switch between very different required responses (e.g. the differentiation between a military ship and unarmed merchant or hospital ships [570]). The change in the farmer's identity also informs how the platoon reacts to this act.

To understand this shift of interpretation, let us add multiple system-level information fields, as a human would. Initially, an observer will not be alarmed if they see a farmer holding a farming tool. After the tool has been used to kill a person, however, the bystander is now in a new context of conflict, and the tool represents something different. The object is no longer a farmer's piece of equipment and is instead a potential weapon. The identity of the tool changes, in spite of being the same object. The shift of context also changes the perceived affordances associated with the tool and the possible responses to it. Its identity as a "normal" artefact depends on the context. The act of striking a soldier has caused the tip from one context to another. The action establishes a new governing ontology.

To make it possible to anticipate this surprising act, let us add the systemic structure shown in Fig. 8.2.5. A history of prior states is added to the reasoning about the situation in which a convoy walks through a village. These could reveal that this particular farmer has a history of aggression, having shouted at passing soldiers earlier that week. At the time, his behaviour was dismissed as "noise" because there was no way to include this information in the model of on-the-ground conditions. However, this non-normal behaviour might make him an agent of interest if commanders of the platoon were attentive to it.

Which anomalous behaviour is important enough to devote attention to? Let us go further and add another system-level structure. Figure 8.2.6 can contribute the notion of "analogy", or particular relationship arrangements that are similar across multiple situations, regardless of syntax. In the case of the farmer, similar situations could be compared across the region. Perhaps this analysis reveals that numerous local residents have shouted or spat at different parts of the occupying force. A regular pattern of aggression would make hostile acts towards troops a norm and indicative of widespread discontent, the kind which can prompt violence. A single instance of aggression is no longer an anomaly. This information will change how a commander prepares for potential risk for their convoy, and the kinds of signals they will be alert to during their transit.

Each of these system-level effects provides structure in a manner that is similar to contextual priming. In contextual priming, noise is identified because it is persistently associated with a larger group of ideas. In human languages, successive layers of context and deviation can create shorthand expressions of potentially complex terms. For example, in Auslan (Australian Sign Language) a sign can be modified according to its immediate "phonological" context (the "signals" occurring around it), with handshapes or location bleeding between signs, yet still clearly retain its meaning [436]. The shorthand signal distributes the complexity of the original structure into the layers and history of that context so that a minimal action can communicate a complex unit of information. Its implicit details are entailed in those invisibly associated dimensions. This phenomenon of distributed intelligence has been observed in relation to distributed cognition in an airline cockpit [264] and narrative [369] and may be used to create a shorthand of signatures for various situations. An awareness of how instances relate to a network of systems means that a small signal can potentially be identified more accurately and an appropriate response determined.

In view of this example, it is evident that the categorisation of a message can be conceived at a variety of levels:

1. experiential/phenomenological context change, in which one situation becomes another
2. context change that permits action of a previously established kind by everything in that system
3. context change that permits action of a previously established kind by everything in that system, but which leads to outcome change and thus a divergence from that original context
4. context change that permits action of a previously established kind by everything in that system, but which leads to outcome change that also revises the organisational paradigm of the original context.

The example of the farmer's attack is an example of (4). Hence, we propose that UAVs and UAAs will require an ability to discover and understand their domains according to this form of domain-specific information analysis. Let us now consider a combination of nesting and precedence, using Back's method of synthetic languages, to consider how this type of complexity could be implemented.

8.6 From Sensing to Synthetic Narratives

The structure of nesting and precedence can be used to build a formal method for the characterisation of noise around a drone, using a machine learning approach to synthetic language by Back.

The fluid flow around the surface of an autonomous aircraft or vessel is dynamic and variable. When interpreted through sensors, this stream of information can indicate whether the robot is translating, rotating or even whether other objects might be nearby because of proximity or intercepted wake. This sensing is a feature of Pounds' drones, which feature embedded rotor force sensors [224] and "vibrissae" [226], which each measure local fluid sheer stresses imparted on the craft as it moves. This is similar to the way wheat in a field moves when the wind blows, except that the motion of the "vibrissae" or whiskers reflects the movement of the drone through its fluid. To accurately interpret these sensor conditions, we must differentiate between anticipated egomotion and more subtle signals hidden within the noisy data stream, which contain valuable novel information (Fig. 8.5).

We propose that it is necessary to characterise collective information at numerous levels: sensing (situated inputs and outputs), signals (composed of multiple inputs/outputs, which gain identity from surrounding signals), history (multiple temporal states), and collaborative behaviour (multiple agents acting in concert).

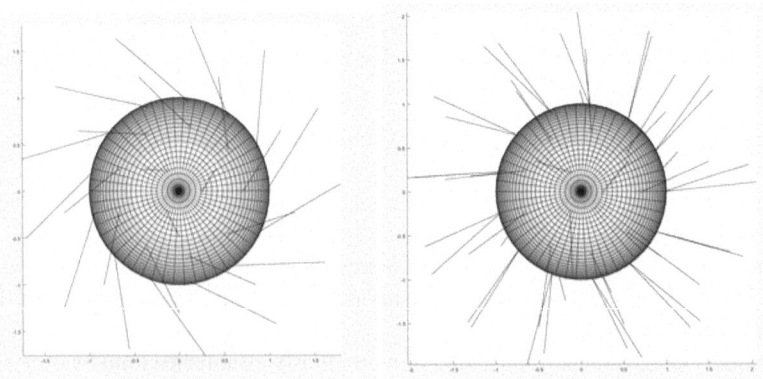

Fig. 8.5 Diagram of collective sensing mechanisms in Pounds' whisker drones. (on left) hairball rotating and (on right) hairball with noise

To handle these levels, data would be passed through multiple stages of information theoretic models to derive meaning across layers [76, 77].

In Back's approach to synthetic language, the change from one field of information to the next is characterised in terms of its altered structure. To interpret these, we treated sensor outputs as symbols within a synthetic language framework so that clusters of outputs provided collections of symbols, namely, "words", which can be understood in combined arrangements and conditions [78]. In narrative terms, this aggregation is similar to the effect of nesting (see Fig. 8.2.2), in which groupings are increasingly assembled to form larger characterisations of a situation. Linking those aggregations from one state to the next is a form of precedence (see Fig. 8.2.4). In both cases, the property of influence, which narrative uses to alter structures towards each other, would not be the same; the closest match is currently an expression of difference from one state to the next.

For this example, at the lowest level, we would derive models for processing the raw signal data. Clusters of signal data are converted into synthetic language primitives (these can be regarded as letters within an alphabet) and then into words using newly derived linguistic information theoretic algorithms [78]. These synthetic language words are then used as the basis for a higher level analysis, using the structures of narrative by linking contexts in different ways.

At the input signal level, a drone's whisker sensors will each detect and reflect the passing situation. Understanding the movement of the vehicle and its surroundings depends on reading the whisker activity collectively—a single signal alone cannot reveal the required situational information. Hence, one approach is to characterise turbulence in a whisker array (or "hairball") as a series of states and then use this as the basis for progressively higher levels of information theoretic models. An additional consideration is that groups of drones will produce simultaneous signals in concert with each other, adding another layer of information to the system.

Processing begins with symbolisation, in which we attach known situations to recurrent states. Collective whisker positions may correspond reliably to rotation, turbulence, and the proximity of known surfaces or the passing of a known agent. Each of these situations will display repeatable features that can form the basis for probabilistic behavioural symbolic events, that is, repeated events or behaviours that can be defined in terms of a probability of occurring. These can be characterised in a reference library of general phenomena, an "elemental" vocabulary, or symbolic alphabet. In narrative terms, these "library" reference situations would operate like analogy (see Fig. 8.2.3), in which situations that recur, such as a hostile farmland community, would have features that can be recognised in new situations and the types of risk that go with them would be anticipated accordingly.

The second stage builds synthetic "words" in accordance with these elemental symbols; these may be considered to be composed situations. In narrative terms, this equates to the building of a situational state (see Fig. 8.2.1). To achieve this, the continuous stream of information will be understood collectively, so that, for example, one kind of wind turbulence around a drone (from the general library above) may frequently follow another, creating a local situation model of air conditions that recur locally. Another example could concern the situational layout, which could be revealed by a tour around a particular territory. This would be recorded as a series of synthetic language "words" that correspond to local events. Such events would include proximity indications for local features. These situated examples will be composed from multiple physical elements, as wind corridors and bluffs, which emerge in response to the architecture and topology of a particular place. Although the above elemental library is generic, these composed situations are attributable to known situations and places. Composed situations therefore perform a different function—they are assembled and called up as a reference for particular situations. In narrative terms, this would be similar to the composed situations ("narrative") of Fig. 8.2.6.

The third stage develops a framework for using the synthetic language output. This involves detecting novel signals within the "noisy" continuous flow. These signals may reveal the presence of entities or conditions that have not been accounted for, indicating unexpected circumstances, activities, or emerging agents. Anomalies can be detected using synthetic language and information topology techniques, such as information topology models, which provide a mechanism to detect deviations from established patterns, regardless of the phenomenon. Unlike classical models, which need to be trained on specific novelties, this approach can be used to detect any anomalies, whether anticipated by a system or not.

The fourth stage composes identified anomalies into combined system structures from narrative. Here, the goal would be to use one or more of these relational structures to correlate unexpected signals across situations. These would reveal larger and possibly unexpected conditions in a situation. Each can establish what kinds of changes are routine and which are not, in different ways. Given the way these structures combine information from multiple times and sources, this will also enable some disambiguation when only partial information is available [869].

8.7 Layers Expressed as Synthetic Languages

Discussing synthetic language in relation to a swarm of UAVs might be initially viewed as there being a single language associated with each UAV. However, we suggest that there can be multiple types and hierarchical layers of language used by each UAV, and each may play a different role. Although limitations of computational power, memory, and energy consumption will inform this scope, setting this aside (e.g. quantum computing devices may overcome these limitations), it is useful to consider how multiple synthetic languages can be understood in this context.

In the first instance, we can consider symbolising the input space and forming synthetic language words. Is there a single way to do this, or will any method do? A symbolisation algorithm has been proposed which optimises intelligibility [79], and another adaptive algorithm for symbolisation has been proposed which is useful when the input space is changing [79]. This is an area requiring further examination;

however, we expect that there will be multiple ways to symbolise the input space depending on usage contexts.

The concept of layers of symbolisation is useful in that it can also provide various resolutions of information. A coarse-grained symbolisation has been demonstrated in other studies to provide rapid assessment of speaker identification in terms of tone and attitude [506]. A similar approach here could be useful to make a rapid assessment of friend or foe: is the speaker aggressive or just worried?

However, a finer grain symbolisation might be useful to include at another level to detect subtle changes in performance. A UAV operating in a swarm might know the identity of all nearby agents, but at the same time, be attuned to subtle changes in behaviour which could indicate a problem. Is the field of farmers listless in a way that is unusual for a group of people tilling a field?

Over the aforementioned general concept of layers, we thus lay a range of parameters that could be part of any drone swarm. These will also be managed as information fields, or "contexts". Some useful suggested contexts are:

1. sensing capabilities and limitations

    (a) limitations of the agent (cost, energy, physical constraints)
    (b) capabilities required (mission, skills)
    (c) symbolisation levels and language extraction

        (i) routine language for "everyday" common functions (mission, energy, communications, situation)
        (ii) self-language to monitor "personal" health and functionality (long term changes, short term changes, alertness)
        (iii) relationship language to monitor nearby agents
        (iv) danger language to monitor for major threats
        (v) narratives language for long-term learning (experience, memory, stories)

2. higher level meaning and integration

    (a) level of authority and autonomy (expectations, culture, ethics)
    (b) integration of all languages (priority, planning)
    (c) human interaction (translation, explainability)
    (d) machine interaction (relationships, friend/foe, discovering, learning new languages)

3. learning and memory

    (a) capacity (to what extent is the agent designed to learn from experience?)
    (b) algorithms
    (c) ageing (replication, death, do we want the agent to change?)

Different types of language can provide the means for understanding and actioning these various forms of situation. We propose that synthetic language systems can be derived for various functionalities within UAVs.

8.8 Discussion

Given our swarming structure of intersecting information fields, several interesting opportunities and issues arise.

Swarm Control and Communication A vehicle equipped with synthetic language unlocks the potential for more than just observation; it also has swarm control capabilities. There is a question of how swarms of agents should behave and how they ought to be autonomously responsible for their other teammates within a swarm. Consider a future scenario of UAAs that have swarm awareness and monitoring capabilities provided by processing the synthetic language signals they detect and the signals they are relayed by the swarm members. In this case, the vehicles or agents will monitor nearby agent vehicles for not only mechanistic problems but also context detection and "understanding" problems. For example, if one agent is "understanding" a situation differently than others in its swarm, it may be being lured away or under duress, or it could be equipped with a form of sensing that makes it better able to detect particular forms of threat. It is important to determine which of these is the case. This could be extended further, to give an individual agent characteristic behaviours that are tuned by the experiences it has, further enabling within-swarm understanding and situational awareness. Hence, it can be observed that synthetic language has a potentially significant role to play in the implementation of such systems.

Distributed Computation The distributed nature of swarms adds complexity to the computational tasks of robot agents and raises several questions: Does every drone need to process every other drones' signals, or is it enough for a few key drones to do so? Should drones pool their aggregate data and understanding of context, either one-to-one or via nominated leaders? Philosophically, if communications and processing capability were infinite, the location at which processing occurred would be arbitrary. However, given there are limitations, optimisations must be made. Processing to extract symbolic, contextual, and narrative meaning may each happen at different parts of a swarm. The weighted path between where data are generated, where they can be computed, and where they must be consumed should be minimised, for every process in the swarm. This is a "travelling salesman" non-deterministic polynomial hard problem to solve; approximate solutions must be adequate in practice.

A Narrative-Based Prospective Provides a Clue Localised information to extract symbols may happen locally (or among proximal peers) to extract symbol candidates. Broader knowledge from the swarm may then be fused locally or at a nexus to create salient context. That context can be interpreted as narrative, at the agent relying on the context to perform the action. An actual narrative would alternate the parts of the inference network that bear most of the interpretive load, according to context and available resources. With a reliable narrative-based system, perhaps those load areas in a UAV swarm could be anticipated. In this way, processing would occur at the point at which the needed information is first collated and thus communication flows are reduced.

Human-Swarm Interaction Designing sensory swarms, and the way they interact with humans, will require careful consideration to ensure safety for both humans and the swarm agents. This means considering not only the obvious cases of humans controlling or teaming with the swarm (i.e. stakeholders who will be affected and also have influence) but also the interactions with bystanders (i.e. stakeholders who will be affected but do not have influence). This parallels the fact that developers of autonomous vehicles have found that they must consider not just the interaction needs of the "driver" or passengers but also those of pedestrians and other road users (e.g. [185, 555]). Drone swarms may need to have "body language" so that bystanders (or civilians) are forewarned of sudden changes in movement, or activation of actuators. This body language could be informed by the rationale of system-level narrative structures so that bystanders can anticipate an artificial agent's behaviour. This can be important in building trust in a swarm.

Exactly how these considerations come into play will be influenced by the domain of use of the drones. In a military setting with hostile actors, it may be less desirable for a swarm to enact bystander communication patterns. By contrast, clear communication or "body language" may be crucial in search-and-rescue missions to avoid causing anxiety to those being rescued and (possibly traumatised) bystanders.

Learning Capabilities Traditionally, we consider machines to have set and specified capabilities. However, sensory swarms in long-term use are likely to require learning abilities, so they can learn what is "normal" in each new scenario they encounter. Given the way narrative enables the classification of novel, familiar, and similar situations, a swarm that is able to add to its library of known situations and their appropriate responses could also be able to learn how to derive normalcy and divergence. To return to the farmer example, a swarm may initially "know" that tools in the hands of people other than those in the uniforms of its teammates and allies represent a potential threat under certain circumstances. However, it may learn that civilians holding tools represent a lower risk when the context involves farm work. This learning could occur through repeated exposure to data from the low-risk situation, or from human team members or controllers sharing their insights or narratives with the system. A swarm's capabilities could thus grow and change beyond its original specifications which may or may not be desirable.

In a defence context, it is important for machines and systems to have known capabilities and behaviours. From a legal perspective, it is a requirement of certifying new technologies, especially weapons technologies, according to the laws of armed conflict and international humanitarian law (and specifically Article 36 of Additional Protocol I to the Geneva Conventions [15]). On the positive side, a learning ability also means that the swarm could be trained in advance of deployment, in the same way that humans are trained, to respond to situations appropriately. On the negative side, it may introduce ambiguity, which may not be able to be accounted for in the fraught situation of a contested environment.

Legal Considerations A swarm of autonomous agents capable of "understanding", learning from, and reacting to its environment completely autonomously falls outside current legal definitions, which assume that there is a human decision-maker responsible for the actions of (semi-) autonomous agents [64]. This means that for a swarm in a military context, there may be a need for a human on the loop, even if humans are not able to react fast enough to be in the loop for swarm decision-making, especially in situations requiring urgency of (re)action. There may also be a need for an autonomous swarms' decision-making to be trained repeatedly, paralleling the ways both human commanders and soldiers operating weapons systems are trained.

Although classical models are invaluable, there is a growing awareness that future technology will need ways to deal with these subtleties [861].

8.9 Conclusion

This chapter explored how "noise" could contribute to the intelligent processing of information in a swarm, using the system-level structures of narrative. At their core, narrative operations are geared to represent anomalies against the norms of an information field and integrate them by adjusting that information field. This integration occurs using the system-level structures of narrative, which can link information in a variety of ways. An overview of some of narrative's system-level structures was provided, grouped into three fundamental forms: states, sequences, and properties. Individually or combined, these structures can provide a means of linking information from multiple ontologies and perspectives.

Two example situations were considered: assessing the situational space around a whisker drone and assessing the risk in a farmland situation through which a convoy is passing. The second example was considered using the affordances of Back's synthetic language and information topology approach, which could be used in conjunction with the narrative structures of nesting and precedence to build an increasingly larger situation model around an anomaly. This distributed system of information processing has the potential for intelligence.

One goal was to lay out conceptual foundations for an approach that could allow swarms to be more reactive in an open world than traditional, highly structured, pre-programmed control systems. Another goal was to develop a means of building models centred around actual situations as they unfold. A desirable outcome was to support the interpretation of unexpected conditions. Hence, this approach provides a potential new way of introducing capabilities for autonomous vehicles as a basis for more general artificial intelligence and is proposed as a foundation for autonomous sensing and decision-making.

Acknowledgments This research received funding from the Australian Government through the Defence Cooperative Research Centre for Trusted Autonomous Systems (TAS-DCRC). TAS-DCRC receives funding support from the Queensland Government.

Open Access This chapter is licensed under the terms of the Creative Commons Attribution 4.0 International License (http://creativecommons.org/licenses/by/4.0/), which permits use, sharing, adaptation, distribution and reproduction in any medium or format, as long as you give appropriate credit to the original author(s) and the source, provide a link to the Creative Commons license and indicate if changes were made.

The images or other third party material in this chapter are included in the chapter's Creative Commons license, unless indicated otherwise in a credit line to the material. If material is not included in the chapter's Creative Commons license and your intended use is not permitted by statutory regulation or exceeds the permitted use, you will need to obtain permission directly from the copyright holder.

Chapter 9
Regulating Maritime Autonomous Swarms

Rachel Horne

Maritime swarming, in which multiple vessels organise themselves as a team to achieve a common objective, is a relatively new phenomenon. Development of maritime swarming capabilities in Australia is primarily driven by defence investment, particularly in the areas of mine countermeasures and persistent surveillance. There are also potential commercial applications for maritime swarms, for example, in reef monitoring and search and rescue. Developers and operators must comply with existing regulatory frameworks, but these were written for traditional crewed vessels and create complexity and uncertainty for emerging technology. This chapter draws on the existing literature to identify and address two key issues relating to the regulation of maritime swarms. The first issue, how to characterise maritime swarms and the impact on their regulatory treatment, is addressed by outlining the three categories available for commercial vessels (domestic commercial vessels, regulated Australian vessels) and foreign vessels and touching on the way defence vessels are regulated both in peacetime and wartime. The second issue, the relationship between regulation, explainability, and trust, is addressed by explaining the meaning of each term and then identifying how they interconnect. The chapter concludes by considering the types of regulatory reform required to support safe and trusted operations into the future, including legislative change, a focus on risk-based regulation, and the introduction of technical standards led by industry.

R. Horne (✉)
Trusted Autonomous Systems Defence Cooperative Research Centre, Brisbane, QLD, Australia
e-mail: DrRachelHorne@Outlook.com.au

9.1 Introduction

Maritime autonomous swarms, in which multiple autonomous vessels (seabots in [41]) work together to achieve a common objective [842], are attractive because of the extended reach, effect, and redundancy they could provide. Swarms enable multiplication of effect, in the context of surveillance and logistics, and possibly lethality in a defence context, in ways that are significant, and comparably low cost. Although the earliest public video purporting to demonstrate maritime autonomous swarms was released in 2014 by the US Office of Naval Research [803], maritime autonomous swarms still appear to be primarily in the realm of research.[1] Some defence projects, for example, in China [66] and the USA, [286], purport to include maritime autonomous swarming, but limited public videos or information have been released. In Australia research on maritime autonomous swarms is primarily driven by defence investment, particularly in the areas of mine countermeasures[2] and persistent surveillance,[3] with additional research being undertaken in the context of reef monitoring.[4] Research is also underway in relation to multi-domain assets such as submersible seaplanes [551, 773] and amphibious bottom/ground-crawling robots [526, 734] capable of operating in littoral (i.e. coastal) environments. It is already challenging to characterise and identify the regulatory treatment for single-domain swarming capability, but when multi-domain elements are introduced, these challenges will multiply.

---

[1] For example, SWARMS: Smart and Networking Underwater Robots in Cooperation Meshes, 2015-1018 collaborative project with 10 European countries and 30 partners (http://swarms.eu/), Oceans 2020 funded by the European Defence Fund (https://ec.europa.eu/defence-industry-space/ocean-2020-eus-largest-collaborative-defence-research-project-under-padr-successfully-completed_en) [842].

[2] Minister for Defence, Senator the Hon Linda Reynolds CSC and Minister for Defence Industry, The Hon Melissa Price MP, Joint Media Release 25 August 2020, Revolutionising future mine countermeasure technology (https://www.minister.defence.gov.au/minister/melissa-price/media-releases/revolutionising-future-mine-countermeasure-technology); [453].

[3] Minister for Defence Industry, The Hon Melissa Price MP, Media Release 16 June 2020, Growing Australia's defence industry through innovation (https://www.minister.defence.gov.au/minister/melissa-price/media-releases/growing-australias-defence-industry-through-innovation).

[4] For example, the ReefScan CoralAUV, developed in partnership between the Australian Institute for Marine Science (AIMS) and Queensland University of Technology, is a potential testbed for swarming purposes (more information available here: https://www.aims.gov.au/research/technology/reefscan/CoralAUV).

9 Regulating Maritime Autonomous Swarms 153

Although there is a fast-growing body of literature regarding technical elements of swarming,[5] also described as "multiple unmanned surface vessels",[6] there is little to guide developers and operators on how their technology is characterised and regulated. These practical matters influence the time, cost, and risk of putting the technology into the water and need to be understood early to guide decision-making. Similarly, there is little to guide developers and operators in understanding the relationship between regulation, explainability, and trust. Understanding these core issues enables more informed consideration of whether changes are needed to better facilitate safe and trusted swarming operations into the future and, if so, what principles should guide those changes.

This chapter considers each of the two issues identified above and argues that more research is needed to support developers and users of emerging technology and assist in the transition from research to operationalisation. This chapter is presented in five sections. The Introduction provides contextual information, introducing the reader to the key topics and arguments contained in the chapter. The second section, "Maritime Swarms", outlines what maritime swarms are and why they are growing in popularity as a research area related to both commercial and defence applications. The third section, "Available Literature", explores the literature available relating to maritime swarming, in three main topic areas—technical, international legal treatment, and domestic legal treatment. Two key issues relevant to the regulation of maritime swarms are identified in the literature: (1) how to characterise maritime swarms and the impact on their regulatory treatment within both commercial and defence regulatory frameworks and (2) the relationship between regulation, explainability, and trust. The fourth section, "Exploration of Key Issues", explores these key issues identified in the literature and provides an analysis of where maritime swarms fit into the Australian maritime regulatory framework. The fourth section reflects on the broad concepts of regulation, explainability, and trust, and considers how they fit together. Concluding the chapter is the fifth section, "Where to From Here", which draws on the outcomes of the previous sections to identify the types of regulatory reform required to support safe and trusted operations into the future, including legislative change, a focus on risk-based regulation, and the introduction of technical standards led by industry.

---

[5] For example, a Google Scholar search of the term "maritime swarm" resulted in 28,100 results, and when the limiter "since 2021" is added, there remained 4,330 results. Examples of articles include Jones. A, Lin. A, Zhang. M, Tall. M, Development of a Steering behaviour for decentralized unmanned underwater vehicle swarm shape formation, Oceans 2021: San Diego 2021, Ieeexplore; Ma. Y, Zhao. Y, Incecik. A, Yan. X, Wang. Y, Li. Z, A collision avoidance approach via negotiation protocol for a swarm of USVs, Ocean Engineering Journal, Vol 224, 15 March 2021.

[6] G Wu, T Xu, Y Sun, J Zhang. Review of multiple unmanned surface vessels collaborative search and hunting based on swarm intelligence, International Journal of Advanced Robotic Systems, 27 April 2022.

9.2 Maritime Swarms

Maritime swarms are attractive because of the extended reach, effect, and redundancy they could provide in contrast to the use of single vessels, which contributes to lethality (where relevant) and survivability. It has been reported that "UUV swarming is already being developed, such as exploration of the ability of multivehicle systems to self-arrange in various swarm formations and simultaneously collect data" [548]. Although no tangible Australian examples are available for analysis, it is expected that commercial use cases will include reef monitoring and search and rescue, and defence uses will include intelligence gathering, mine countermeasures, and torpedo countermeasures. Some defence projects, for example, in China [66] and the USA [286], purport to include maritime autonomous swarming, but limited public videos or information have been released.

The lack of tangible examples of maritime swarming exacerbates the difficulty in forming a cohesive understanding of what "swarms" and specifically "maritime swarms" are, which challenges technology, regulatory, and policy development. Guihen [342] highlighted the importance of identifying common language to describe swarms, noting that "a communicated understanding of swarming objectives will make progress of each underpinning technology more cohesive to the overall vision of autonomous underwater swarms" [342]. In addition to the impact on developing technology, "... a lack of agreement on key terms has hampered discussions, for instance in relation to the ethical and technical challenges of autonomous weapon systems" [685]. The "explainable artificial intelligence", or XAI, movement also focuses on creating an ontology able to express the reasoning for decisions, which is necessary for building trust [30]. In a regulatory context, without language to describe maritime swarms, seeking to use them and develop suitable regulatory approaches is difficult.

The concept of multiple unmanned surface vessels [842] does not have a single agreed definition or description, and there is large variance in the complexity and autonomy expected of the technology involved. One description, from Wu, Xu, Sun, and Zhang, relates to multiple autonomous vessels that organise themselves as a team to achieve a common objective [842], with Lundquist adding "collaboratively".[7] Other descriptions do not include the centralised control element, instead relying on multiple independent agents to execute a task, taking inspiration from biomimicry. The Australian Army Research Centre considers definitions and

---

[7] RAND Australia, authors L Slapakova, P Fusaro, J Black, P Dortmans, Supporting the Royal Australian Navy's Campaign Plan for Robotic and Autonomous Systems: Emerging Missions and Technology Trends: Research Report, 2022 (https://www.rand.org/content/dam/rand/pubs/research_reports/RRA1300/RRA1377-1/RAND_RRA1377-1.pdf, pg. 12); E Lundquist, 'NATO's Autonomous UUVs Are Working Together to Find Mines', The Maritime Executive, 29 January 2021.

descriptions of swarms,[8] and helpfully identified that "a homogeneous swarm is three or more UxVs [uninhabited vehicles] of the same type undertaking the same task in cooperation. A heterogeneous swarm contains independent UxVs delivering alternate effects in unison. UxVs can also be deployed in small manoeuvre units, called 'clusters', which can surge to swarm and provide effects in time and space" [71]. From a non-domain-specific perspective, the Australian Defence Glossary defines "swarming" as "the large mass of autonomous systems interoperating collectively to act and respond in a coordinated effort to provide an overwhelming effect" [71].

The variety of definitions and descriptions related to swarms, including non-domain specific, or specific to the maritime domain, make discussion of the regulation of the technology difficult and necessitates being clear upfront on what we mean when we say "swarm" and "maritime swarm" (for further discussion on definitional issues, see [388]). For the purposes of this chapter, a maritime swarm is a group of three or more autonomous vessels acting as a team to complete a mission.

Autonomous technology development continues to advance in Australia and globally, with three major clusters of research identified: autonomous systems and society, operationalisation of autonomous systems, and networks of autonomous systems [471]. There is little corresponding exploration of fundamental questions such as how is a maritime swarm characterised, what is the role of regulatory frameworks in ensuring minimum levels of trust and explainability, and are changes needed to better facilitate safe and trusted swarming. In a report commissioned for the Royal Australian Navy's Warfare Innovation Navy branch, RAND highlighted that [746] "Evolving regulatory, legal, policy and ethical frameworks (including sensitivities around autonomous uses of force) might significantly determine when RAS-AI technologies are employed in particular missions or whether they are employed at all".

The risk of a continued focus on the technology rather than its regulation is that even if the technology advances to a state of operational readiness, it may be unable to be deployed.

9.3 Available Literature

There is a fast-growing body of literature on the technical elements of swarming,[9] ranging from Abdelli, Amamra, and Yachir's work surveying swarm robotics [23]

---

[8] Australian Army Research Centre also sets out some minimum requirements for a robot swarm, being (1) three or more robots, (2) which have limited or no human control, and (3) which perform tasks cooperatively, with the following reference [61].

[9] For example, a Google Scholar search of the term "maritime swarm" resulted in 28,100 results, and when the limiter "since 2021" is added, there remained 4,330 results. Examples of articles include [438, 553].

to Gregory and Vardy's consideration of the use of micro unmanned surface vehicles [337], expediting the recovery of vessels [453], and developing autonomous docking technology [75].

This technology-focused literature helps to establish the state of the art and the key areas of attention, but it does not assist developers and operators to know how to actually use the technology lawfully.

There is a smaller body of literature considering the legal implications of autonomous vessels from an international law perspective, in both defence and commercial contexts, but this does not yet contemplate swarms. For example, McKenzie [571, 572] addressed the question "when is a ship a ship" in the context of military use of autonomous vessels in international waters, and Chadwick [180] considered similar issues with a focus on whether the international regulatory framework is ready. Klein et al. [475] considered the use of autonomous vessels for commercial purposes in international waters and the requirements that arise. The unique features of maritime swarms, including the potential to determine sequences of action independent of a human operator, and the increased (and different) risk inherent in increased numbers of vessels in use in proximity, mean that specific legal analysis is required.

A smaller body of literature has considered the legal implications of autonomous vessels from a domestic law perspective, particularly under Australia's domestic law. Only one grey literature (i.e. non-academic) resource considering the regulation of swarms was identified in the author's search. For example, Judson and Horne [444] addressed the general regulatory approach for autonomous vessels, Horne et al. [387] analysed the domestic legal framework and considered whether a new technical standard is required to support it, and Horne, in grey literature, set out regulatory requirements and pathways.[10] Devitt, Horne, and Assaad et al. considered the challenges of assurance and regulation for autonomous systems and proposed new approaches [235]. Trusted Autonomous Systems published a "swarming primer" which summarised the current state of swarm regulation, including, specifically, in the maritime domain [796].

The Trusted Autonomous Systems swarming primer was the only literature available that specifically considered the regulation of maritime swarms. The document aimed to "increase awareness and understanding, from both a commercial and defence perspective, of swarming and its potential associated issues" [796]. It identified swarming as requiring three elements: a multi-agent system compromising three or more individual agents working together, cooperative coordinated

---

[10] R Horne, Regulatory requirements and pathways for autonomous and remotely operated marine equipment, 26 July 2022 RAS-Gateway (https://www.rasgateway.com.au/resource-hub/regulatory-requirements-and-pathways-for-autonomous-and-remotely-operated-marine-equipment); Regulatory requirements and pathways for survey-exempt vessels, 26 July 2022, RAS-Gateway (https://www.rasgateway.com.au/resource-hub/regulatory-requirements-and-pathways-for-survey-exempt-vessels); and Regulatory requirements and pathways for vessels in survey, 26 July 2022 RAS-Gateway (https://www.rasgateway.com.au/resource-hub/regulatory-requirements-and-pathways-for-vessels-in-survey).

task completion, and decentralised, autonomous decision-making [796]. It identified regulation as a key enabler of swarming, because without a suitable regulatory framework, the technology cannot be operationalised, but also noted it is a complex topic. The primer summarises the regulatory approach from a commercial and defence perspective for maritime swarming. In short, although swarms are not specifically catered for in existing regulatory frameworks, they are subject to those frameworks and operators will need to ensure their activities are compliant. The primer states:

> One key enabler of swarming is the regulatory framework supporting it. While there are separate regulations for the land, maritime and air domains, across Defence and commercial industry, in general, swarming is not precluded by regulation. However, the process for obtaining approvals to operate is not well tested and is therefore time consuming and difficult to achieve. There are also several other technological and operational limitations to the more widespread adoption of swarm technology, such as assurance of autonomy and the development of ethical frameworks. [796]

Some consistent themes can be identified from the available legal and regulatory literature, including the following: (1) the importance of characterising a technology to appropriately regulate it, for both commercial and defence applications, and (2) the need to understand the relationship between regulation, explainability, and trust for autonomous technology. This chapter explores these core issues and identifies that regulatory changes are needed to better facilitate safe and trusted swarming operations into the future. The chapter concludes by identifying the principles that should guide regulatory development across domains.

9.4 Exploration of Key Issues

9.4.1 Key Issue 1: How Do We Characterise Current and Likely Future Maritime Swarms, and What Is the Impact on Their Regulatory Treatment?

It is important for designers and operators to understand how their specific maritime swarm, and the individual vessels that comprise it, are characterised in a regulatory context as early as possible because this will affect decisions and options throughout the lifecycle of the technology. This section identifies how maritime swarms can be categorised and the implications for their regulatory treatment.

In Australia, vessels are categorised and regulated according to their use. Vessels used for commercial, research, or government purposes are regulated by the Australian Maritime Safety Authority (AMSA) as either "domestic commercial vessels", "regulated Australian vessels", or "foreign vessels".[11] Vessels that are

---

[11] For more information see Horne. R, Vanderkooi. M, Guihen. D, Autonomous vessel regulation in Australia: Why an Australian Code of Practice is required, IndoPacific 2022 International

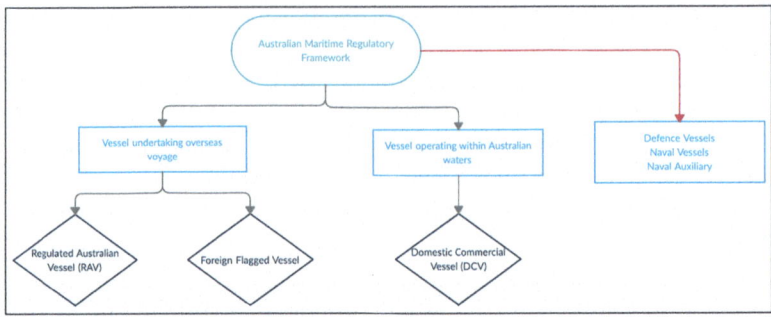

Fig. 9.1 Categorisation of vessels in Australian maritime regulatory framework by R Horne

"defence vessels"[12] are regulated through the Defence Seaworthiness framework.[13] In general, autonomous and remotely operated vessels, which are being developed by commercial industry for defence, are considered domestic commercial vessels until they are actually in use by uniformed members of the Australian Defence Force. This means that it is critical in the defence context to understand both the commercial and the defence regulatory environment.

There are examples of autonomous and remotely operated vessels operating in Australia as "domestic commercial vessels", "foreign vessels", and "defence vessels". Although there are no examples of "regulated Australian vessels", this chapter still includes them for completeness, and because they do offer opportunities for swarms that should be explored. There are no publicised examples of maritime autonomous swarms, in the general understanding of the phrase, operating in Australia under either commercial or defence regulation (Fig. 9.1).

According to the existing regulatory framework, it is likely that a "maritime swarm" would not be a regulated entity; instead, the individual vessels that comprise it would be regulated. For swarms that are homogeneous, meaning that the vessels are the same type, this should be more straightforward than for heterogeneous swarms, in which the vessels are different types [71] and may have different risk profiles. A new approach to regulating groups of vessels will be needed to manage the complexity of swarms as an entity. Until that new approach is designed and

---

Maritime Conference (Sydney, Australia) (https://www.indopacificexpo.com.au/IMC2022/papers/papers.asp or by emailing info@tasdcrc.com.au).

[12] Noting that the vessel must meet the definition of "defence vessel" in the *National Law Act 2012* (Section 6 Marine Safety [domestic commercial vessel]) (https://www.legislation.gov.au/Details/C2018C00484) or the definition of "naval vessel etc" in the *Navigation Act 2012* (Section 10) to be "carved out" of those respective laws and therefore not subject to the AMSA's remit.

[13] For more information see the Defence Seaworthiness Management System Manual available at: https://defence.gov.au/publications/docs/SeaworthinessMgmtSystemManual.pdf and the Defence Seaworthiness website at https://www.defence.gov.au/business-industry/seaworthiness; for more information see [796].

implemented, designers, manufactures, and operators will need to understand how the vessels that constitute their swarm are characterised and the corresponding regulatory requirements.

9.4.1.1 Domestic Commercial Vessels

Domestic commercial vessels, that is, vessels for use in connection with a commercial, government, or research purpose within Australian waters, are regulated by the *Marine Safety (Domestic Commercial Vessel) National Law Act 2012*.[14] Relevantly, a "vessel" is defined as meaning "a craft for use, or that is capable of being used, in navigation by water, however propelled or moved, and includes an air cushion vehicle, a barge, a lighter, a submersible, a ferry in chains and a wing in ground effect craft"("Definition of vessel", Marine Safety (Domestic Commercial Vessel) National Law Act 2012, Section 8). This broad definition captures a wide variety of autonomous vessels, even very small autonomous underwater vessels (AUVs), but excludes physically tethered vessels such as remotely operated vessels (ROVs).

Domestic commercial vessels, unless subject to an exemption, are required to have a unique vessel identifier, have a certificate of survey, be listed on a certificate of operation, and be crewed by persons holding the required certificate of competency, and the owner, master, and others interacting with the vessel must comply with general safety duties.[15] The size, intended use, and intended area of operation determine the exact technical requirements with which to comply, noting that the larger the vessel and the further from shore it operates, the higher the risk it is considered.[16] Autonomous vessels must seek a specific exemption from the AMSA to operate lawfully.[17]

This model is fit for purpose for traditional vessels, for example, fishing or passenger vessels; however, for small autonomous and remotely operated vessels, which are often subject to iterative development, have short life spans, and present as low risk, it can be unduly cumbersome, confusing, and slow to navigate [235]. Flexibility mechanisms, such as specific exemptions, enable operation without meeting all standard regulatory requirements, for example, having a certificate of survey, subject to conditions,[18] but still generally require an application for each vessel. At time of writing all commercial autonomous vessels operating lawfully in Australia have a specific exemption issued by the AMSA related to, at a minimum,

---

[14] The law is accessible on the Australian Government Federal Register of Legislation: https://www.legislation.gov.au/Details/C2018C00484.

[15] *Marine Safety (Domestic Commercial Vessel) National Law Act 2012 (sections 12-70)* [72].

[16] Starting a legal analysis by identifying the vessel, what it is doing, and where it is operating, applies equally when considering domestic law and international law. See [474] and RAS-Gateway.com.au for more information on regulatory requirements for vessels.

[17] Information on the regulatory requirements for domestic commercial vessels can be found on the AMSA website, www.amsa.gov.au.

[18] *Marine Safety (Domestic Commercial Vessel) National Law Act 2012 (section146)*.

crewing requirements, and subject to conditions intended to ensure a minimum level of safe operation is maintained.

It would be very cumbersome for the operator of a heterogeneous, or mixed, maritime autonomous swarm[19] to work through the existing regulatory requirements and process for each vessel within the swarm as an individual vessel, and equally cumbersome for the AMSA to assess the required applications. Although the AMSA has not yet had to grapple with the regulation of a true maritime autonomous swarm, in the general understanding of that phrase, it will undoubtably be called on to do so in the coming years.

9.4.1.2 Regulated Australian Vessels

Regulated Australian vessels are generally Australian vessels that operate outside of Australian waters, and are regulated under the Navigation Act 2012 (Cth). "Vessel" is defined as meaning "any kind of vessel used in navigation by water, however propelled or moved, and includes the following: (a) a barge, lighter or other floating craft; (b) an air cushion vehicle, or other similar craft, used wholly or primarily in navigation by water.[20]" Individual vessels are required to meet the full range of regulatory requirements imposed unless an exemption has been granted by the minister or the AMSA or if the AMSA has issued an equivalence. Requirements are generally based on the tonnage (i.e. weight), length, and purpose of the vessel. Failure to comply with the requirements set out in the *Navigation Act 2012*, which are usually imposed on the vessel owner and master, could result in fines or imprisonment. Vessel owners must engage a recognised organisation (i.e. class society such as Lloyds Register or DNV GL) to identify the specific requirements that will apply, and to survey the vessel and provide the required certification.

To date, no regulated Australian vessels are autonomous vessels. This is likely because there are few large commercial autonomous vessels in use, and applicable regulatory requirements do not convert well to small iteratively designed vessels. It is difficult to identify the requirements that do apply, primarily because doing so requires reference back to multiple international conventions and Marine Orders 1–98,[21] which have different applicability and flexibility criteria. There are also no examples to follow, and no published guidance from the AMSA on how the *Navigation Act 2012* applies to autonomous vessels.

For maritime swarms, an option to consider within the regulated Australian vessel category is the concept of a "mothership" and "ships equipment" deployed

---

[19] Heterogeneous means mixed. A heterogeneous swarm means a swarm comprising various types of assets, for example, which have different technical specifications and capabilities.

[20] The *Navigation Act 2012* (Section 14 "Definitions").

[21] Marine Orders are regulations that provide the technical detail underpinning legislative requirements. An index of Marine Orders is available here: https://www.amsa.gov.au/about/regulations-and-standards/index-marine-orders.

from it. Essentially, if an operator intends to deploy a swarm from a "mothership" that is already a regulated Australian vessel, then those smaller vessels comprising the swarm could possibly be considered "ships equipment"; therefore, rather than being individually regulated, they are considered part of that mothership for the purposes of survey and safety management systems. Issues such as maintaining line of site, communication, and displaying the relevant vessel names remain relevant for individual autonomous vessels or vessels operating in a swarm. This option is likely very attractive to operators who have access to a regulated Australian vessel because they could avoid the requirement to address the regulation of each small autonomous vessel in a prospective swarm. A second option would be for one vessel within the swarm to become a regulated Australian vessel (if this is possible), and then, possibly, all other swarm members could be considered "ship equipment" and be somewhat interchangeable. It is unknown whether the AMSA would be supportive of either option raised, noting that the concept of "ships equipment" is not explicitly stated in the *Navigation Act 2012* framework, and a maritime autonomous swarm presents new risks that other small vessels using that category do not.

9.4.1.3 Foreign Vessels

Foreign vessels operating in Australian waters are subject to the *Navigation Act 2012* and are routinely inspected in Australian ports by the AMSA Port State Control marine surveyors [74]. Autonomous vessels are generally unable to comply with the international law conventions implemented in the *Navigation Act 2012*, or in the corresponding legislation of its flag state [73] (i.e. its home government). Where a foreign vessel has an exemption or equivalence enabling it to operate by its flag state, it is generally allowed to continue operating under that document in Australian waters following appropriate dialogue with the AMSA. This approach has been used to enable foreign vessels that are autonomous vessels to operate in Australian waters. However, where the subject vessel does not have an exemption or equivalence from its flag state, for example, a flag state determines that they are not "vessels" but perhaps "flotsam", that is, marine rubbish, it makes identifying and enacting the correct regulatory pathway more difficult.

It seems likely that if the operator of a maritime swarm, composed of individual autonomous vessels, had an exemption or equivalence from its flag state, and the AMSA had assurances regarding the safe operation of the vessels (e.g. by viewing the safety management system and safe operating history of the operator), then it would be allowed to operate in Australian waters.

9.4.1.4 Defence Vessels Carved Out from Commercial Regulatory Framework

Vessels that fit the definition of "defence vessel" in the *Marine Safety (Domestic Commercial Vessel) National Law Act 2012* or "naval vessels" in the *Navigation*

Act 2012, are not subject to regulation by those Acts, and are instead regulated by defence through the Defence Seaworthiness framework. This requires that the vessel is (a) a warship or other vessel that (i) is operated for naval or military purposes by the Australian Defence Force or the armed forces of a foreign country, (ii) is under the command of a member of the Australian Defence Force or of a member of the armed forces of the foreign country, (iii) bears external marks of nationality; and (iv) is manned by seafarers under armed forces discipline, or (b) a government vessel that is used only on government noncommercial service as a naval auxiliary. The AMSA does not "determine" whether a vessel fits this definition but can provide guidance to defence on specific scenarios. Ultimately, only a court can determine the correct interpretation of the application of legislation against specific circumstances.

It is considered generally possible for autonomous vessels to meet the definition of "defence vessel" and "naval vessels" described above. Livoja, Massingham, and McKenzie explored the critical issue of whether an autonomous vessel is under "command", determining that "the command requirement does not necessitate direct oversight by a (human) commander for every decision made, but rather requires asking whether the system is fulfilling the intent of the commander" [532]. The defence exercises Autonomous Warrior 2018[22] and Autonomous Warrior 2022[23] provide examples of Defence (through Warfare Innovation Navy) working with the AMSA to determine which participating vessels were defence vessels, foreign vessels, or domestic commercial vessels and implementing appropriate regulatory approaches to ensure the exercise occurred consistently with the requirements of both the *Marine Safety (Domestic Commercial Vessel) National Law Act 2012* and the *Navigation Act 2012*.

Whether it is preferable for a vessel used by Defence to be considered a domestic commercial vessel, a regulated Australian vessel or carved out as a defence vessel or naval vessel is circumstance-dependant and a matter for defence. The benefit of applying the civilian regulatory framework is the certainty and resulting efficiency and trust generated; however, this is offset by the potential delay and expense involved in confirming requirements and seeking suitable exemptions. The benefit of applying the Defence Seaworthiness framework is that it is tailored to Defence's need to prioritise mission completion while considering the safety of persons and the environment. Similar to the civilian regulatory framework, there is likely to be uncertainty in some cases about how to apply the seaworthiness framework to autonomous vessels, whether acting individually or in a swarm, which can cause delay and jeopardise project timeframes.

---

[22] Stephen Kuper, "Defence to highlight future of unmanned operations at Autonomous Warrior 2018", Defence Connect.

[23] Australian Government: Defence, "Exercise Autonomous Warrior testing new technologies to meet emerging maritime security challenges", media release issued by Defence Media, 16 May 2022.

9.4.1.5 Defence Vessels: Defence Seaworthiness Framework

Vessels used by Defence, across all services, are regulated by the Defence Seaworthiness Authority, under the Defence Seaworthiness Management System (DSwMS) [796] "[796]" ([p.15)]. The DSwMS aims to ensure that the operation of maritime materiel achieves the desired tasking while eliminating or minimising risk so far as reasonably practicable. The DSwMS achieves this through an outcome-based approach, using both governance and management compliance obligations (GMCOs) and activity- and condition-based compliance obligations (ACCOs).

Under the DSwMS the safety and capability risk associated with maritime materiel must be assessed by a suitably competent and authorised person. To conduct the assessment, the assessor must have a thorough understanding of the item, its intended use and the risks associated, and sufficient evidence that both the governance and management compliance obligations and activity- and condition-based compliance obligations have been met [796].

The successful undertaking of Autonomous Warrior 2019 and Autonomous Warrior 2022, which both included defence vessels, is evidence that autonomous vessels are capable of operating under DSwMS, although the complexity and time taken to achieve this is unknown. It is evident that the completion of a regulatory safety and environmental framework is part of meeting compliance obligations, and also assists in communicating risks and controls to other relevant regulators, whether that be AMSA or a local council.

The Trusted Autonomous Systems swarming primer [796, p.15] suggests that "the DSwMS may not suitably address swarm systems where the boundary of the system is not clear, as in the case of a swarm with adaptable/flexible size where individual agents may be added or removed". Further work is required to identify how a maritime swarm would fit within the Defence Seaworthiness framework, and the adjustments to language, intent, and risk management that may be required. Conducting events that include autonomous vessels, and maritime swarms, is a good way of continuing regulatory development, specifically, by creating tangible examples for legal analysis, and regulatory and policy development, to follow.

9.4.1.6 Defence Vessels: International Law in Peace and Conflict

If a maritime swarm is to be used by Defence outside of Australian waters, or during times of armed conflict, it is critical to consider the international law obligations that would apply. There is an important distinction between a maritime swarm used for defence purposes and a maritime swarm used as a weapon. Legal obligations also differ depending on the characterisation of the swarm and between operations in peace or conflict. In times of peace general international law applies, for example, the United Nations Convention on the Law of the Sea. Where there is armed conflict, the requirements of international humanitarian law (IHL) also apply [696]. Where a swarm has lethal capability, the legal framework for assessing new means and methods of warfare apply. This includes the Article 36 requirements in Additional

Protocol I to the Geneva Conventions regarding ensuring the ability of new weapons, means, and methods of warfare comply with international humanitarian law.[24]

When a maritime swarm is developed, a legal analysis needs to occur to identify the exact requirements that apply under the various international conventions that are in force. The first critical question in that analysis is how to characterise the swarm: Is it considered a single legal entity, or a collection of individual legal entities that work together? The answer to that question affects the legal requirements and processes that need to be complied with [338]. Given that this is the central topic of a later chapter, this chapter does not address this beyond highlighting it as an important consideration for developers and funders to consider well in advance.

Research is underway considering a range of questions related to the use of swarming for defence purposes, including how the technology will work; how it will comply with existing legal requirements in peacetime, for example, related to navigation and safety; and its ability to comply with the legal requirements that apply in times of armed conflict, including international humanitarian law [180, 571, 572]. Given that real-life examples of autonomous vessels are being weaponised and used in armed conflict, there is increasing urgency to understand these issues and to identify and enact any regulatory reform needed to ensure ongoing safe and trusted operations [591].

Summary of Key Issue 1: How Do We Characterise Current and Likely Future Maritime Swarms, and What Is the Impact on Their Regulatory Treatment?

This section highlighted the importance of characterising maritime swarms and outlined the regulatory treatment for vessel categories, ranging from domestic commercial vessels and regulated Australian vessels to defence vessels operating in peacetime or during armed conflict. Empowering developers and operators with this knowledge as early as possible in the design process enables greater flexibility and more streamlined regulatory processes. This section demonstrated the breadth of categories and regulatory treatments available to vessels operating as part of a maritime swarm or for commercial or defence applications while identifying the opportunities or limitations presented specifically for swarms.

---

[24] "All states are required to undertake legal reviews of new weapons either because of an express obligation under Article 36 of the Additional Protocol I to the Geneva Conventions, or to give effect to a broader International Humanitarian Law requirement to ensure the lawful use of weapons in armed conflict. While international law does not require a particular review methodology to be used, the rapid changes to technology that enable autonomy—such as Artificial Intelligence and machine learning—raise the question of how states can practically conduct legal reviews of weapons that are enhanced by such technology" [696].

9.4.2 Key Issue 2: Discussion of the Relationship Between Regulation, Explainability, and Trust for Maritime Swarms

Trust[25] in technology, including by operators, regulators, and the public, is critical for its successful use in public spaces. In a general sense, trust in technology is based on a combination of compliance with applicable regulatory frameworks, a history of successful use, familiarity, reliability, and trust by association. When new technology is developed, it must become trusted before it becomes an accepted inclusion in operations, and an accepted presence in public spaces, including harbours and waterways. The danger and uncertainty of new technology, to a large degree, determines the level of trust needed before that technology is accepted. Specific to autonomous underwater vessels (AUVs), Keane et al. identified that reliability and resilience, being the ability to [453] (referencing [801]) "recover from unexpected disturbances in the operating environment", are critical factors in confidence or trust. Keane et al. also identified that more research is needed on how to build operator trust in autonomous technology [453], which highlights the need to consider trust from multiple perspectives, including operators as well as regulators and the public.

Understanding the concept of regulation in the context of autonomous technology is critical, as is its relationship to trust and explainability. Regulation can be described as "sustained and focused control exercised by a public agency over activities that are valued by a community" [453]. The word regulation implies some sort of influence by government, but it[26] [94] "may be carried out not merely by state institutions but by a host of other bodies, including corporations, self-regulators, professional or trade bodies, and voluntary organisations". Blaxland et al. identified that although regulation can be restrictive, it can also be facilitative or enabling [94]. From an Australian Government perspective, regulation includes "any laws or other government-endorsed 'rules' where there is an expectation of compliance. In Australia, regulation is made at the federal level as well as by the states and territories, in the form of legislation and subordinate legislation and at a local government level as regulations and by-laws" [232].

It is generally accepted that regulation is necessary to ensure a minimum level of safety throughout society and to maintain stability for our economy and that regulation being in place generates trust [232]. Regulation also contributes to explainability because it provides a standard against which a product, service, or operation can be measured. Typical examples of regulation include road safety rules,

---

[25] "Trust, the firm belief in the reliability, truth, or ability of someone or something, underlies all social and economic relations and is central for the acceptance and adoption of autonomous vessels both by the maritime community and the general public. Trust requires explanations but is a much broader concept facilitating interaction among people and between people and technologies" [325].

[26] Reference note 8: see [343]. Note: It is difficult to find a single precise definition of regulation; see [479].

product liability, and food safety regulation. Products and businesses that comply with applicable regulatory frameworks are typically trusted by the public, and advertising campaigns sometimes include reference to having specific certification or compliance levels to capitalise on that.

> A challenge for the Government or applicable regulator is to properly balance risk with regulatory overlay. If there is insufficient regulatory overlay to meet the risk of the operation, it will jeopardise trust in the technology and its regulation, but if the regulatory overlay is too heavy, it can stifle innovation and jeopardise trust in the intentions of the regulator. For this reason, adopting an appropriate regulatory approach is critical to ensuring innovation is supported, the benefits of autonomous systems are realised, and trust is maintained between all stakeholders involved. [386]

Appropriate regulatory frameworks foster trust, particularly when a history of compliance is established.

A core component of regulation is assurance, which is a broad term for the process of demonstrating compliance with applicable requirements. Trusted Autonomous Systems explained that "safety assurance of ... autonomous vessels, aircraft, vehicles and associated operations ... are all the processes, checks, requirements etc. that are undertaken to confirm that the specific regulatory requirements are met, and therefore minimise uncertainty related to an operation being able to achieve safe outcomes" [795]. In a maritime context, inspection by accredited marine surveyors and sea trials are common parts of the assurance process [795]. Assurance for autonomous technology is particularly difficult. Three key issues are as follows: (1) the need to identify and use novel technical approaches because of the unpredictable and changeable nature of algorithms and software [795], (2) the need for third parties providing assurance services to possess sufficient technical expertise and experience [795], and (3) the lack of tailored regulatory framework and technical standards to inform assurance requirements [235]. A fourth issue, related specifically to AUVs in the maritime domain, is the complexity of the operating environment in which the vessel needs to operate safely, and how assurance activities can be built to address that. The subsea environment is changeable and complex, and depth, temperature, and salinity all affect operations while communication options are extremely limited. These factors mean that AUVs need higher levels of automation than their surface equivalents, and there is lower certainty about factors such as their speed and position [453]. Assurance is a core component of trustworthiness, but until the aforementioned issues are addressed, it will remain a problematic part of the regulatory process.

In a maritime context, vessels operating lawfully, specifically, in compliance with applicable legal requirements, including relevant assurance requirements, are generally trusted or considered "trustworthy",[27] to operate in public waterways. This trust and how it is built differs between operators, regulators, and the public. For example, operators may build trust in a technology through compliance with

---

[27] "Trustworthiness is a property of an agent or organisation that engenders trust in another agent or organization" [237].

accepted design, construction and survey processes, and operating experience; a regulator may build trust through legal compliance, successful inspections, and a history of safe operations, and the public builds trust through seeing them in use, seeing respected/trusted people use or endorse them and knowing that they are being lawfully used. Where new technology is developed, there is no baseline level of trust between operators, regulators, and the public, and that needs to build up with time, experience, and appropriate regulation [386].

Explainability is a principle that requires actions to be explainable in a way a human can understand, and it is a critical foundation for trust.[28] When applied to AI, the principle is described as XAI [756].

> Explainable artificial intelligence (XAI) is a set of processes and methods that allows human users to comprehend and trust the results and output created by machine learning algorithms. Explainable AI is used to describe an AI model, its expected impact and potential biases. It helps characterize model accuracy, fairness, transparency and outcomes in AI-powered decision making. Explainable AI is crucial for an organization in building trust and confidence when putting AI models into production. AI explainability also helps an organization adopt a responsible approach to AI development. [411]

Explainability builds trust in autonomous systems enabled via AI because it enables humans to understand the reasoning behind specific actions. It is clear that for autonomous vessels that comprise a maritime swarm to be trusted, their behaviour needs to be explainable, in a way that the recipient of the information can understand.[29] The emerging area of "human-centred XAI" is working to address the need for explainability outputs interpretable by a wide range of end users, for example, passengers or other vessels [814].

Noting the criticality of explainability in building trust in autonomous systems, it would be reasonable to expect that it was a requirement under the applicable regulatory framework. However, in a maritime context, the applicable regulatory framework does not contemplate autonomy at all, and so there are no specific requirements or standards related to issues such as explainability. As described earlier in this chapter, regulations are still based on those for traditional vessels, such as fishing vessels, and have not yet been updated to include specific requirements relevant for autonomous vessels. This inhibits the development of trust in autonomous technology for all parties, including operators, regulators, and the public. Whether this needs to change is considered in the following section.

---

[28] Explainability is included in Australia's AI Ethics Principles (https://www.industry.gov.au/publications/australias-artificial-intelligence-ethics-framework/australias-ai-ethics-principles) and NATO's Principles for Responsible Use of Artificial Intelligence in Defence (for more information see [612]).

[29] For a comprehensive discussion on explainability for various audiences, see [325].

9.5 Where to from Here

As maritime swarms research progresses closer to operationalisation, a focus on how to characterise the technology and the corresponding regulation will become increasingly important. As demonstrated in this chapter, there are numerous vessel categorisation options for developers and operators to consider, for both commercial and defence applications, and an early appreciation for these will support successful projects. Understanding the relationship between explainability, trust, and regulation will support the development of new technical standards and practices that embed trustworthy practices early and support the use of the technology in a way that builds trust among all stakeholder groups.

It is essential that regulation is understood, and areas for improvement identified, to support appropriate regulatory change. To better facilitate autonomous technology in the maritime domain, whether it be single vessels or swarms, legislative change is needed to enable the AMSA to determine requirements for vessels rather than relying on exemptions from automatically applying requirements. The AMSA needs to introduce a risk-based regulatory approach, enabling it to better match the risks of a vessel and its proposed operation with the regulatory overlay required. Consideration of facilitating a single operator to supervise multiple vessels, and of a human supervisor supervising a maritime swarm, should occur, as well as ensuring any regulatory change is flexible enough to enable technological developments not yet thought of. Changes to be considered should include enabling a "system of systems" approach, in which a system rather than a single vessel is regulated [235],[30] providing more flexibility in deciding which regulatory requirements apply, and incorporating tailored technical standards led by industry [17, 235].

Further research is needed on issues such as how to apportion responsibility and liability, how to manage multi-domain operations, and how to require explainability in a way that is achievable by operators and interpretable by third parties. Policy makers need to learn from the experience of other nations and other domains to ensure a fit-for-purpose, practical approach is taken and enables rather than inhibits innovation. Conducting regulatory sandbox activities to trial new approaches and gather data to inform risk-based regulation should also occur.[31]

---

[30] Further discussion on a "systems of systems approach" is available, for example, [386]); and Rachel Horne, Caroline Law-Walsh, Zena Assaad, and Keith Joiner, "Ten regulatory principles to scaffold the design, manufacture, and use of trustworthy autonomous systems, illustrated in a maritime context" (Conference Paper, TAS '23 First International Symposium on Trustworthy Autonomous Systems, July 2023).

[31] Note: Sandboxes were a recurring theme at the Autonomous Vessel Forum 2022, reference: R Horne, Reflecting on the Autonomous Vessel Forum 2022, Trusted Autonomous Systems Blog, 12 Nov 2022 (https://tasdcrc.com.au/reflecting-on-the-autonomous-vessel-forum-2022/). Note Regulatory Sandboxes have been implemented by the AMSA at the ReefWorks Test Range at the Australian Institute of Marine Science in Townsville, Queensland, Australia, and at the AMSTEC Test Range at the Australian Maritime College Search at Beauty Point in Tasmania, Australia.

The chapter on Trust and Safety in the Australian Robotics Roadmap report highlighted the link between trust, safety, and regulation and identified how autonomous technology is changing traditional paradigms:

> In order for the robotics industry to thrive, a regulatory response that adapts, protects, and engenders community trust, while accounting for rapidly evolving industrial environments, is essential. This is an area where investment can lead to improved outcomes for the whole robotics sector, and Australia can lead the world. [235]

To transition the existing regulatory framework from somewhat static and human-centred to something that exhibits the values called for in the Australian Robotics Roadmap, a regulatory reform process is required.[32] To support this future process, a series of principles have been identified that are relevant to the maritime domain and air and land domains.

9.6 Principles Arising Relevant to Other Domains

Considering regulation, how it applies to maritime swarms, and the changes needed has produced several principles to help shape regulatory reform. These are relevant to all maritime, air, and land domains:

- Establishing common language is crucial. Establishing a common language regarding autonomy and swarming is crucial for productive discussion and change. Without this, repeatable legal analysis is not possible and meaningful progress is more difficult.
- Establishing why regulation is necessary and critically considering whether it needs to be changed, how the change will occur, and what success will look like is necessary to inform successful regulatory change.
- Identifying fundamental questions that need to be answered to inform appropriate regulatory pathways, and seeking to address them early, is critical.
- Establishing objective ways to assess risk is crucial.
- Building trust in technology among the community and regulators is critical.
- Regulators across domains and across commercial and defence sectors must collaborate on regulatory change to ensure smooth regulatory and assurance pathways for multi-domain and dual-use technologies.
- Industry and third parties, such as government-funded research centres, will play a critical role in advocating for regulatory change and in informing what it should look like.

See R Horne, An Australian Maritime Regulatory Sandbox, Trusted Autonomous Systems, https://tasdcrc.com.au/an-australian-maritime-regulatory-sandbox/.

[32] Note that further discussion on the regulatory reform required to support the safe and trusted use of autonomous and remotely operated vessels is available, for example, [385].

- Regulatory sandboxes should be used to enable testing of regulatory and policy settings and to assist developers in testing and trialling their projects.
- Multidisciplinary collaboration is crucial to progress suitable regulatory frameworks to enable the safe and trusted use of autonomous vessels, as individuals or in a swarm, into the future.

9.7 Conclusion

Maritime swarming is on the horizon, and consideration must be given to how this technology will be characterised and regulated, how that regulation relates to explainability and trust, and whether regulatory changes are needed. This chapter set out to investigate these key issues, drawing on the relevant literature and highlighting areas of importance for further research, in addition to proposing several principles to guide future regulatory reform. As autonomous vessel technology continues to advance, and maritime swarms move closer to operationalisation, it is critical that regulatory reform occurs to ensure a regulatory framework is available that supports safe and trusted operation into the future.

Acknowledgments The author would like to acknowledge the support of her PhD supervisory team, Prof Kieran Tranter (QUT), Associate Professor Felicity Deane (QUT), and Dr Keith Joiner (ADFA), in addition to the input of the UQ Future of War team, Dr Eve Massingham (UQ), Dr Simon McKenzie (Griffith University), and Dr Tim McFarland (UQ).

Open Access This chapter is licensed under the terms of the Creative Commons Attribution 4.0 International License (http://creativecommons.org/licenses/by/4.0/), which permits use, sharing, adaptation, distribution and reproduction in any medium or format, as long as you give appropriate credit to the original author(s) and the source, provide a link to the Creative Commons license and indicate if changes were made.

The images or other third party material in this chapter are included in the chapter's Creative Commons license, unless indicated otherwise in a credit line to the material. If material is not included in the chapter's Creative Commons license and your intended use is not permitted by statutory regulation or exceeds the permitted use, you will need to obtain permission directly from the copyright holder.

Chapter 10
An Approach to the Legal Review of Autonomous Swarms

The Legal Implications for Designing Thinking Swarms in Armed Conflict

Damian Copeland, Philip Sammons, and Lauren Sanders

Autonomous weapon systems (AWS), employed in swarms, promise enhanced military effectiveness. However, with that efficiency comes complexity in understanding how those swarms operate and how their use might comply with international law. States have agreed that the use of AWS must conform with international law and, more relevantly, the laws of armed conflict. Prior to the use of weapons, means and methods of war, States must ensure whether as a legal obligation under Additional Protocol I or as a matter of practicality that new weapons, means and methods of war are capable of complying with their international legal obligations. This chapter explains how the use of simulation testing and evaluation methods can be used to support the conduct of these legal reviews. It identifies which agents of an AWS, and each agent's constituent parts, must be the subject of review and concludes that to properly undertake an assessment of the lawfulness of AWS swarms, reviewers must take a risk-based approach. This risk-based approach can be informed by simulation testing techniques, such as Monte Carlo methods, and will aid in assisting reviewers to understand what limits must be placed on AWS swarms to make them lawful, such

---

D. Copeland
Director, Article 36 Legal, Canberra, ACT, Australia

TC Beirne School of Law, University of Queensland, St. Lucia, QLD, Australia
e-mail: damian@article36legal.com

P. Sammons
Contour Advisory, Canberra, ACT, Australia

University of New South Wales, Sydney, NSW, Australia
e-mail: philip.sammons@contouradvisory.com.au

L. Sanders (✉)
TC Beirne School of Law, University of Queensland, St. Lucia, QLD, Australia

International Weapons Review, Fortitude Valley, QLD, Australia
e-mail: l.sanders@uq.edu.au

as limitations on swarm size or mandating of detection, classification, recognition and identification confidence level to undertake particular combat actions.

10.1 Introduction

The challenges relating to the development and use of autonomy in military weapons have now been the subject of international debate for over a decade.[1] Although the prospect of new international laws regulating autonomous weapon systems ("AWS") remains uncertain, researchers, governments and industry from around the world are working towards the use of autonomous systems in coordinated swarms to achieve a military effect [354]. The use of AWS in military swarms, a tactic promising greater military advantage, involves converging massed physical effects and concentrating combat power through autonomous technology [833]. This tactic allows for less complex computing in the autonomous individual agents and allows the swarm to conduct more complex tasks than an individual agent. However, it provides redundancy in cases of destruction of individual agents within the "smart, small and many" autonomous systems that comprise the swarm as a whole. Unlike investing in 'few and exquisite' high-end military capabilities, there are obvious tactical and acquisition advantages in adopting such technologies [355].

Despite the advantages, AWS swarms pose several legal challenges concerning an individual States's obligation to certify these systems' compliance with international law obligations. States that are signatories to Additional Protocol I to the 1946 Geneva Conventions ("AP I"), adopting new methods of warfare, such as swarming autonomous systems, are obliged to undertake a review for compliance with international law, and international humanitarian law (IHL) in particular.[2] This review must occur before they are used in situations of armed conflict.

When used as a method of warfare, swarming challenges defence paradigms of human control and necessitates a re-evaluation of the traditional approach to the AP I Article 36 weapons review (hereafter referred to as a "weapons review" or "Article 36 weapons review").[3] Specifically, when assessing the intended tactics for deploying military swarming systems, the review must assess whether the normal and anticipated use of the swarm complies with the IHL rules governing an attack. This specifically includes the capacity of the system (and its operator) to comply with the principles of precautions, distinction and proportionality. The unique technological features of autonomous swarms used for military operations

---

[1] See UN Office for Disarmament Affairs, Background on LAWS in the CCW (Website) https://www.un.org/disarmament/the-convention-on-certain-conventional-weapons/background-on-laws-in-the-ccw/.

[2] Additional Protocol I to the 1946 Geneva Conventions ("AP I"), Article 35(3).

[3] AP I, Article 35(3).

therefore necessitate specific legal consideration not required in the review of other AWS capabilities.

Building on the traditional approach to Article 36 weapons review, this chapter considers normative, risk-based approaches to the review of swarms to enable their lawful use in armed conflict. It commences, in Sect. 10.2, with a description of the common features of existing "traditional" Article 36 weapons review processes, to enable these steps to be applied, and modified as required, to meet the specific review requirements of swarm technology.

An analysis of the common rules governing the swarm's behaviour and the various roles that individual swarm agents may play in the use of force must be undertaken to enable its weapons review. Section 10.3 will outline which technical features of military swarms necessitate additional review for Article 36 compliance. The specific parts of the swarm that require review may vary depending on the swarm's architecture. This means that the review process may need to consider the individual agents and their functions or the potential effects of the entire swarm.

The implications for weapons review processes related to these specific design features are considered in Sect. 10.4. Subsequently, Sect. 10.5 assesses the key legal issues relevant to the use of AWS swarms in armed conflict. Notably, this does not cover all legal issues related to the regulation and use of swarms, nor does it survey all *jus in bello* issues. Instead, it focuses on those legal aspects relevant to weapons regulation within the Australian context, thereby assessing the potential adoption of drone swarms as Australian Defence Force capabilities. The regulation and governance of drone swarms warrant separate detailed analysis.[4]

Following this, Sect. 10.6 contains a brief survey of existing policy and ethical landscapes to ascertain what additional considerations may affect the conduct of a legal review of a drone swarm's capability. Using the Martens Clause and adopting the Australian interpretation of this clause, a limited analysis of ethical considerations is included.[5] It is to be noted that this chapter does not propose to focus specifically on the ethical issues related to the adoption of swarm technology. The brief canvassing of ethical considerations is limited to those adopted by the current Australian Defence Force ethics doctrine. Further, the chapter does not propose to incorporate a broader ethical debate about the use of AWS swarms, seeking only to analyse limited governance and assurance considerations that may affect the review of drone swarms for introduction into military service (for more information on this issue, see, e.g. [685]).

---

[4] See, for example, "Regulation of Maritime Swarms".

[5] The Martens Clause first appeared in the preamble to the 1899 Hague Convention II (Hague Convention (II) with Respect to the Laws and Customs of War on Land (adopted 29 September 1899, entered into force 4 September 1900) Annex: Regulations concerning the Laws and Customs of War on Land art 22) and was subsequently reaffirmed in the 1907 Hague Convention IV (Hague Convention (IV) respecting the Laws and Customs of War on Land (adopted 18 October 1907, entered into force 26 January 1910) (hereafter "1907 Hague Convention IV")); and AP I, above (n 5), Article 1(2); International Court of Justice, Requests for Advisory Opinions on the Legality of Nuclear Weapons—Australian Statement, 1996 AUSTL. Y.B. INT'L L. 685, 699, cited in [578].

Finally, in Sect. 10.7, a methodology to assess risk related to the above legal challenges is discussed, thus identifying a practical methodology to enable the weapons review of AWS swarms. Specifically, this section discusses how simulation systems can undertake edge case testing to assess whether a swarm can meet minimum review standards relating to two issues: first, how a swarm can be tested to meet minimum risk standards for IHL task compliance, and second, how a military swarm can be tested to articulate the legal risk associated with adopting swarms to capitalise on the efficiencies presented by swarm technology compared with nonswarming autonomous systems.

10.2 The Traditional Article 36 Weapons Review Process

AP I expressly requires States to legally review any new weapon, means or method of warfare, before its use in armed conflict, for compliance with that State's international legal obligations. Article 36 states:

> In the study, development, acquisition or adoption of a new weapon, means or method of war, a High Contracting Party is under an obligation to determine whether its employment would, in some or all circumstances, be prohibited by this Protocol or by any other rule of international law applicable to the High Contracting Party.

This provision is intended to enable the creation of national mechanisms to "prevent the use of weapons that would violate international law in all circumstances, and to impose restrictions on the use of weapons that would violate international law in some circumstances" [414]. The determination is based on the normal or expected use of the weapon under the circumstances in which it is expected to be deployed.

Given that weapons reviews are implemented at the national level, the approach varies between States.[6] However, it is possible to identify four common steps from existing State practice and distil what is legally necessary to discharge the requirements set out in Article 36, in what this chapter describes as the "traditional weapons review" process [414]. The first, preliminary step confirms the applicability of Article 36 as a matter of law by assessing whether the subject is determined by the reviewing State to be a "new weapon, means or method of warfare". Given that AP I does not define these concepts, the reviewing State can rely on its own interpretations. This step also identifies the normal or expected use of the weapon, anticipated at the time of the review, which serves to limit the scope of the legal analysis. Effectively, this step asks: "Must the State review this system, and if so, for what kind of use?".[7]

---

[6] All references to "States" in this chapter connotes a government with rights and obligations, subject to international law.

[7] "AP I", Article 36(5).

The second step determines whether the type of weapon is specifically prohibited or restricted by treaty or customary law binding the reviewing State. That is, does the law ban this kind of weapon outright? As not all States have signed up to all international law treaties regulating the possession and use of weapons, this determination is unique to the reviewing State. The analysis in this step focuses on the type of weapon or the mechanism designed to cause harm to persons or damage to objects.

If the weapon is not caught by a specific international law prohibition or restriction, the third step in the traditional weapons review process considers whether the weapon causes prohibited effects. That is, can it be used in accordance with the applicable law? This step applies international law rules of general application that apply to all weapons.

First, giving effect to the IHL principle of distinction, the weapons review considers whether the subject weapon is, by nature, indiscriminate. This means that it would be unable to be directed at a specific military objective. To be considered discriminate, a weapon must be capable of being directed at a specific military objective and, therefore, distinguishing between lawful targets and unlawful targets. For States that are part to AP I, this determination is informed by Article 51(4). In practice, the threshold for finding a weapon to be indiscriminate by nature is high, because almost any weapon can theoretically be used in circumstances that do not cause harm or destruction to civilians or civilian objects.

The second general prohibition is against weapons that are designed to cause superfluous injury or unnecessary suffering to combatants. This reflects the cardinal IHL principle of humanity[8] and requires consideration of the military advantage anticipated from the use of the weapon in comparison with the harm it causes to combatants. In practice, this requires multidisciplinary input from medical experts and military commanders to inform a determination of whether the anticipated weapon effect causes "harm greater than that unavoidable to achieve legitimate military objectives".[9]

Although the prohibitions against indiscriminate weapons and weapons that cause unnecessary suffering are customary in nature, the final general prohibition against weapons that cause long-term, widespread harm to the natural environment only bind States party to AP I.[10]

The final step in the traditional weapons review process is quasi-legal in nature in that it includes the consideration of national policy in the analysis. In some cases, although a State may not be prohibited or restricted by international law from the possession or use of a specific weapon, means or methods of warfare, national policy

---

[8] Additional Protocol I to the 1946 Geneva Conventions ("AP I"), Article 51(4).

[9] Legality of the Threat or Use of Nuclear Weapons (Advisory Opinion) [1996] ICJ Rep 226 [260] ("Nuclear Weapons Case").

[10] "AP I", Article 35(3) states, "It is prohibited to employ methods or means of warfare which are intended, or may be expected, to cause widespread, long-term and severe damage to the natural environment".

may impose restrictions for ethical, moral or political reasons. These may reflect interpretations of international law requirements as contemplated by the Martens Clause, which was first expressed in the Preamble to the Hague Convention II of 1899, containing the Regulations on the Laws and Customs of War on Land.[11] The contemporary Martens Clause is found in Article 1(2) of AP I.[12]

These traditional weapons review steps focus on the legality of the weapon per se and do not consider the contextual uses of the weapon. This reflects the traditional dichotomy between the legal analysis in Article 36, designed to determine the legality of a weapon prior to its use in armed conflict, and the individual determinations of lawful use in armed conflict, informed by the advice and training of a legal adviser deployed in compliance with Article 82 of AP I. The challenge for the development and use of AWS swarms is whether the traditional weapons review steps are sufficient to adequately determine their legality for use in situations of armed conflict.

Before considering this challenge, it is important to consider the characteristics and design of AWS swarms to identify whether the traditional approach is sufficient, or whether there is a need for additional review processes. Swarms are not only novel in their design but also may enable new, collective methods of warfare that challenge existing interpretations of IHL.

10.3 Technological Features of Military Swarms That Impact Weapons Review Considerations

Reduced to its most basic description, a swarm is a group of individual systems or agents that operate as a collective [817]. Although this volume engages with the difficulties of properly defining and articulating exactly what swarming is, in this chapter, to the extent possible, the issues raised about the review of AWS swarms will be definition-agnostic, noting that this issue would require resolution in the capability description prompting the weapons review itself.

The specific AWS swarms that would be the subject of the weapons review, and those that this chapter is interested in, are that of a team of armed, autonomous mobile drones capable of dynamic allocation of resources and intercommunication. This is a specific distinction from swarming as a tactic, although the drone swarms described are capable of, and likely to be tasked to, employing swarming tactics when deployed during armed conflict. The item being reviewed is the swarm drone

---

[11] 1907 Hague Convention IV.

[12] Additional Protocol I to the 1946 Geneva Conventions ("AP I"), Article 1(2) states, "In cases not covered by this Protocol or by other international agreements, civilians and combatants remain under the protection and authority of the principles of international law derived from established custom, from the principles of humanity and from the dictates of the public conscience".

and swarm drone agent. The action that may require separate review (specifically, the method of warfare) is the use of AWS to undertake swarming tactics.

Drone swarms differ to drones employed en masse, in which large numbers of individual drones are individually tasked to complete a mission [446]. This is also distinct from systems that use multiple, centrally controlled, or pre-programmed agents, to achieve an outcome [446]. For example, the use of drones to create spectacular light shows is not swarms; the individual drones used for such displays follow a set, pre-planned path and are not designed to make individual decisions or alteration to the set plan.[13] AWS swarms rely on dynamic individual and group behaviours to complete the provided high-level objectives. This is achieved through internal communication and monitoring with other agents. The dynamic and uncertain nature of this process makes it particularly difficult to determine their outcomes.

Swarming as a military tactic is designed to overwhelm or saturate the target's ability to adequately defend. When describing the act of swarming as a tactic, the Australia Defence Force defines it as "[t]he large mass of autonomous systems interoperating collectively to act and respond in a coordinated effort to provide an overwhelming effect".[14] Emulating the swarming behaviour of different animal species, military swarming permits low-cost systems to operate as a collective, without a central controller, using common algorithmic rules to achieve military goals.

Two key features of swarming technology trigger additional legal considerations when undertaking a weapons review:

1. A key feature of the use of autonomous technology capable of employing this tactic is that the agents that comprise the swarm are typically smaller, with less computational power, and thus component parts of the swarm contribute to a greater overall task capability by allocating specific tasks to specific swarm agents.
2. A second feature of the use of swarming tactics is that there is capacity for decentralised control of tasks and agents to independently reorganise tasks depending on the mission objectives and requirements (see generally [368]).

In relation to this first issue, the agents of the swarm may be capable of performing a list of specific tasks, and may be tasked to complete them sequentially, having limited capability to conduct them concurrently. The swarm, when operating as a whole, may be tasked with specific missions that combine discrete tasks, such as surveillance, target acquisition and delivery of a munition. When combined, however, these tasks can form complete missions when operating in a swarm.

The recent employment by the Russian Federation of the Iranian-made Shahad 129 to attack Ukrainian targets is an example of how existing drone technology is

---

[13] Barrett, B., "Inside the Olympics Opening Ceremony World-Record Drone Show," Wired, 9 February 2018, www.wired.com/story/olympics-openingceremony-drone-show, cited in [446].

[14] Australian Defence Force, Australian Defence Force Glossary.

being used in armed conflict. The Shahad can be programmed to loiter above a target area and conduct kamikaze-styled attacks on infrastructure such as munitions stores and power stations [67]. In its current configuration, this loitering drone requires human control in the conduct of the target identification and selection, and only limited parts of the targeting cycle are conducted by this capability.

When combined with a swarm of drones, agents of a swarm could be used to support all tasks associated with the targeting cycle. Using a simplified example, adopting the find, fix, finish, exploit, analyse and disseminate (F3AED) targeting process, this entire cycle could be undertaken by a drone swarm, and the computational power in a particular agent could be allocated to a particular task within the F3AED cycle [275]. Parts of the swarm are allocated to munition carriage and delivery, while others are tasked to conduct the analysis, identification and surveillance functions. By being allocated only one part of the overall task, the agent can be smaller in terms of computational processing, and size, and centralised control, which enables better performance when combined with a swarm of coordinated agents. US military research using simulated swarms has suggested that swarm tactics are more efficient and more lethal than conventional tactics [833]. For example, a swarm of 800 agents was able to identify and attack 8% more targets than the 1,000 agents using nonswarm tactics [833].

In relation to the second key feature, the capacity for task reorganisation among agents of the swarm to conduct those individualised component tasks supports redundancy but also a differential approach to standard setting when completing different tasks. For example, target identification may require a certain number of sensors slewed to corroborate the identity of a target, but once positively identified to whatever standard is articulated in the system of control dictated for use in the particular operating environment, fewer sensors may be required to continue monitoring that target up until strike. As a basic example, it might require ten sensors to identify a target, and then five of them may be able to be re-tasked to observe civilian patterns of life in and around the identified target while the remaining five sensors maintain sensor identification (and "lock") on the target. This task reorganisation not only enables the swarm to be dynamic in response to its deployed environment but also enhances mission conduct by the dynamic internal re-tasking of swarm agents.

The use of drones in the battlefield to conduct planned attack missions is generally not considered true swarm behaviour. The use of drones in this way is effectively the automation of drones. Swarm behaviour involves the implementation of high-level behaviours that govern the actions of the drone swarm members rather than direct instructions. The emergent swarming behaviour drastically increases the scalability and effectiveness of the individual agent; however, it eventually comes at the cost of not knowing exactly how or why the swarm is completing the set task in the way it is because the data becomes too large and too complex to appropriately manage. It is the unpredictability and massively scalable aspects of drone swarms that warrant careful consideration of the existing Article 36 review process.

The use of swarms to perform military functions is still in the early stages of research and development [422]. Many countries, such as the USA, Russia, the

UK, France and India, recognise the utility of swarm tactics and are currently developing these capabilities for military use [354]. However, noting this scalability and unpredictability, it is not yet clear how the adoption of such technology might meet key legal standards when being used in armed conflict.

10.4 Legal Considerations Attached to Weapons Review of Military Swarms Used in Armed Conflict

Consequent on the unique technological features of AWS swarms, the approach to the weapons review of individual swarming agents, as well as the review of the method of swarming adopted by groups of agents, it is anticipated to require adjustment to adequately assess the lawfulness of these capabilities. This adjustment is in addition to the adjustment adopted for the review of individual autonomous capabilities [210].

An important, initial weapons review question is whether the AWS swarm design and use requires unique weapons review considerations or processes. This is a question that is currently being considered in the broader context of AWS [696]. There are a number of factors that are relevant to considering this question. First, a distinguishing feature of a swarm is that its control is decentralised by design. It relies on individual agents acting in compliance with an internal behaviour model rather than on an external control mechanism. This means that an AWS swarm is likely to be highly autonomous, having little direct or supervisory human control. This means that a weapons review must be capable of determining the ability of the AWS swarm to perform its role in compliance with those rules of IHL allowing certain tasks to be conducted autonomously. For example, if a system is expected to undertake a proportionality assessment, in which the system increases or decreases the lethality of effects, this requirement becomes even more complex according to how decentralised the system is and how removed it is from centralised decision-making. Thus, including an ability to self-moderate the lethality of effects will require an assessment of the certainty of the system's ability to choose the correct response in such a situation.

Second, to avoid random or completely unexpected behaviour, the rules programmed into individual swarm agents to perform swarm tactics are assessed to have to be deterministic in nature with the inputs to the system being sufficiently structured. A deterministic algorithm will, given a certain input, always produce the same output [804]. A number of components of an artificial intelligence (AI) system can be deterministic. Deterministic programming enables essential swarm behaviour, such as separation, alignment and cohesion, which allows swarms to move to a destination or avoid obstacles [354]. However, this is not sufficient to enable the execution of complex tasks given variability of the input space and the potentially large number of state space variables [714]. This is particularly important where IHL rules regulating methods of warfare require the exercise of human judgement. For example, some targeting rules, including distinction, proportionality and precautions in attack, contemplate the application of distinctly human cognition, such as "do everything feasible" or specify certain actions "in cases of doubt".[15] Where an AWS swarm is designed to conduct swarm attacks, the weapons review process must be designed to consider its ability to do so lawfully in the range of operational and environmental circumstances with a given level of assurance.

Third, although the concepts of swarming AWS are typically associated with the use of drones in the air domain, the approach to Article 36 weapons review is not limited to any one domain.[16] The weapons review must consider all domains and accept the possibility of multi-domain platforms, where future drone swarms or combination of agents across multiple domains could be acting as a swarm. It is the parts of the system that contribute to its kinetic (or harm-causing) functionality that are of particular relevance to the review. Thus, when conducting a review, all agents that can form part of the swarming system must be subject to a weapons review.

As previously noted, a separate approach may require the consideration of drone swarms as a method of warfare. The process for reviewing a method of warfare, rather than a weapon, requires different considerations during the review process. This may require the review to be made in the context of an AWS capability such as a single drone. It is possible that the AWS is in-service with the reviewing State and has already been the subject of a weapons review. However, it is possible that a State may re-program an existing AWS to employ swarming tactics and, in doing so, trigger the requirement for an additional weapons review to determine the legality of the new AWS swarm methods. In light of increasing autonomy in weapon systems, the component(s) of the swarms that has(have) "an impact on [an] offensive capability of the force to which it belongs" will require weapons review, even if individually they would not be considered a weapon ([568] cited in [140]). This effectively expands the componentry and parts of the system that are subject to review when assessing the legal compliance of AWS swarms.

Finally, the weapons review obligation is one of result rather than process. International law does not require States to undertake a particular process or methodology in determining the legality of new weapons, means or methods of warfare [137]. It simply requires States to have made such a determination in advance of a weapon's use. The State is then able to rely on the weapon being used lawfully by the members of its military, who are subject to international law accountability mechanisms, such as war crime prosecution for the misuse of weapons they control [210]. However, this approach does not enable a State to determine whether those functions may be performed by the AWS swarm autonomously and without direct human input in compliance with IHL. This is particularly pertinent to consider where those functions are critical to the targeting

---

[15] "AP I", Article 52(3) and 57(2).

[16] See The Australian Defence Force, Concept for Multi-Domain Strikes (2020) which defines five domains as land, air, maritime, space and cyber, as described in [423].

process and specifically regulated by IHL rules, such as distinction, proportionality and precautions in attack.

These factors suggest that the traditional weapons review process requires expansion to enable consideration of those aspects of swarming that are governed by IHL and performed without individual human control. This requires an additional process focused on the lawful use of the AWS swarm, in addition to the legality of the AWS. The former requires consideration of IHL targeting rules, whereas the latter concerns the international law prohibiting or restricting weapons.

10.5 Additional Review to Address IHL Rules That Impact Use of Military Swarms

Having established that additional review processes are required to account for the technical features of AWS swarms, this section seeks to propose how specific IHL rules can be assessed in these additional processes. The focus of the additional review process is an analysis of the IHL rules regulating AWS swarm's functionality enabling the use of force. For example, if an AWS' swarm agents are designed to identify persons and objects on the battlefield, the additional review process should consider the AWS swarm's ability to distinguish between lawful and unlawful targets as required by the rule of distinction in Article 48 to AP I, and then how this information is passed to other members of the swarm. This requires the individual swarm agents to operate in reliance on IHL definitions that are relevant to the distinction rule (e.g. the definitions of civilians, civilian objects, combatants and military objectives).

Alternatively, an AWS swarm may be required to demonstrate through performance during appropriate tests that it will only engage lawful targets, if one can describe its processes as amounting to "understanding" [140]. Such programming is likely to be complicated and require careful analysis to ensure those responsible for the AI system's development and programming understand these rules, definitions and their application to enable them to be accurately translated into code and to reflect the State's legal obligations. In addition, the Article 36 weapons review must consider the data used to train an AI system's neural network to identify persons and objects (should the swarm use neural network technology). This includes consideration of the specific indicia or characteristics that are intended to enable an AI neural network to accurately identify and classify persons and objects defined by IHL. Although object classification is a relatively simple task, determining legal definitions of objects once classified is context and conflict specific and thus more complex.

To achieve this additional review, we propose that a functional review, and a use review, is conducted, rather than the traditional review described above. The conduct of a functional review is a departure from the traditional Article 36 weapons review methodology because it would consider those IHL rules that regulate the conduct of an attack, beyond just those that prohibit the causing of unnecessary suffering or are by nature indiscriminate. The focus of the use review would be the AI system and its ability to interpret data from its environment and apply the IHL rules programmed in its system. The use review would depend on several State policy decisions. These include which IHL rules the AI system may advise on and the compliance standards required before humans can act on these AI recommendations. By combining these additional review processes, the uniquely complicating issues of AWS swarms can be mitigated, to the extent that they will be capable of informing a sufficiently robust legal review that will support a State to be reasonably assured that the weapon can be used in accordance with its legal obligations.

The timing, currency, multidisciplinary and iterative nature of these review processes is beyond the scope of this chapter. It is worth noting, however, that in addition to the technical considerations of systems containing AI or autonomy, which would prompt special consideration of these factors, the unique nature of swarming between agents would require further analysis as to whether the adopted review process might be adopted in relation to these factors. It is sufficient to note that in addition to implementing these additional review processes, it is also useful to consider having regard to the needs of the particular system (noting also that this issue has been specifically addressed in Australia's updated weapons review policy) [331].

10.6 Policy Considerations Impacting Weapons Review of a Military Swarm

Given the requirement to consider how a swarm might operate, to enable its review for compliance with international law, it is likely the weapons review will contemplate the associated system of control used to deploy the swarm in accordance with the intention of its operators. The method of controlling an AWS agent must therefore be known at the time of the weapons review, and some understanding of the control measures put in place during its fielding in specific operational environments will also require assessment as part of the weapons review. It is more likely, however, that this information will only be known as a particular armed conflict or mission evolves and thus a review will consider general policies and processes related to the deployment of swarming systems. This requires the review to contemplate existing policies, tactics, techniques, procedures, doctrines and control measures. Specifically, the Article 82 obligation to provide legal advisers "when necessary" to advise military commanders about compliance with the requirements of the Geneva Convention law will thus become relevant.[17] That is, the ability to ascertain compliance with a State's international legal obligations cannot be determined during the design and development of the swarming agents

---

[17] AP I, Article 82.

and swarm, but, rather, requires determination by a legal adviser during the use or proposed use of the swarm during a situation of armed conflict.

For example, the Australian systems of control would require consideration during the weapons review process, but the particulars of the policies put in place to control the swarm in a particular mission profile on a specific operation would require consideration separately [70]. The system of control proposed by Australia articulated the process by which a military ensures sufficient control over capabilities to ensure that it is able to "operate in a lawful and regulated manner" [70]. It has been described as "an incremental, layered approach to applying control, covering all aspects of a weapon system from design through to engagement" [70]. In the case of AWS swarms, the required controls would be identified during the weapons review as an output of the review, while the specific limits and standards of those controls applied to a particular mission or operation would be in accordance with the relevant legal profile applicable for that mission or operation.

Although rules of engagement are often misconstrued in shorthand as being the method for legal compliance by military forces and their capabilities during armed conflict, they combine legal obligations for the specific mission and incorporate other policy-driven limitations or permissions pertaining to who may authorise the use of the capability [390]. For example, where it might be authorised for use by law, a swarm may not be authorised to operate in a particular location or to the full extent of its technical capability according to a policy (or political) constraint. In this sense, it is likely that a weapons review will identify what limitations or controls must be considered for the AWS swarm to operate, and those controls will be moderated and further defined and refined during the operational fielding of the capability.

Similar to the concept of kill switches being built into AWS, or design-led interventions to prevent misuse of AWS, such enhanced control options can provide control measures that safeguard the legal compliance of the AWS swarm when factoring in its general reduced levels of human control [379]. Methods to achieve these control measures could include enhanced notification of specific behaviours, to require the system to trigger human intervention, or control of particular agents or classes of agent within the swarm. In addition to the requirement to articulate control measures that reflect the specific operational and legal profile of the swarming mission, it is likely additional control measures will be required. This is as a result of the decreased capacity to control swarms given their mass and the decentralised method of control in true swarming capabilities. It is also likely that this greater level of attention to control measures will reflect the decreased level of human intervention in swarms, as potentially compared with other autonomous systems.

The combination of these policy considerations, although in some cases inter-linked with legal requirements, provides an opportunity through the process of review to create an assurance framework for the use and deployment of AWS swarms.

10.7 Methodology to Test Military Swarms to Support Article 36 Weapons Review

Simple Article 36 reviews may be conducted as a single final review in which the technical specifications, test and evaluation data and intended employment literature are reviewed to determine the legality of the weapon, as a gateway of sorts before its introduction into service. A drone swarm review would likely require a multistage review culminating in a final Article 36 review, to account for the complexity of compliance issues discussed above and the need to augment certain design characteristics to meet legal specifications through the design and testing and evaluation process. The review methodology must therefore enable the reviewer to confirm that the swarm has met the specified level of reliability for its deployment in its articulated use case. This necessarily requires testing in edge case situations and testing the swarm's ability to dynamically re-task agents to meet the required standards for each IHL subtask that the swam undertakes.

The review and analysis of Monte Carlo computer-based risk analysis simulations, which support operational analysis and real-world testing, are an example methodology. This approach could ensure that behaviour models perform as expected and meet review criteria [535, 544]. Monte Carlo methods are often used as a data analysis tool to introduce variability and randomness to the test environment. Risk analysis methods could be used to simulate and verify the safety factors and operationalised tolerances. Data gathered from Monte Carlo simulations support the effectiveness, management and safe operation of the swarm given the full range of swarm agents deployed and variation in scenario configuration [535, 544]. By replaying and capturing the simulations with varying scenario and environmental parameter data, key areas of inflexion and areas of diminishing returns are identified [695]. This enables the identification of the minimum number of swarm agents to achieve a task, according to the acceptance of legal risk. Put simply, the testing allows the reviewer to determine when the allocation of agents ensures that the anticipated swarm tasks will meet the legal standard, and identifies when higher compliance through allocation of more swarm agents represents a diminishing return. A practical outcome of the review of AWS is the ability to then dictate the sensitive issue of articulating a data-match standard, by reference to a general level of risk acceptance grounded in the edge case testing conducted using this methodology [89].

Specifically, critical outcomes may be the recognition of a minimum number of simultaneous target tracking agents to reach the required confidence level for detection, classification, recognition and identification (DCRI) or the maximum recommended swarm size to ensure adequate levels of control. Testing may also support the identification of additional emergent behaviour. For example, there may be a relationship identified between the collateral damage that occurs during a strike compared with the number of swarm agents. In cases in which a swarm strategy that uses multiple simultaneous engagements or obfuscation methods to ensure engagement with the target is allowed, the result may be many munitions landing on unintended targets. In this case, this testing methodology provides an ability to record and predict the risk associated with the swarm behaviours as they relate to specific IHL rules, noting the simulated emergent behaviours will be tied to a particular operational environment and use case.

The conduct of simulations to test AWS swarms provides a meaningful way to support the weapons review of these capabilities. However, the completed review must reflect the limitations of such testing methodologies, namely, that the results produced allow the reviewer to accept compliance of the AWS in line with a level of risk and within limited testing parameters (i.e. using particular use cases, environments and articulated specifications for systems of control). It also provides a mechanism to identify the acceptable level of risk in terms of swarm numbers, from a task reorganisation perspective, as well as from a massed effect and command and control perspective. These results can provide meaningful inputs to the weapons review of an AWS swarm, which will necessarily require repetition as use cases change or behaviours of the swarm continue to learn and adapt to the deployed environment. The level of repetition of such tests will require consideration in light of the other weapons review considerations discussed throughout this chapter. Further, introduction of a multi-tiered testing protocol, which combines simulations with controlled field testing and post deployment monitoring, could facilitate the continual assessment of such systems to enable compliance with legal standards.

10.8 Concluding Observations

Although the assessment of thinking swarms for legal compliance is not dissimilar to the requirements of any autonomous capability for the purposes of an Article 36 weapons review, the disaggregation of the system and the interactive nature of the parts create a number of specific challenges. First, the capabilities must be individually capable of complying with the international law obligations attached to their specific functionality. That is, the swarm can achieve what the individual agent cannot because each agent has specific roles and capabilities; however, each of those roles and capabilities must be legally compliant as far as they assume IHL functions. Second, depending on the IHL rules that the swarm triggers, there is a requirement to consider the aggregate standards created by the many, as compared with the one. That is, the standard to meet legal obligations, such as those of feasible precautions, necessitates a separate review of how the individual capabilities interact. In essence, swarms must be reviewed once for compliance of each of its constituent parts and again for its action as a swarm.

The use of swarming technology can have significant effects in terms of legal compliance. The capacity to corroborate single-source sensor observations can ameliorate some of the errors seen in the use of single-source intelligence, surveillance and reconnaissance capabilities. However, the number of sensors required to demonstrate that all feasible precautions have been taken requires a risk-based certification approach to the weapons review of such military swarm capability.

Despite this legal challenge, there are existing simulation and testing methodologies that can be undertaken to assist a weapons review to establish whether the swarm can meet these legal obligations. They will, however, require adjustment to account for each use case and deployed environment, and thus will require regular review and potentially reiteration of testing as the swarm's behaviour learns or adapts.

Despite the existence of these testing methodologies, further consideration as to the level of risk acceptance (or additional testing obligations) is necessary to enable military swarms to deploy and reach their full potential. Novel testing or review regimes to support the use of military swarms in novel environments is warranted to enable the full potential of military swarms to be lawfully and ethically deployed.

Open Access This chapter is licensed under the terms of the Creative Commons Attribution 4.0 International License (http://creativecommons.org/licenses/by/4.0/), which permits use, sharing, adaptation, distribution and reproduction in any medium or format, as long as you give appropriate credit to the original author(s) and the source, provide a link to the Creative Commons license and indicate if changes were made.

The images or other third party material in this chapter are included in the chapter's Creative Commons license, unless indicated otherwise in a credit line to the material. If material is not included in the chapter's Creative Commons license and your intended use is not permitted by statutory regulation or exceeds the permitted use, you will need to obtain permission directly from the copyright holder.

Part III
Thinking Swarm: Topology and Architecture

Chapter 11
Cognitive Architecture of Aware System of Systems

Peter Bernus, Beth Cardier, Ovidiu Noran, and Glen Smith

The systems engineering community is often faced with complexity barriers that develop because modern systems must be agile and resilient. This flexibility requires dynamic changes to the system so it can adapt to changing missions and changes in the internal and external environments. We are particularly concerned with architectural solutions applicable for creating a system of systems (SoS), such as a community of (human, machine or hybrid) agents, which cooperates and collaborates, sharing situation awareness and assessment and creating an emergent virtual agent that acts as one aware individual. We present a concept related to the long-standing evolution of cognitive architectures. The novelty is the demonstration of how Pask's "conversation theory" can be used to achieve two important traits of emergent general intelligence: communication and learning. The relevance is in the intent to support the thinking swarm initiative, in which the swarm is considered a virtual organisation tasked to perform a mission.

11.1 Introduction

The systems engineering community is often faced with complexity barriers that develop because modern systems must be agile and resilient. This flexibility requires dynamic changes to the system so it can adapt to changing missions and also

---

P. Bernus (✉) · B. Cardier · O. Noran · G. Smith
Institute for Integrated and Intelligent Systems, School of Information Communication and Technology, Griffith University, Nathan, QLD, Australia
e-mail: P.Bernus@griffith.edu.au; B.Cardier@griffith.edu.au; O.Noran@griffith.edu.au; Glen.Smith@griffith.edu.au

changes originating from the internal and external environments.[1] Thus, there is a need for guidance on the architectural development of agile and resilient systems.

Traditionally, architectural models are produced in the initial stage/s of a system's life (model-based systems engineering but models are increasingly being used for model-based control, and for supporting various management, control and service functions during system operation, partial or total decommissioning or during maintenance, repair and overhaul). Accordingly, the term "digital twin" [439, 615] has been popularised as the set of models collectively mirroring the system of interest in real time during its operation [533] to support management decisions and business intelligence and serve a range of other purposes.

Process management systems have also become popular [809], allowing the enterprise to maintain a real-time mirror of some process instances (and history traces), and these could be used for improved management and control, as well as to inform and support process improvement practices.

Today, the concept of digital twin potentially covers the entire enterprise in scope, including physical equipment, humans and software and hardware. Importantly, this coverage also includes the products and product lines as entities of interest, as well as the relevant elements of the environment, and the relationships among all of the above.

After the concept was thus generalised, it found applications in product life cycle management [533]], logistics [644], model-based control [634], predictive maintenance [538] and supply chain management [421]), to name a few. Technologies that support this development include the Internet of things (sensor networks, actuators, etc.) and various computational techniques, such as machine learning, data analytics and associated technologies capable of handling large amounts of heterogeneous data (e.g. data lakes [563]).

To illustrate this guidance, the authors make use of the Generalised Enterprise Reference Architecture and Methodology architecture framework, which is part of the ISO standard ISO15704:2019 on "enterprise modelling and architecture" [420]. This chapter is therefore about the underlying architectural models, their dynamics and evolution.

11.2 Architecture Models and Their Relationships

This architecture framework was selected in preference to other frameworks because:

---

[1] This chapter relies on and extends our recently published paper [114], but recasts it in light of (1) its relevance to the thinking swarm initiative, (2) our proposal's relationship to cognitive architectures and (3) the communication, learning and interpretations that occur in the conversation among agents.

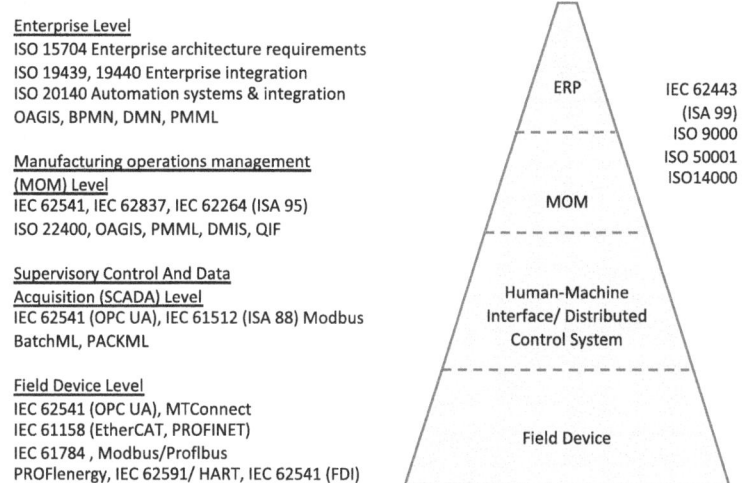

Fig. 11.1 National Institute of Standards and Technology's standards landscape for smart manufacturing (redrawn after [542])

1. according to the National Institute of Standards and Technology's "Standards Landscape for Smart Manufacturing" [542], ISO15704 is the top-level enterprise architecture standard for smart manufacturing (see Fig. 11.1)
2. it has a built-in modelling framework Generalised Enterprise Reference Architecture and Methodology (GERAM) (Fig. 11.2), and
3. it has an explicit representation of relationships among the life cycles of multiple entities of interest.

   For example, the "supporting entities" in ISO15288 [419] refer to systems that contribute to the life of the enterprise in various stages of its life. The relationship between the life cycles of entity A and B can be represented (as a shorthand) by drawing an arrow from the operation of A to those life cycle activities of B, to which A contributes (see Fig. 11.3).

   As opposed to this, all other architecture frameworks are based on architecture descriptions that can be derived as views from underlying models.

ISO15704 differentiates between various types of entities (such as a sociotechnical SoS, such as an enterprise), with each having its life cycle, such as entities including products, projects, programmes, organisations and various virtual enterprises.

The concept of the life cycle of an entity uses an abstraction from time and consists of life cycle activity types, while the temporal aspect is covered by the concept of life history (being a collection of life cycle activity instances grouped into stages). As the entity's life progresses, some entities are created, some are decommissioned and each has a life cycle and life history of its own.

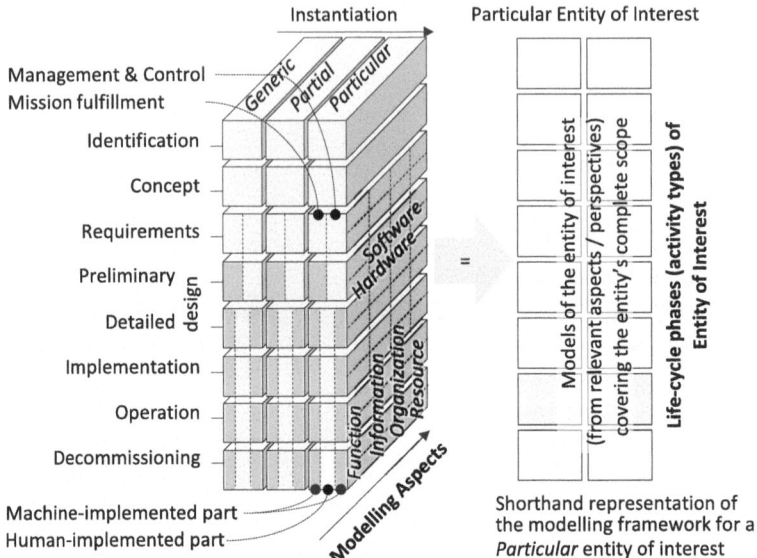

Fig. 11.2 ISO15704/GERAM's modelling framework. The framework may be populated by models for a particular entity, as well as reference models ("partial models") and generic models (expressed as ontological theories or just metamodels) of the domain

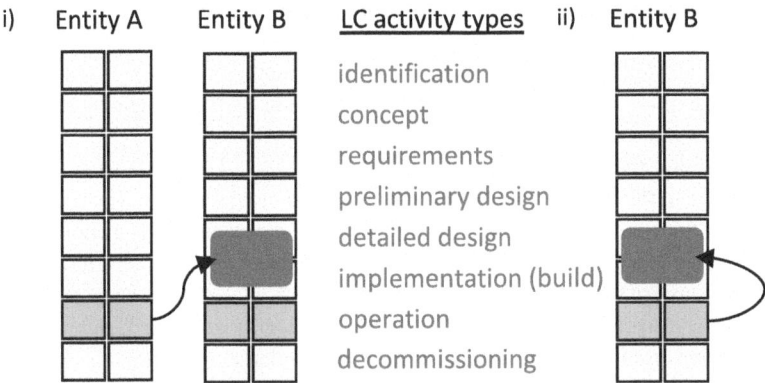

Fig. 11.3 Example of life cycle relationship between two entities: either (1) "A" performs the detailed design and building of "B" or (2) "B" performs its own detailed design and building activities and maintains the models of itself (displaying a form of adaptation and agility)

This history includes the formation by enduring entities (headquarter, forces, supporting entities, networks) of virtual entities (such as programmes, projects, operations and missions). Thus, complex coordinated operations can be architected as separate entities (systems) in their own right.

11 Cognitive Architecture of Aware System of Systems

Although agility (as a fast adaptation capability of both mission fulfilment and command and control) of all entities is important, as an example we use missions performed by collaborative agents (humans and machines).

Typically, a long chain of life cycle relationships exists (see Figs. 11.4 and 11.5).

Figure 11.6 arranges (along the atemporal life cycle dimension, from "top to bottom") the models collectively constituting the "digital twins" as created in the design, building, deployment, (re)configuration and operation of the system of interest.

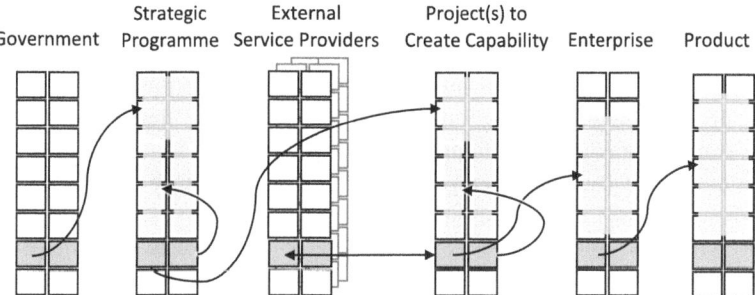

Fig. 11.4 Chain of life cycle relationships

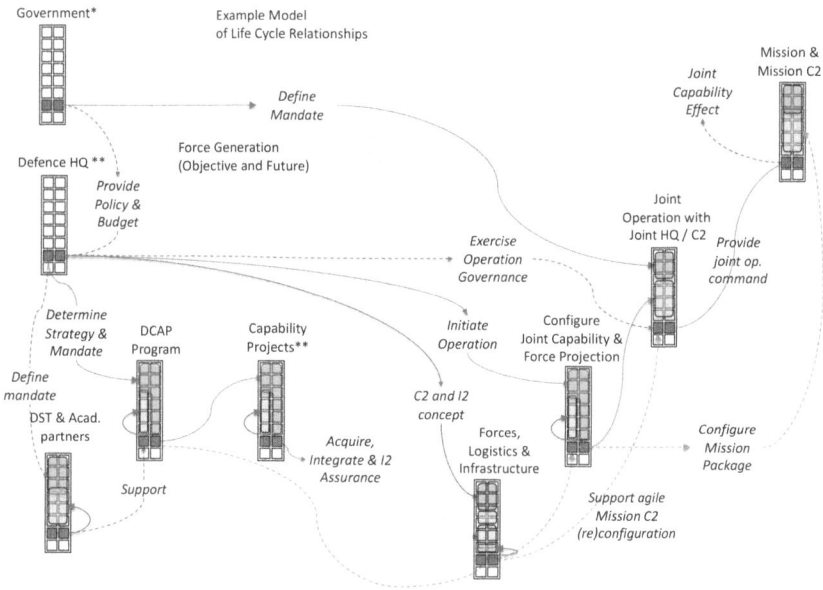

Fig. 11.5 A long chain of life cycle relationships (note the self-generating nature of programme and project entities)

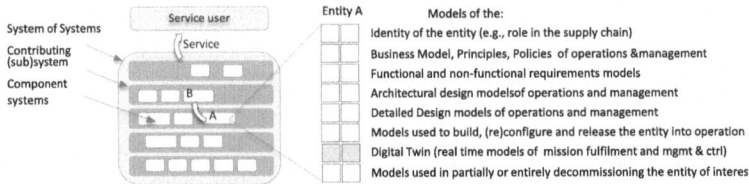

Fig. 11.6 Models populating the modelling framework during the creation versus during the operation of an entity (in a layered system of systems each component may be operated by a different entity and has its own life cycle)

In the life of any entity, there are models capturing the so-called static architecture that is immutable, because modifying that would constitute a major change, essentially changing the entity's identity. Conversely, the dynamic architecture results from adaptively (re)configuring the entity as needed.

In the case of such self-configuring, the entity must have the requisite configuration processes as a capability. This capability is part of the entity's static architecture, to be able to generate the needed dynamic structures. (This is not to say that there cannot exist even higher-level processes that in turn change the entity's abovementioned capability set.)

11.3 Patterns for Architecting Aware Systems

Intuitively, an entity maintaining its own models may be construed as a system that is aware of itself. However, to claim to have created an architecture of aware systems, a theoretical underpinning is necessary. Here, we apply a theory that explains the fundamental elementary building blocks of aware systems, whether these systems are natural, artificial or hybrid.

This theory was originally proposed by Pask [645] for systems that can learn. Pask demonstrated a minimal set of processes necessary for a system to display awareness and represented these in a generic model called a conversational skeleton (see Fig. 11.7). The two sides of the model illustrate the participating processes in a so-called "conversation", and both sides have a minimum of two levels.

$Level_0$ is the set of processes (P_0^a and P_0^b) that participants a and b use to perform some actions in the domain of interest, and which are connected through an "interface" or "externalisation platform" that both sides can observe and manipulate. Any form of such externalisation is admissible (observable demonstration, or communication using one or more language modalities, collectively forming an abstract Language L_0).

Note the similarity of this abstraction to the synthetic languages discussed in Chap. 7, namely, that interaction may occur via observable events, language tokens and various other signal modalities, whether persistent or not.

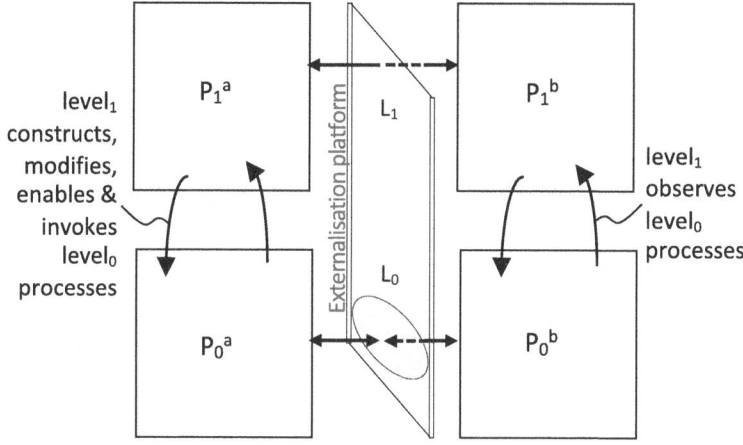

Fig. 11.7 Conversation theoretic process pattern (called a "P-individual")

The structure of this cybernetic model is orthogonal to the models of intelligent agents proposed in so-called cognitive architectures [482], as exemplified by ACT-R [49, 138], SOAR [494, 495], CLARION [770, 771] and others, such as various more recent large language models. This is because as we shall see, each set of processes itself has agency, for example, $level_1$ processes acting as sentinel agents for $level_0$.

The intent of this chapter includes illustrating of the two major features of intelligent cooperative agents, learning and communication [582], as well as the emergence of higher-level agenthood through cooperative communication.

For example, $level_0$ processes may collaborate by performing tasks that achieve a common goal, or one side (using L_0) may demonstrate to the other the performance of a task while the other side's $level_0$ observes.

However, $Level_1$ includes two sets of contextual processes P_1^a and P_1^b, which have three roles:

- Using the interface, they establish context in a conversation, which includes the common understanding of a goal, or concept, and a commitment to follow a course of action.
- They communicate while acting cooperatively to progress the two sides' $level_0$ processes.
- Further, $level_1$ may be used by one side to instruct or teach the other side, or to use "teach back" to prove that common understanding has been achieved. Therefore, $level_1$ includes the set of processes able to create, invoke, modify and observe the progression of $level_0$ processes of their respective sides.

Pask called this set of four processes a "psychological individual" (henceforth P-individual).

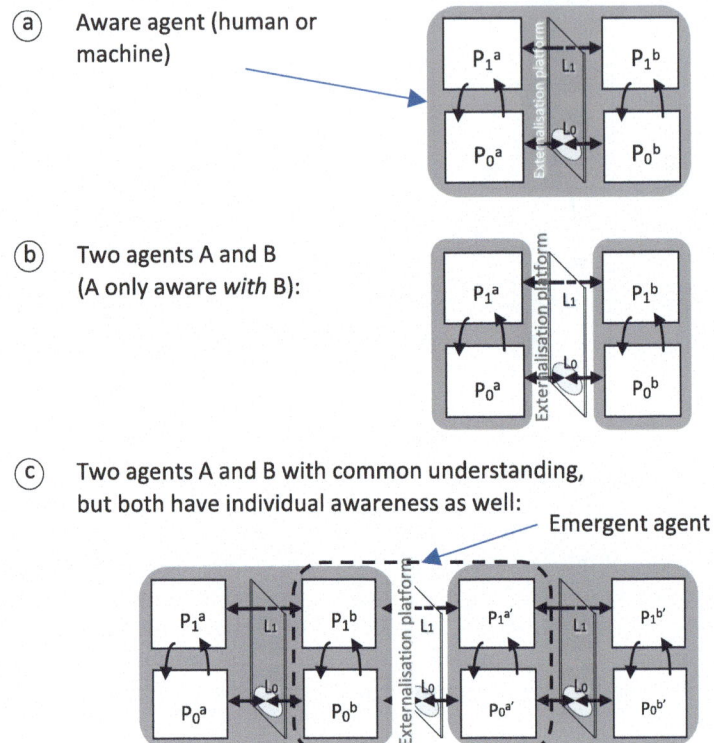

Fig. 11.8 Multiple ways of physically manifesting P-individuals (grey box represents physical individuals; dotted box represents an emergent agent that in itself is a P-individual)

There exist multiple possibilities of embodying a P-individual in physical individuals (human, machine or hybrid agents; see Fig. 11.8), the simplest one being where all four sets are implemented in one agent or in two agents (see Figs. 11.8a and 11.8b, respectively), or even four.

Further, these individuals can be dynamically combined into emerging individuals:

- by composition and abstraction (creating an emergent agent out of multiple cooperating agents; see Fig. 11.8c, or
- by dynamically creating and recreating a temporary control hierarchy (holarchy) that suits the situation [588, 808].

There is an opportunity here that the literature does not seem to have explored yet, namely, that there exists a relationship between models of P-individuals and physical individuals (agents), and models of situation awareness (SAW).

We used Endsley's early work, which established a functional model of processes involved in creating SAW [269], as well as that of [623], who extended the scope

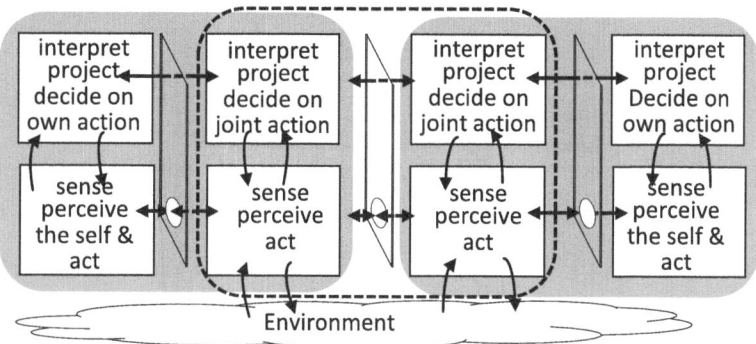

Fig. 11.9 Sharing situation awareness through conversation between two physical individuals (agents), both implementing their own P-individual (and together implementing two sides of an emergent agent)

of situation analysis to the theory of shared SAW. We did this because these models can be further elaborated using conversation theory.

For example, it can be demonstrated what processes are necessary for a P-individual to create individual SAW on the one hand, and a shared SAW in conversation with others, as well as necessary additional conditions for physical individuals to achieve the same.

Figure 11.9 illustrates the sharing of SAW using a conversation. The terminology "being aware with" (see Fig. 11.8b) may sound unusual; it is an expression (coined by Pask) stressing that (1) if the two sides are implemented by two agents, then neither will have SAW without the other; however, (2) if the P-individual is implemented in one agent, then that agent can make sense of the situation on its own (see Fig. 11.8a) [113, 381, 716].

It is to be noted that some or all of these processes can be tacit (performed without awareness and very efficiently), even though for the external observer the entire set of processes appears to be performed in an aware way (i.e. awareness emerges).

Each of the eight "white boxes of processes" P_i^x (where i is the *level* and x is running across the two sides of processes for each agent) can be subdivided into three more levels (see Fig. 11.10) according to the nature of implementation [448].

Ostensibly the bottom (tacit) level in Fig. 11.10 is the only efficient way to implement processes in humans, whereon in machines some efficient algorithms (pattern recognition, neural nets) are candidates while procedural/symbolic/algorithmic implementations are often not good candidates.

However, if the efficient tacit process produces ambiguity or unexpected outcome, the aware level must "take over" to explain, interpret, project and decide. This is the reason we suggest that a hybrid symbolic-sub-symbolic implementation of the cognitive architecture would be necessary, similarly to the duality principle in some notable cognitive architectures, such as CLARION [770].

One may assume that this requires the existence of an explicit situation model that would be maintained on $level_1$; however, this is not necessarily the case. The

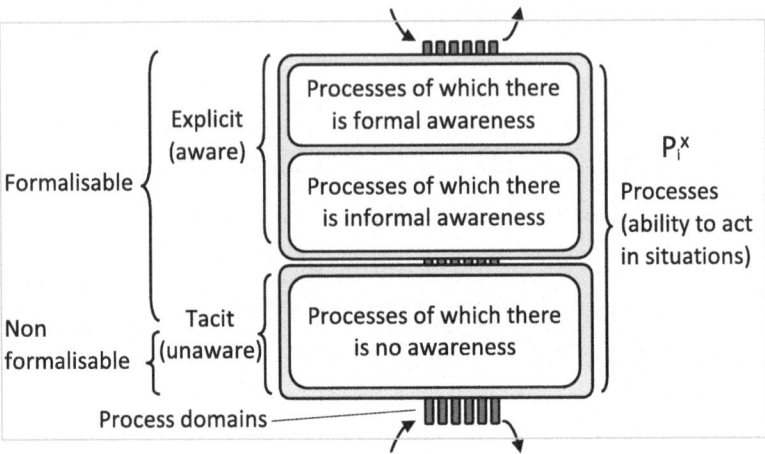

Fig. 11.10 Internal levels of knowledge (according to level of awareness) [448]. The bottom (tacit level) process type is the aim for efficiency reasons, but the model accounts for the need to learn new processes on $level_0$, and new situations on $level_1$

narrative derived from the signals emanating from $level_1$ flow of tokens and the information from $level_0$ (about the status of $level_0$ processes) together may form an internal interpretation or narrative (see Chapter 10) of the situation type at hand: this may be as simple as a "qualitative danger" signal (common to all dangerous situation types). This would then be sufficient for $level_1$ to change the controls to $level_0$, which, in turn, invokes canned responses on that level ("flee").

This model is recursive, that is, it repeats on all aggregation levels. The aggregation does not have to be static: it can be dynamically (re)created as an adaptation or optimisation measure in light of changing internal or external events. This is because through conversation, the two agents naturally create a higher-level virtual agent that in turn may participate in further higher level aggregates (see Fig. 11.11).

In terms of the hierarchy of control, the P-individual can shift levels, for example, the (until now) implicit "intent" of cooperative action and the implicit knowledge of a range of shared situation knowledge is because of a conversation that occurred prior to action (see Fig. 11.11).

The (two or more) agents on the left side are the cooperating agents (cf. Fig. 11.9) tasked with completing a mission. The agent on the right side is the agent (like a commander) that establishes common intent with them but typically does not directly participate in the cooperative action (mission fulfilment). The common understanding of intent maybe achieved using various conversational techniques, including "teachback", as experimentally validated by Pask [646] and extensively used ever since in many disciplines, for example, the medical field [775]. This fact requires that the agents engage in a conversation that is on $level_2$ (for them), for example, training prior to action and debriefing, and accounts for learning new situations.

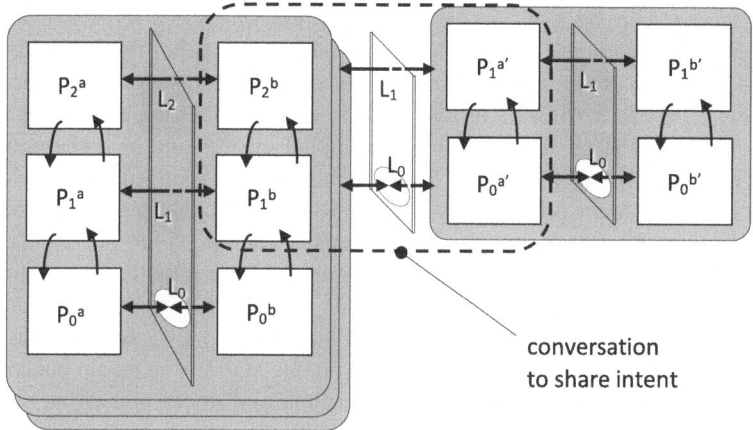

Fig. 11.11 Recursive process-pattern for aware entities

Further, at the same time, a $level_2$ conversation within an agent (such as between processes P_2^a and P_2^b) has a meta-cognitive role, such as determining whether the right reasoning processes are being used on $level_1$. This is in line with Cox's remarks that what is required for self-awareness is the integration of the meta-cognitive knowledge and processes in a system [212, 213].

Another notable structure is when agent A does not perform its $level_0$ process, but the required process is implemented by (can be "outsourced" to) another agent B.

Note that agents never directly control one another but instead negotiate in a conversation and agree to cooperatively perform a task (see Fig. 11.9). The benefit is that additional resources and capability become available, but the cost is the need to communicate.

The present discussion assumes that cyber-physical systems [507] are widely distributed, and in competition with others, therefore acting in a manner akin to military operations and missions to remain adaptive, competitive, resilient and agile.

11.4 A Hierarchy of Interpretations

Conversations provide a method by which an agent can summarise its knowledge and then transfer that abstraction to another agent in a SoS. The goal is the faithful transfer of information (such as SAW) from one part of the SoS to another.

However, in a swarm of autonomous agents, parts of the system may need to report and receive different information. How can we establish a reporting chain that enables knowledge to be reported from the local to global level or among peers while from time to time including any new information detected by individual members of the swarm?

The answer may be found in narrative structure, which is a form of everyday reasoning. The structures of storytelling have features that differ from logic, in that the goal is not to reproduce information exactly when it is transferred from one state to the next. Instead, each state of a story adds some new information to the prior state, simulating the process of gaining additional information as a consequence of moving through the open world. Some of this additional information could not have been known in the prior (or lower) state. To include it, a metastructure may be needed to use as a guide, or an adjustment to enable it to fit.

Examples of higher-level devices that can guide the integration of new elements into a conversation (that may be one way or two way) so they can produce a valid interpretation are "analogy" [313] and "framing" [370]. Both enable new and unexpected information to be connected among states to maintain coherence. "Framing" is particularly relevant to Pask's structure of interaction through information exchange.

Cognitive narratologist David Herman's notion of framing occurs when a key story is reported by a character in another situation. This literary device can be seen in classic literature, such as Mary Shelley's *"Frankenstein"* and Bronte's *"Wuthering Heights"* [832], as well as modern films, such as *"The Usual Suspects"* and *"The Princess Bride"* [284]. Framing sacrifices some of the fidelity and details of an initial state in exchange for offering a different perspective, new details or perhaps even a different underlying ontology. In addition, it often reveals information about the agent performing the translation. A story consists of many frames, each contributing different information to the overall interpretive fabric. To become integrated in that fabric, contributing frames must both include and exclude details. Determinations of inclusion are made in relation to context, which in this work is the dialogue paradigm and goals shared by conversing agents.

The device of framing enables a distributed system to aggregate and record knowledge from a wider scope. In this sense, Herman conceived of narrative as a "distributed system of intelligence" [370, p358], which makes possible shared thinking about "past events and about one's own and other minds". For example, a novel may switch between many situations or points of view, accumulating a situation model that seem more robust from having verified information from multiple angles [871]. Herman's description of a "frame" thus elaborates Pask's conceptualisation of a "transfer of knowledge" from one agent to another inasmuch as narrative transmission is a translation rather than a transfer.

In every jump of frame, information is re-contextualised. The frame represents a knowledge structure that may differ from the incoming information. As incoming information is registered there, its structure may change to accommodate its new housing. To replicate this narrative feature, the system would need the ability to partially match structure, so that alignments could be found even if the match was not perfect. Those alignments register the two contexts against each other so that relevant information can be exchanged or re-positioned correctly. This basic mechanism would assist with the problem of managing the variety of information that is produced by a swarm.

Many agent-specific interpretations can emerge throughout the swarm, and some may seem contradictory (nevertheless, each supports the respective individual agent's cooperative behaviour). However, on the SoS level, these individual interpretations have to integrate into the SAW of the virtual emergent entity that maintains relevant interpretation so the many can function as one to achieve intent.

11.5 Scope of Models

For the processes discussed above to generate useful information for decision-making, they need sensory input and must selectively perceive that input. This is the condition for developing (and maintaining) situation assessment and sensemaking/SAW (*sensing → perceiving → comprehending → projecting*) [623, 628]. In a SoS, that situation assessment would include an assessment of what contextual frame within which one is operating so that information and communication could be positioned correctly. The frame could determine the conversational paradigm of the other agent, or be as complex as deciding how to characterise the situation through which one is physically moving to understand what kinds of responses are appropriate.

The model of life cycle relationship of a cyber-physical system contains multiple entities of interest and their models (see Figs. 11.4 and 11.5), each comprising many agents, but also every entity (ideally) acting as an emergent aware agent. The "digital twin" of the physical system is comprising these models, but with a caveat: the digital twin cannot be statically designed; it must develop through learning. Specifically, the agent implementing the aware P-individual must be directly invoking, enabling, constructing or modifying and sensitising the set of sensory and perception processes.

The awareness of the cyber-physical system boils down to not only having processes in the operation (that fulfil the mission of the system) and associated processes that implement the needed control functions but also having processes that reason about these processes in context (and as a result change them in some way). Such change may involve improvement, reorganisation and reassignment of tasks.

11.6 An Example Scenario

What follows is an illustration of the types of conversations that agents in a swarm may enter into to establish individual and shared intent, individual and shared awareness of the situation the swarm is in, the status of the swarm and the individual members of the swarm. A condition placed on the example is that the communication among members of the swarm needs to exhibit proof of common understanding, which means that some of the conversations involve learning. This is because the open world nature of situations requires the conceptualisation of situations not encountered before, thus lacking common a priori reference.

The scenario is based on the real-life situation that arose, for example, in the 2022 devastating Lismore floods (NSW, Australia) and relies on numerous details reported in the NSW Independent Flood Inquiry [301].

Some findings of the inquiry include reports of:

- inadequate flood models or inadequate data on flood gauges
- loss of telecommunications (Internet, mobile phones)—for extended periods (days)
- failure to communicate in an unambiguous way (between multiple levels of emergency services and the community)

We attempt to demonstrate how a swarm of helper drones may augment the awareness of the situation at all levels.

The hypothetically required functions of individual drones are a subset of the swarm functions listed in Table 1.2 of Chap. 1 in this book.

11.6.1 Disaster Management Preparedness Stage

A large number of flood gauges had been supplying false readings because of various technical and environmental factors. In the preparedness stage, individual drones could be used to inspect and report on the state of these devices, so that during a flood event, adequate data can be supplied for decision-making. However, during flood events the same drones could supply additional data and, with coordinated planning, eliminate false information. This would contribute to system-level situation awareness, with each individual contributing to an overall model, even if individual drones did not have such an overarching model.

The issue of trust extends to both the gauge infrastructure and the sensors, as well as the communication functions of the drone agents. This is why the sensing and reporting function would be constructed using a M-individual superstructure (see Fig. 11.8a).

While drone agent A's P_1^a level processes invoke and monitor the sensing and reporting processes are performed at P_0^a level, P_1^b can interrogate P_1^a about this performance, while at the same time P_0^b is observing P_0^a's inputs and outputs. If, for example, the sensor reports data that are inconsistent with the temporal projection of the situation model maintained by P_0^b (such as water levels reported by the gauges vs those reported by the P_0^a sensors), this discrepancy is reported back to P_1^b processes to disambiguate the situation.

If that is not possible, P_1^b could call on another drone agent B to check out the same and compare the findings. In narrative terms, this operates as a common context that shares terms with and can therefore at least partially align with both of the original agents. This would involve a $level_1$ conversation between A and B, establishing whether the discrepancy is due to A's sensor faults or faults with the gauges that report on a separate channel.

The above discussion is to demonstrate drone agent A's self-awareness, as well as the ability of the swarm to adaptively resolve swarm behaviour that, unless the swarm checks out A's functionality, would cause a situation model that is not self-consistent.

A similar self-healing scenario would unfold if for some reason agent A identifies that its GPS-sourced geolocation is inconsistent with previous locations, as well as the direction and speed of travel.

11.6.2 Disaster Management Response Stage (During Flood Event)

Loss of telecommunications is a major (and typical) problem in flood events, and this exacerbated the situation in Lismore. How could a thinking swarm of drones have helped with this issue?

Again, individual drone agents could be tasked to participate in monitoring communication facilities, identifying sudden or creeping failures, predicting a situation in which intervention is necessary and taking remedial action to the extent that is possible.

Members of the swarm, with the capability to participate in performing the above tasks, would be initially informed of the intent of the operation (in a briefing conversation; see Fig. 11.11). This could happen immediately at the onset of the flood event, or prior to that.

Although the general intent to monitor communication facilities (such as mobile carriers) would be programmed into the drone agents, some specifics would only be available for sharing much later. Drone agents would need to learn the terrain, weather forecasts and other relevant parameters necessary to maintain a shared model that could be used to locate facilities and find communities or individuals in need of help.

This learning is not static, because individual agents in action are likely to find themselves in ambiguous situations. If the alignment between frames across any level is ambiguous, this must be realised first by the individual agent (call it agent A). This ambiguity is in a sense similar to the previous flood gauge example, even if the actual processes operate in a different domain. Nevertheless, A can escalate the problem and communicate with other agents, for example, the need for information by A required to resolve the ambiguity, and the explanation this ambiguity exists.

Examples include, "Is it a fault of mine that prevents communication with mobile phone towers?", and "Is the mobile phone tower malfunctioning?". Other agents in the swarm may designate an agent (call it agent B) to seek the needed information and, in a conversation with A, establish a new shared narrative. This may be possible because B has knowledge of the fact that neither it nor others in the swarm can receive signals from the tower.

The ensuing conversation would then unfold by designating a temporary leader of the swarm (cf. managed levels of swarm governance; see Chap. 1). This leader would collect information from all relevant frames and prepare a new intent and plan, communicating the same with the rest. For example, agent C senses that in an affected area, mobile phones are trying to connect to a tower but are not able to.

The swarm, however, has the capability of creating a temporary chain of relays to connect the distressed callers to appropriate authorities or to send a signal (text message) to them, informing that help is on the way. A similar architectural choice for heterogeneous distributed underwater swarms is discussed in Chap. 13.

Here, a chain of relays is an emergent agent; thus, another level ($level_2$) needs the same kind of self-awareness as available on the individual agent level. This ensures that the chain remains unbroken, because faults or performance degradation on the individual drone agent level is monitored and the swarm can reconfigure; however, swarm configuration requires reasoning about both the domain effectiveness of individual drones and their individual abilities to self-monitor, adding $level_2$ processes to agent C that can reason about the swarm as a whole.

11.7 Conclusion

The chapter followed a conceptual-analytical methodology, aiming to develop a high-level model of cyber-physical systems that have awareness as a distinguishing feature. The model was inspired by conversation theory applied to cooperative multi-agent systems (potentially including a mix of human, machine and hybrid agents).

Such systems are intrinsically more robust than systems managed using conventional control methods, and could be experimentally implemented using existing artificial intelligence programming methods of two kinds (see below), or by humans, or in a hybrid human-machine symbiosis:

1. Tacit processes implemented by humans or by machine learning/pattern recognition techniques, or purpose-built hardware: these are efficient, are fast and do not rely on any sort of reasoning. Although they are to some extent vulnerable, their efficiency makes them prime candidates for deployment.
2. Explicit processes that are performed in a rule-based/procedural manner: although they may be less efficient, they can be verified and used for training to eventually make the process tacit. Such robust, verifiable processes guard against the deployment or continued use of tacit processes if the situation is not appropriate for their use.

There is a danger of possible misinterpretation here. We stress that both (1) and (2), tacit and explicit, may exist on each level of the conversation, including $level_0$ and $level_1$. However, given the overwhelming need to act in the domain using tacit processes (for efficiency), it is expected that explicit processes on $level_0$ will predominantly be used during training, while $level_1$ will require explicit processes for reasoning about situations (to create explainable situation awareness) [330].

Thus, $Level_1$ processes deploy, invoke, train and supervise $level_0$ processes, and this could be done in a tacit way, but in terms of situation interpretation or projection, explicit processes are needed.

The other novelty of the proposal is the combination of the architectural models of cyber-physical systems and of conversation theory with the considerations of knowledge domains, such as tacit, informal explicit and explicit.

In the future the authors aim to demonstrate the above model using multi-agent environments, with the dual goal of validating it as a reference architecture and to demonstrate how agents could reason about forming emergent agents to maximise effectiveness and efficiency.

Open Access This chapter is licensed under the terms of the Creative Commons Attribution 4.0 International License (http://creativecommons.org/licenses/by/4.0/), which permits use, sharing, adaptation, distribution and reproduction in any medium or format, as long as you give appropriate credit to the original author(s) and the source, provide a link to the Creative Commons license and indicate if changes were made.

The images or other third party material in this chapter are included in the chapter's Creative Commons license, unless indicated otherwise in a credit line to the material. If material is not included in the chapter's Creative Commons license and your intended use is not permitted by statutory regulation or exceeds the permitted use, you will need to obtain permission directly from the copyright holder.

Chapter 12
Trochoids

Spirograph, Multi-agent Formation and Beyond

Jerome Moses Monsingh, Hoam Chung, and Arpita Sinha

As a child, everyone probably had the experience of playing with a toy called a Spirograph. This geometric drawing toy can create intricate patterns by combining two circular motions. These patterns fall into the class of mathematical curves known as trochoids. The trochoidal motion has been observed in planetary movements, biological spiral waves, electron behaviours and many other naturally occurring phenomena. Engineers have used such patterns for applications including atomic force microscopy, rotary pumps and cam gears, among others. One of the main characteristics of trochoids is that they are contained in a limited area, which is finely swept through by the curves. Suppose that a team of mobile robots wants to trace these patterns, for example, for surveying an area. What kind of control laws will ensure that the robots trace trochoidal patterns? Would it be possible to design such a control law only using locally exchanged information? In this chapter, we use the distributed consensus law to achieve a coordinated trochoidal formation among a team of mobile agents. The mobile agents are either modelled as single or double integrators, and a general communication topology is considered. We modified the standard consensus protocol and design the gains to create trochoidal motion through local information exchange. Several examples are presented that show complex trochoidal patterns. We discuss our current results and the potential application of our findings across various civilian and military applications.

J. M. Monsingh (✉)
Robert Bosch Centre for Cyber Physical Systems, Indian Institute of Science, Bangalore, India
e-mail: jerome.monsingh.c2022@iitbombay.org

H. Chung
Department of Mechanical and Aerospace Engineering, Monash University, Melbourne, VIC, Australia
e-mail: hoam.chung@monash.edu

A. Sinha
Systems and Control Engineering, Indian Institute of Technology, Bombay, India
e-mail: arpita.sinha@iitb.ac.in

12.1 Introduction

Swarm robotics has emerged as an increasingly popular field, offering a promising approach to tackling complex engineering challenges by using robotic systems that collaborate with one another. Consider a scenario such as a search and rescue operation following a natural disaster, in which the environment is chaotic and potentially dangerous for human intervention. In such situations, relying on a small team of highly equipped autonomous agents may not be practical, because the team's effectiveness would diminish rapidly in the face of adverse conditions. Alternatively, a well-organised swarm of simple mobile agents can offer a viable solution, thanks to its inherent characteristics such as scalability, robustness, adaptability and parallelism. These qualities align closely with the core objectives of swarm robotics research, making swarm robotics an increasingly compelling area of study [145, 693].

Swarm robotics presents several advantages over single-robot systems. First, although the global mission objective may exceed the capabilities of a single unit, the collective behaviours of individual units enable the goal to be achieved. In addition, it is more robust than a single-robot system, because even if one or more units fail, the remaining units can adapt and complete the global objective, thereby avoiding single-point failure. Further, each unit in a swarm can access information only from its local neighbour, eliminating the need for a costly global communication infrastructure [613].

Despite the desirable features mentioned, the challenge lies in designing algorithms that ensure specific collective behaviours within operational limitations. Coordinated motion is a prime example of such swarm behaviours, often observed in nature among flocks of birds, schools of fish and swarms of ants. The models used to characterise these collective motions have inspired numerous studies, especially in the fields of control and robotics [251, 252].

To tackle this challenge, cooperative control problems, in which multiple agents collaborate autonomously to achieve common tasks, have become a popular area of research. These agents represent physical entities, such as ground robots, satellites, quad-copters and other mobile machines. The objectives of these multi-agent systems can vary widely, including achieving consensus, forming specific formations, assigning tasks, ensuring coverage, flocking behaviour and distributed estimation. Multi-agent systems have diverse applications, including rendezvous, area coverage, surveillance, patrolling, target tracking and environmental monitoring, among others [418, 607].

Consensus is a fundamental problem in cooperative control, which has garnered significant attention since the early 2000s in systems and control research. The goal is to devise control laws for a group of agents to converge on a common agreement. Consensus protocols have been extensively studied and come in various forms to address various scenarios [523]. These protocols can be classified according to agent dynamics, such as leader-follower or leaderless approaches, as well as network topologies, including homogeneous and heterogeneous agents. They also vary in terms of communication mechanisms, collective behaviours and the transition from theoretical concepts to practical implementation. Formation control is another well-studied problem within multi-agent systems, in which agents must maintain a predefined geometric shape dynamically in space [194]. Trochoidal formation control is a specific subclass of formation control that has received significant attention in research.

Trochoids have intrigued the scientific community in the past, for example, when it was discovered that the planetary motion was not elliptical but trochoidal trajectories [477, p. 450]. Physicists observed the motion of Mercury to be trochoidal and Newtonian mechanics failed to explain the motion. The Perihelion precession of Mercury was explained using general relativity. In biology, the tip of a spiral wave traces a pattern that resembles a trochoid [837]. These spiral waves are generated when an abnormal generation of excitation waves traverse through the cardiac tissues. The electron motion in the presence of magnetic and electric fields leads to a trochoidal motion [547]. These are exploited in the mass spectrometer, in which mutually orthogonal electric and magnetic fields separate ions [128]. This particular arrangement is called a prolate trochoidal mass spectrometer, widely used in chemical analysis. Further, it seems to improve the focusing properties.

Mathematicians have also studied trochoids for their intrinsic geometric properties. For an excursion refer to [277, 352, 742, 743]. Artists have also explored these trochoids for various art forms [740]. This class of curves appears to be the solution to the shortest path problem given some constraints [711]. Even Bernoulli's brachistochrone problem in the calculus of variations, the curve of fastest descent between two points is a cycloid rather than a straight line. These trochoids can also be obtained by the popular drawing tool called the Spirograph.[1] It was developed by English architect and engineer Peter Hubert Desvignes.

Trochoidal patterns emerge when colloidal particles are illuminated by a diverging laser beam [602]. Double eyewall hurricanes or typhoons create trochoidal paths, and details of these paths are carried out in [380]. Pattern-based scanning methods are designed and analysed in [624]. These designed patterns are subsequently used in atomic force microscopy. These trochoidal patterns also find their application in many mechanical devices, such as rotary pumps [206], rotary engines [297] and cam gears [848].

Trochoid, in general, is defined as the locus of a point attached to a circle that rolls over another circle. An epitrochoid is generated when a point P attached at a radial distance d from the centre of a circle of radius r rolls over a circle of radius R exteriorly and $d < r$. When $d > r$ and the circle of radius r rolls over the interior of the circle of radius R, the locus is called a hypotrochoid. However, d cannot exceed R. The trochoidal family includes a rich class of curves, namely, hypotrochoid, epitrochoid, epicycloids, hypocycloids and cardioids. The common curves, such as ellipses and circles, also fall into this class. For a more general

---

[1] https://en.wikipedia.org/wiki/Spirograph

treatise of such curves, refer to [540, Chapter 17]. Examples of some trochoidal curves are given in Fig. 12.1.

The general parametric form of the trochoids is given as

$$x(t) = x_0 + a_1 \cos(\omega_1 t + \phi_1) + a_2 \cos(\omega_2 t + \phi_2)$$
$$y(t) = y_0 + a_1 \sin(\omega_1 t + \phi_1) + a_2 \sin(\omega_2 t + \phi_2),$$
(12.1)

where a_i, ω_i and ϕ_i, $i = 1, 2$, are constants and t is the independent variable time. x_0, y_0 is the centre of the trochoid. When the centre is the origin, these trochoids are called centered trochoids.

The generation of such trochoidal motion by a single agent or multiple agents has received some attention in the control theory literature. Traditionally, the collective behaviours obtained under the linear consensus protocol are rendezvous [744] and geometric formations, with a principal focus on circle formation [534]. Pavone and Frazzoli [649] provided control policies for single-integrator agents interacting through ring topology. Depending on the critical angle, three collective motions are observed, namely, rendezvous, circles and spirals. Ren [674] extended the idea to a more general interaction, namely, weighted digraphs. The author introduced Cartesian coupling in the consensus algorithm to produce circles and logarithmic spirals over and above the rendezvous of the multiple agents. Extending the behaviour of the agents to the realm of aesthetic curves was done by Tsiotras and Castro [798]. In the case where agents are modelled as single integrators, [797] studied an extended consensus law with a skew-symmetric matrix appended to the standard consensus protocol. The underlying interaction among the agents was captured using an undirected graph. The case where the interaction topology is a ring was dealt with in [441]. They introduced a stabilisation term, in addition to a coupling term in the consensus strategy. Conditions are derived for epicyclic pattern formation. Ansart and Juang [55] used model reference adaptive control to trace the epicyclic motion to overcome the parameter variations in the above control policy. Essentially the model reference adaptive control serves as a tracking law. Ansart and Juang [56] used a linear quadratic estimator to estimate the absolute position, in addition to using model reference adaptive control to sustain the epicyclic motion. An integral sliding mode approach was used in [57]. The integral sliding mode controller rejects the uncertainties in the gains and ensures that the network maintains its formation.

Pattern generation using unicycle agents is relatively more prevalent. Using range-only information, a guidance law was discussed in [793]. Conditions are derived for pattern formation for a single unicycle agent, and the effects of controller gains and initial condition are discussed. Further extension of this work using relative heading information was discussed in [31]. Tripathy and Sinha [794] proposed a control law for a single unicycle agent where the control input was a continuous function of range. Moreover, a switching control strategy was proposed to obtain the desired pattern from any initial conditions. Galloway et al. [306, 307] studied constant bearing cyclic pursuit in low dimension. The authors derived the conditions on the control parameters that result in hypotrochoid-like trajectories

12 Trochoids

Fig. 12.1 Representative examples of a trochoid. (**a**) Epitrochoid (when $d < r$). (**b**) Hypotrochoid (when $d > r$)

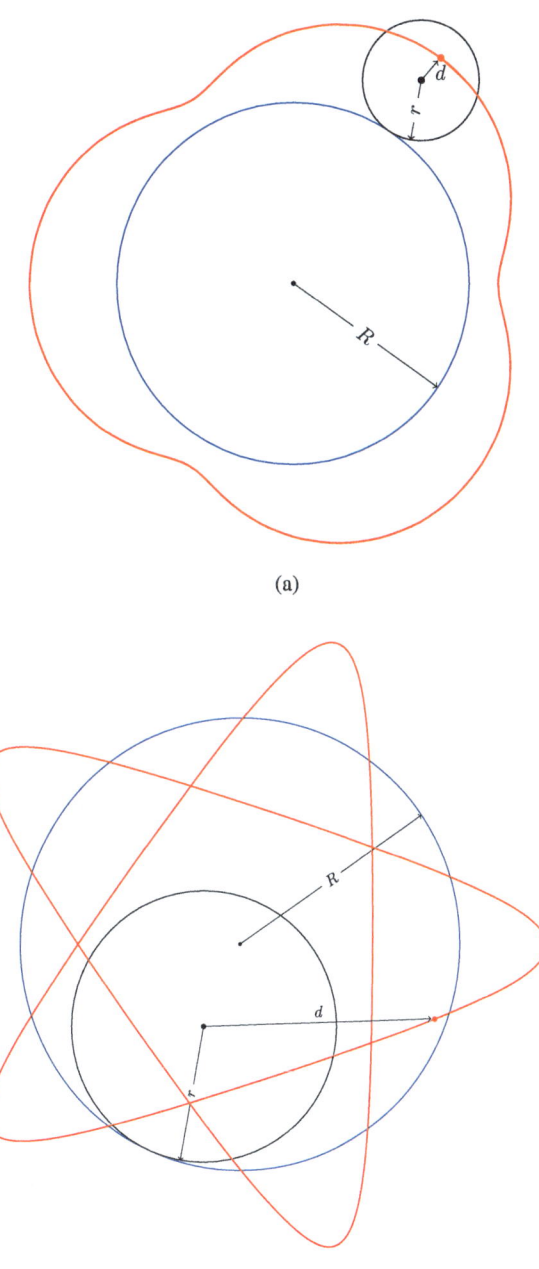

(a)

(b)

under certain rectilinear equilibria. Parayil and Ratnoo [642] proposed a Hopf bifurcation-based control law for a single unicycle agent that leads to a trochoidal motion. Further, the authors classified the generated patterns according to the geometric characteristic to design the parameters. Spirograph-like trajectories were generated using coupled oscillator steering control in [467]. The authors showed that these trajectories are possible with a general coupling function, and a detailed analysis was performed for a three-vehicle system. Table 12.1 summarises various pattern formation control algorithms in the literature.

These trochoids are not only aesthetic but also have multifaceted applications which includes area coverage [443], coverage path planning [586], hurricane sampling [239], orbit designs for satellites [513] and aerial surveillance in the presence of wind [209].

The other collective motion, in addition to circles, spirals and trochoids, was obtained in [241]. The author proposed control laws for hierarchical cyclic pursuit for single integrators. In addition to rendezvous and circles, complex circular motion and concentric circular motion were obtained. Simulations are provided to validate the theory. Ding et al. [242] proposed a pursuit-based approach for various formations. The agents follow either single-integrator kinematics or double-integrator dynamics, and the underlying interaction topology falls under directed acyclic graphs. Various collective behaviours are obtained, which include V-formation, vortex motions and tornado motions.

There are two main advantages in using trochoids for coverage applications: radial and azimuthal coverage. In a coverage or a sampling mission, agents are used to collect temperature, convection, salinity and precipitation measurements. Typically either circular (radial coverage) or folium trajectories (angular or azimuthal) are used to sample various points. However, when trochoidal trajectories are used, it amalgamates the advantages of both circular and folium trajectories, thus improving the sampling density. Further, the trajectories are dense, and the points of a revisit of the agents can vary for each agent, making it suitable for applications such as target containment and target tracking. These factors motivated our research on the distributed consensus law for trochoidal formations.

We attempted to address a few research gaps in our research. The existing control laws for pattern formation with linear consensus strategy have been more focused on interaction topologies that are either undirected or cyclic. Therefore, for a more general interaction topology, namely digraphs, the consensus law must be further generalised. Moreover, under the cyclic graph, the agents traced the same pattern (epicyclic) with an identical radius. Nevertheless, in a pragmatic scenario, each agent in the team may have different capabilities and hence is required to cover different regions, which means that the region covered by each agent can differ. We show that with general graph topologies, the agents can cover regions of different annular radii. The control laws existing in the literature focus primarily on either single-integrator models or unicycle models. However in the real world, many system behaviours can be approximated by a double-integrator model that represents an inertial system. Thus, we propose a control law for trochoidal pattern generation with double integrators under a more general interaction topology. In the

12 Trochoids 213

Table 12.1 Overview of pattern formation in the literature

Author	Year	No. of agents	Kinematics/dynamics	Underlying topology	Hardware implementation	Type of control law	Remarks
Tsiotras et al. [798]	2011	Multiple	Single integrators	Undirected graph	No	Consensus-based	A skew symmetric matrix was appended to the consensus law
Juang [441]	2013	Multiple	Single integrators	Cyclic graph	No	Cyclic pursuit-based	Final formation was epicyclic
Juang [442]	2012	Multiple	Double integrators	Cyclic graph	No	Cyclic pursuit-based	Epicyclic-like motion was obtained
Galloway et al. [306, 307]	2013	Multiple	Unicycle	Cyclic graph	No	Cyclic pursuit-based	Hypotrochoid-like motions were observed under certain rectilinear equilibria for a three-particle system
Twinkle et al. [793]	2015	Single	Unicycle	None	No	Range-based	Hypotrochoid-like motions were achieved
Shashank et al. [31]	2017	Single	Unicycle	None	Yes	Relative heading angle-based	Experimentally validated using a differential drive robot
Twinkle et al. [794]	2018	Single	Unicycle	None	No	Range-based	Control input is a continuous function of range
Parayil et al. [642]	2019	Single	Unicycle	None	No	Bifurcation-based	Different trochoidal patterns were observed depending on control parameters
Kimura et al. [467]	2008	Multiple	Dubins-type	Complete graph	No	Coupled oscillator steering control	A general coupling function was used for phase controller

current literature, pattern formation has been discussed only in the 2-D plane, which can be a serious limitation when applied to real scenarios. We extend the consensus algorithm for agents moving on a 3-D space keeping aerial applications in mind.

Given this perspective, we address the following research problems in this chapter.

- design a consensus strategy for a group of agents having single-integrator kinematics, interacting through a more general network topology and moving on a 3-D plane such that the trajectories of the agents resemble trochoidal patterns.
- extend from single-integrator kinematics to double-integrator dynamics without the knowledge of relative velocities for agents interacting through a general network topology.

The chapter unfolds as follows. In Sect. 12.2, mathematical preliminaries and background are provided. In Sect. 12.3, the distributed control laws for single integrators along with the conditions for pattern formation are presented. In Sect. 12.4, the distributed control laws for double integrators along with the conditions for pattern formation are presented. Section 12.5 demonstrates pattern formation through simulations and experimental validation. The study is concluded in Sect. 12.6.

12.2 Mathematical Preliminaries and Background

In this section, we present a short review of graph theory notions that are used to abstract the interaction topology among agents (adapted from [674]). Suppose there are n agents in the network. The interaction between the agents is represented by a weighted directed graph or weighted digraph $\mathscr{G}$. A weighted digraph $\mathscr{G}$ can be characterised mathematically by a node set $\mathscr{V}$, an edge set $\mathscr{E} \subseteq \mathscr{V} \times \mathscr{V}$ and a weighted adjacency matrix $\mathscr{A} = [a_{ij}] \in \mathbb{R}^{n \times n}$, where the set of nodes represents the agents and the edges indicate the interaction between the agents. If an edge (i, j) exists, it implies that the agent j can obtain information from agent i. In the case of an undirected graph, $(i, j) \in \mathscr{E}$ implies $(j, i) \in \mathscr{E}$. The weighted adjacency matrix $\mathscr{A}$ is defined such that a_{ij} is positive if $(j, i) \in \mathscr{E}$ and 0 otherwise.

Analogously, the neighbours of agent i are denoted by $\mathscr{N}(i) = \{j \in \mathscr{V} : (i, j) \in \mathscr{E}\}$. A directed path or dipath of a graph $\mathscr{G}$ is a sequence of edges that connects a sequence of vertices. A directed spanning tree exists in a digraph if at least one node has a dipath to all other nodes. Several definitions exist for the Laplacian matrix in the literature. It is defined as $\mathscr{L} = \Delta(\mathscr{G}) - \mathscr{A}$, where $\Delta(\mathscr{G})$ is the degree matrix, which is a diagonal matrix defined as $\Delta(\mathscr{G})_{(ii)} = d_{in}(v_i)$ for all i. $d_{in}(v)$ is the weighted in-degree of the vertex v defined as $d_{in}(v_i) = \sum_{\{j \mid ((v_j, v_i) \in \mathscr{E})\}} a_{ij}$. For an undirected graph, $\mathscr{L}$ is symmetric, whereas this property does not hold for a digraph ([580], Chapter 2).

12 Trochoids

Notations: We assume that $\mathbb{Z}$, $\mathbb{R}$ and $\mathbb{C}$ represent the set of integers, real numbers, and complex numbers, respectively. The set of all positive real numbers is denoted by $\mathbb{R}^+$. For $\mu \in \mathbb{C}$, we represent $\arg(\mu)$ as the phase of μ, $\overline{\mu}$ as the complex conjugate of μ and $|\mu|$ as the modulus of μ. i denotes the imaginary unit. $\mathfrak{Re}(\cdot)$ and $\mathfrak{Im}(\cdot)$ represent the real and imaginary parts of a complex number, respectively. conv$(\cdot)$ denotes the convex hull. We present the following lemmas that are used in the analysis conducted out in this chapter.

Lemma 12.1 ([503], Chapter 13) *Let $P = [p_{ij}] \in \mathbb{R}^{m \times m}$ have eigenvalues α_i associated with eigenvectors $u_i \in \mathbb{C}^m$, $i = 1, \ldots, m$ and let $Q = [q_{ij}] \in \mathbb{R}^{n \times n}$ have eigenvalues β_j associated with eigenvectors $v_j \in \mathbb{C}^n$, $j = 1, \ldots, n$. Then, the Krionecker product of P and Q is $P \otimes Q = [p_{ij}Q] \in \mathbb{R}^{mn \times mn}$. The mn eigenvalues of $P \otimes Q$ are $\alpha_i \beta_j$ with associated eigenvectors $u_i \otimes v_j$, $i = 1, \ldots, m$, $j = 1, \ldots, n$.*

Lemma 12.2 ([671]) *Let $\mathscr{L}$ be the nonsymmetric Laplacian matrix associated with weighted directed graph $\mathscr{G}$. Then, $\mathscr{L}$ has at least one zero eigenvalue and all nonzero eigenvalues have positive real parts. Further, $\mathscr{L}$ has a simple zero eigenvalue, and all other eigenvalues have positive real parts if and only if $\mathscr{G}$ has a directed spanning tree.*

Lemma 12.3 ([674]) *Let $R \in \mathbb{R}^{3 \times 3}$ be a rotation matrix with $a = [a_1, a_2, a_3]^\top$ and θ denote the Euler axis (i.e. the unit vector in the direction of rotation) and Euler angle (i.e. the angle of rotation) respectively. Then, the eigenvalues of R are $1, e^{i\theta}, e^{-i\theta}$ with the associated right eigenvectors given by, respectively, $\zeta_1 = a$, $\zeta_2 = [(a_2^2 + a_3^2)\sin^2(\theta/2), -a_1 a_2 \sin^2(\theta/2) + ia_3 \sin(\theta/2)|\sin(\theta/2)|, -a_1 a_3 \sin^2(\theta/2) - (ia_2 \sin(\theta/2)|\sin(\theta/2)|)]^\top$, and $\zeta_3 = \overline{\zeta_2}$. The associated left eigenvectors are, respectively, $\nu_1 = \zeta_1$, $\nu_2 = \overline{\zeta_2}$, $\nu_3 = \overline{\zeta_3}$.*

Lemma 12.4 *Let $A \in \mathbb{R}^{n \times n}$ with eigenvalues γ_i and associated right and left eigenvectors t_i and w_i respectively. Also let $B = \begin{bmatrix} 0_{n \times n} & I_n \\ A & -\tau I_n \end{bmatrix}$, where $0_{n \times n}$ denotes the $n \times n$ zero matrix and τ is a positive real constant. Then, the eigenvalues of B are given by*

$$\sigma_{2i-1} = \frac{-\tau + \sqrt{\tau^2 + 4\gamma_i}}{2}, \tag{12.2}$$

$$\sigma_{2i} = \frac{-\tau - \sqrt{\tau^2 + 4\gamma_i}}{2}, \tag{12.3}$$

with associated right eigenvectors $\begin{pmatrix} t_i \\ \sigma_{2i-1} t_i \end{pmatrix}$ and $\begin{pmatrix} t_i \\ \sigma_{2i} t_i \end{pmatrix}$, respectively and left eigenvectors $\begin{pmatrix} \sigma_{2i-1} w_i + \tau w_i \\ w_i \end{pmatrix}$ and $\begin{pmatrix} \sigma_{2i} w_i + \tau w_i \\ w_i \end{pmatrix}$, respectively.

Proof Refer to [596].

12.3 Proposed Control Law for Single Integrators

Consider n agents in $\mathbb{R}^3$ with single-integrator kinematics:

$$\dot{r}_i = u_i, \tag{12.4}$$

$i = 1, \ldots, n$, where $r_i = [x_i, y_i, z_i]^\top$ is the position and u_i is the control input corresponding to the ith agent. The agents exchange information through a fixed topology represented by a weighted digraph. We assume that the weighted digraph has a directed spanning tree. We propose the control law as

$$u_i = -g_p r_i - \sum_{j \in \mathcal{N}(i)} k_p R(\theta)(r_i - r_j), \tag{12.5}$$

$i = 1, \ldots, n$, where $g_p \geq 0$ and $k_p > 0$ are real constants, $R(\theta)$ is a rotation matrix as defined in Lemma 12.3 and $\mathcal{N}(i)$ are the neighbour of agents i. The components of r_i are coupled with a rotation matrix $R(\theta)$. In addition, a self-position feedback is used with a gain g_p. The control law described in (12.5) can be written in matrix form as

$$\dot{r} = Ar, \tag{12.6}$$

where

$$A = -g_p I - k_p \mathcal{L} \otimes R(\theta), \tag{12.7}$$

with $\otimes$ denoting the Kronecker product, $\mathcal{L}$ being the nonsymmetric Laplacian of the network topology, $r = [r_1^\top, \ldots, r_n^\top]^\top$ and $I \in \mathbb{R}^{3n \times 3n}$ being the identity matrix.

When $g_p = 0$, (12.7) is the same as the control law proposed in [674] where conditions on θ are derived to obtain rendezvous, circular motion and logarithmic spirals. Fixing $\theta = \pi/2$, and keeping $g_p = 0$, [797] achieved trochoidal patterns for undirected graphs. For a cyclic graph, [441] showed that (12.7) results in pattern formation under some conditions on g_p and θ. All the results in [441, 797] are achieved in a 2-D plane. In this chapter, we extend the results in [441] to any graph topology and in a 3-D plane. We consider a directed graph having a directed spanning tree. Since the behaviour of the system depends on the location of the eigenvalues, we find the conditions on g_p and θ to place the eigenvalues at desired locations.

Let μ_i, $i = 1, \ldots, n$, be the ith eigenvalue of $-\mathcal{L}$. Without loss of generality, we assume that

$$\arg(\mu_1) < \arg(\mu_2) \leq \cdots \leq \arg(\mu_n). \tag{12.8}$$

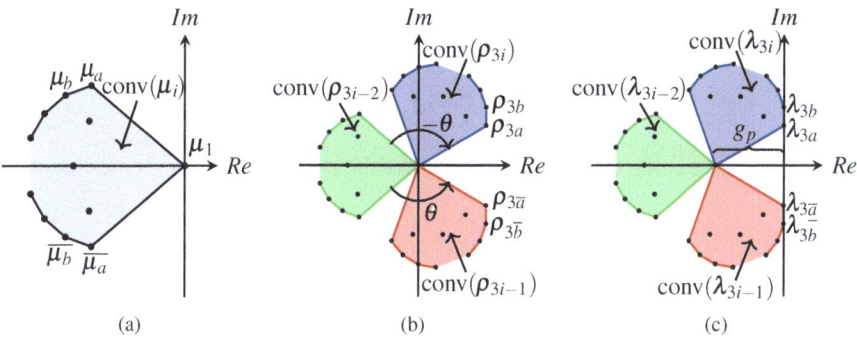

Fig. 12.2 (a) Representation of the eigenvalues of (a) $-\mathscr{L}$, (b) $-k_p\mathscr{L} \otimes R(\theta)$ and (c) A

From the properties of the Laplacian matrix $\mu_1 = 0$ and $\arg(\mu_i) \in (\pi/2, 3\pi/2)$, $i = 2, \ldots n$, since the digraph has a directed spanning tree (Lemma 12.2). Let the eigenvalues of $-k_p\mathscr{L} \otimes R(\theta)$ be ρ_i and the eigenvalues of A be λ_i. Then, from Lemma 12.1 and Lemma 12.3, for each eigenvalue of $-\mathscr{L}$, μ_i, there are three eigenvalues of $-k_p\mathscr{L} \otimes R(\theta)$ given by

$$\rho_{3i-2} = k_p\mu_i, \quad \rho_{3i-1} = k_p\mu_i e^{i\theta}, \quad \rho_{3i} = k_p\mu_i e^{-i\theta}. \tag{12.9}$$

The eigenvalues ρ_j and λ_j are related as

$$\lambda_j = \rho_j - g_p. \tag{12.10}$$

The effect of $R(\theta)$ and g_p on the eigenvalues $-\mathscr{L}$ can be seen graphically in Fig. 12.2. Let $\mathrm{conv}(\mu_i)$ represent the convex hull of the eigenvalues μ_i of $-\mathscr{L}$. The convex hull $\mathrm{conv}(\mu_i)$ is shown as shaded region in Fig. 12.2a. The effect of $R(\theta)$ is presented in Fig. 12.2b. Corresponding to the three eigenvalues of $R(\theta)$, the convex hull shown in Fig. 12.2a either remains the same (for eigenvalue 1), rotates by θ (for eigenvalue $e^{i\theta}$) or rotates by $-\theta$ (for eigenvalue $e^{-i\theta}$). By scaling the three convex hulls by k_p, we obtain the eigenvalues of $-k_p\mathscr{L} \otimes R(\theta)$. The convex hull formed by the eigenvalues ρ_{3i-2}, ρ_{3i-1} and ρ_{3i}, $i = 1, \ldots, n$ are represented respectively by $\mathrm{conv}(\rho_{3i-2})$, $\mathrm{conv}(\rho_{3i-1})$ and $\mathrm{conv}(\rho_{3i})$ and are shown in Fig. 12.2b. The parameter $-g_p$ (with $g_p \geq 0$) shifts the eigenvalues of $-k_p\mathscr{L} \otimes R(\theta)$ to the left as shown in Fig. 12.2c. Therefore, Fig. 12.2c shows the eigenvalues of A with $\mathrm{conv}(\lambda_{3i-2})$, $\mathrm{conv}(\lambda_{3i-1})$ and $\mathrm{conv}(\lambda_{3i})$ representing the convex hulls formed by the eigenvalues λ_{3i-2}, λ_{3i-1} and λ_{3i}, $i = 1, 2, \ldots, n$ respectively (Fig. 12.2).

For every complex eigenvalue in $\mathrm{conv}(\rho_{3i})$, its conjugate eigenvalue is present in $\mathrm{conv}(\rho_{3i-1})$. This can be shown as follows: for every complex conjugate eigenvalue pair, μ_i, μ_j with $\mu_i = \overline{\mu_j}$, we can write using (12.35)

$$\overline{\rho_{3i}} = \overline{\mu_i e^{-i\theta}} = \overline{\mu_i} e^{i\theta} = \mu_j e^{i\theta} = \rho_{3j-1}. \tag{12.11}$$

The same property holds for the eigenvalues λ_i of A, since from (12.10) and (12.39), we can write:

$$\overline{\lambda_{3i}} = \lambda_{3j-1}. \tag{12.12}$$

The following Lemma discusses the location of the eigenvalues of A on the complex plane.

Lemma 12.5 *Given θ, the eigenvalues of A in (12.7) have negative real parts if*

$$g_p > \max_i k_p |\mu_i| \cos(\arg(\mu_i) - \theta). \tag{12.13}$$

Proof Refer to [595].

Let the right and the left eigenvector of $-\mathscr{L}$ associated with μ_i be v_i and w_i, respectively. Then, from Lemma 12.1 the associated right eigenvectors of A are given by $m_{3(i-1)+l} = v_i \otimes \zeta_l$ $i = 1, \ldots, n$, $l = 1, 2, 3$. The associated left eigenvectors of A are given by $p_{3(i-1)+l} = \frac{w_i}{w_i^\top v_i} \otimes \frac{v_l}{v_l^\top \zeta_l}$ $i = 1, \ldots, n$, $l = 1, 2, 3$.
The solution of the system given in (12.6) can then be written as

$$r(t) = e^{At} r(0) = \sum_{k=1}^{3n} e^{\lambda_k t} m_k p_k^\top r(0), \tag{12.14}$$

where $r(0)$ is the initial condition. Also note that, when (12.13) is satisfied, e^{At} goes to 0, since all the eigenvalues corresponding to A have negative real part.

For smaller values of θ, we can have $\arg(\mu_i) \notin (\theta - \pi/2, \theta + \pi/2)$ for all i which implies that there will be no eigenvalues of $-k_p \mathscr{L} \otimes R(\theta)$ on the right half of the complex plane. In that case, g_p can be zero. When the right side of (12.13) is nonzero, we can have a system that is marginally stable as follows.

Corollary 12.1 *For the matrix A in (12.7), there exists at least one pair of eigenvalues on the imaginary axis and remaining on the left half of the plane if the following conditions hold:*

$$\arg(\mu_2) - \pi/2 \leq \theta \leq \arg(\mu_2) + \pi/2, \tag{12.15}$$

$$g_p = \max_i k_p |\mu_i| \cos(\arg(\mu_i) - \theta). \tag{12.16}$$

Proof Refer to [595].

It is shown in [674] that when all the eigenvalues are on the left side except one which is at the origin, we obtain rendezvous. Similarly, when there is a pair of eigenvalues on the imaginary axis, then circular formation is obtained while exactly one pair of complex eigenvalues on the right side generates logarithmic spirals. Therefore, in our case, when conditions (12.15) and (12.16) are satisfied, we will obtain a circular formation, whereas if (12.13) is satisfied, then all the states will go to zero. We can similarly find the conditions on g_p and θ to obtain rendezvous and logarithmic spirals.

When θ and g_p satisfy (12.15) and (12.16), it follows from Corollary 12.1 that there is a pair of eigenvalues on the imaginary axis and the rest on left side of the complex plane. Let $\tilde{\lambda}$ correspond to the eigenvalue of A on the imaginary axis. We use $(\tilde{\cdot})$ for the right and left eigenvectors of $\tilde{\lambda}$ for notational convenience. Then, using (12.48) we can write the state response of (12.6) as $\lim_{t \to \infty} r^*(t) = \lim_{t \to \infty} e^{At} r(0) = (e^{i|\tilde{\lambda}|t} \tilde{m} \tilde{p}^\top + e^{-i|\tilde{\lambda}|t} \overline{\tilde{m} \tilde{p}}^\top) r(0)$, where $r(0)$ is the initial condition. Since $\tilde{\lambda}$ and $\overline{\tilde{\lambda}}$ are conjugate of each other, we can further simplify r^* and write each component as

$$r^*_{3(i-1)+l}(t) = 2 \Re(e^{i|\tilde{\lambda}|t} \tilde{v}_{(i)} \zeta_{3(l)} \tilde{p}^\top r(0))$$
$$= |\zeta_{3(l)} \tilde{v}_{(i)} \tilde{p}^\top r(0)| \cos(|\tilde{\lambda}|t + \arg(\tilde{v}_{(i)} \zeta_{3(l)} \tilde{p}^\top r(0))) \qquad (12.17)$$

for $i = 1, \ldots, n$, $l = 1, 2, 3$. Since $x_i(t) = r^*_{3i-2}(t)$, $y_i(t) = r^*_{3i-1}(t)$, and $z_i(t) = r^*_{3i}(t)$, we observe from (12.17) that at steady state, the agents undergo a circular motion with the centre at the origin and period $2\pi/|\tilde{\lambda}|$. The radius of the circle can be obtained after some manipulation as $2|\tilde{v}_{(i)} \tilde{p}^\top r(0)| \sqrt{a_2^2 + a_3^2} \sin^2(\theta/2)$. It can be verified that at steady state the Euler axis a is perpendicular to $[x_i(t), y_i(t), z_i(t)]^\top$. Thus, whenever Corollary 12.1 is satisfied, the trajectory of the agents is circular.

Instead of just one pair, if there are multiple pairs of distinct eigenvalues of A on the imaginary axis, then the trajectory of the agents displays trochoidal patterns as shown in [797] and [441]. We can find the conditions on g_p and θ to have at least two distinct pairs of eigenvalues of A on the imaginary axis. In the next section, we derive the conditions under which the system will produce trochoidal patterns, which is the interest of this chapter.

12.3.1 Trochoidal Pattern Formation

The trajectories of the agents in a multi-agent system depend on the eigenvalues of the system matrix A given in (12.6). Juang [441] showed that if there are two distinct pairs of eigenvalues of A on the imaginary axis and the rest on the left side, then at steady state the evolutions of the states resemble an epicyclic equation. In our case, we can place the eigenvalues of A by appropriately selecting the parameters g_p and θ. We derive the condition on θ and g_p to have two distinct pairs of eigenvalues on the imaginary axis, while the rest are on the left side.

Let us define:

$$a := \arg\min(\mu_i - \mu_1). \quad (12.18)$$

We assume that the index a satisfies

$$\arg(\mu_a - \mu_1) < \arg(\mu_i - \mu_1) < \pi, \quad (12.19)$$

for all $i \neq a, 1$. From (12.8) it follows $\mu_a = \mu_2$. Let us also define:

$$b := \arg\min(\mu_i - \mu_a). \quad (12.20)$$

We again assume that the index b satisfies

$$\arg(\mu_b - \mu_a) < \arg(\mu_i - \mu_a) \quad (12.21)$$

for all $i \neq a, b, 1$. The assumption in (12.19) implies that there exist no other eigenvalues on the line passing through μ_1 and μ_a. Similarly, the assumption in (12.21) implies that there are no eigenvalues on the line joining μ_a and μ_b. We make these assumptions to develop the theory and relax them in the later part of this section. Let

$$\mu_{\bar{a}} = \overline{\mu_a}, \quad \mu_{\bar{b}} = \overline{\mu_b}, \quad (12.22)$$

then for the convex hull $\mathrm{conv}(\mu_i)$, μ_1, μ_a, μ_b will be the three consecutive vertices in a counterclockwise direction and $\mu_1, \mu_{\bar{a}}$ and $\mu_{\bar{b}}$ will be the three consecutive vertices in a clockwise direction, as shown in Fig. 12.2a. From (12.40), $\overline{\lambda_{3a}} = \lambda_{3\bar{a}-1}$ and $\overline{\lambda_{3b}} = \lambda_{3\bar{b}-1}$. We aim to put the pairs of eigenvalues $\lambda_{3a}, \lambda_{3\bar{a}-1}$ and $\lambda_{3b}, \lambda_{3\bar{b}-1}$ on the imaginary axis. The idea is pictorially depicted in Fig. 12.2. If we select θ as

$$\theta^c = \arg(\mu_b - \mu_a) - \pi/2, \quad (12.23)$$

then the edge between μ_a and μ_b in $\mathrm{conv}(\mu_i)$ will be rotated such that the corresponding edge in $\mathrm{conv}(\rho_{3i})$ becomes parallel to the imaginary axis. This will ensure the edge corresponding to $\mu_{\bar{a}}$ and $\mu_{\bar{b}}$ in $\mathrm{conv}(\rho_{3i-1})$ is also parallel to the imaginary axis with the same real parts. Since $\arg(\mu_b - \mu_a) \in (\arg(\mu_a), 3\pi/2]$, θ^c is bounded by

$$\theta^c \in (\arg(\mu_a) - \pi/2, \pi]. \quad (12.24)$$

Now if we select g_p as

$$g_p^c = k_p |\mu_a| \cos(\arg(\mu_a) - \theta^c), \quad (12.25)$$

which is equal to the $\mathfrak{Re}(\rho_{3a})$, then the edge will be pushed to the imaginary axis. This is mathematically addressed in the following lemma.

12 Trochoids

Lemma 12.6 *For the matrix A in (12.7), there exist at least two pairs of eigenvalues on the imaginary axis and remaining on the left-half of the complex plane if $\theta = \theta^c$ and $g_p = g_p^c$ as given in (12.23) and (12.25), respectively.*

Proof Refer to [595].

Lemma 12.6 does not guarantee that the pairs of eigenvalues of A on the imaginary axis will be distinct. However, for a pattern, there should be at least two distinct pairs of eigenvalues on the imaginary axis.

Lemma 12.7 *Under the conditions mentioned in Lemma 12.6, there will be two distinct pairs of eigenvalues of A on the imaginary axis if and only if*

$$\theta^c \notin \left\{ \frac{\arg(\mu_a) + \arg(\mu_b)}{2}, \arg(\mu_a), \arg(\mu_b) \right\}. \tag{12.26}$$

Proof Refer to [595].

In essence, the critical angle θ^c is the angle made by the projection from the origin to the line joining μ_a and μ_b. If the projection from the origin intersects at the vertex μ_a, it leads to $\lambda_{3a} = \lambda_{3\bar{a}-1}$, which implies $\lambda_{3a} = 0$. If the projection intersects at μ_b, then it leads to $\lambda_{3b} = \lambda_{3\bar{b}-1}$, which implies $\lambda_{3b} = 0$. If the projection bisects the line joining μ_a and μ_b, it leads to the condition $\lambda_{3a} = \lambda_{3\bar{b}-1}$, which, in turn, implies $\lambda_{3a} = -\lambda_{3b}$. This is illustrated in Fig. 12.3. Thus, we will not obtain distinct eigenvalue pairs on the imaginary axis if either one of $\lambda_{3a} = \lambda_{3\bar{b}-1}$ or $\lambda_{3a} = \lambda_{3\bar{a}-1}$ or $\lambda_{3b} = \lambda_{3\bar{b}-1}$ holds.

Remark 12.1 In the results obtained so far, it is not necessary to select μ_a as a vertex immediately left of μ_1 of the convex hull $\text{conv}(\mu_i)$. Any edge of $\text{conv}(\mu_i)$ can be rotated and shifted to ensure that the eigenvalues on this edge will fall on the imaginary axis. Therefore, if condition (12.26) is not satisfied when $\mu_a = \mu_2$, we can select some other vertex as μ_a and check condition (12.26).

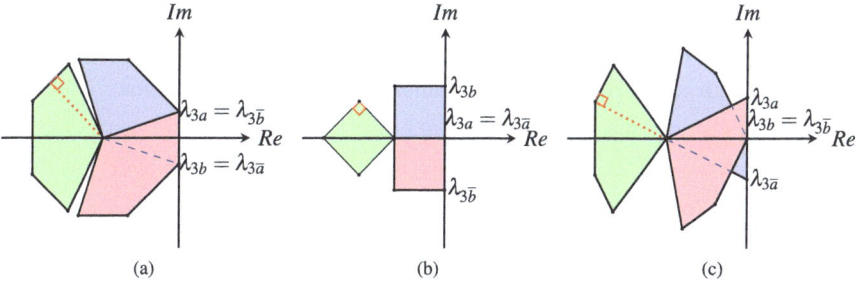

Fig. 12.3 Eigenvalues of A satisfying conditions (a) $\lambda_{3a} = \lambda_{3\bar{b}-1}$ (b) $\lambda_{3a} = \lambda_{3\bar{a}-1}$ (c) $\lambda_{3b} = \lambda_{3\bar{b}-1}$

Remark 12.2 If the assumptions (12.19) and (12.21) are not satisfied, A can still have multiple pairs of imaginary eigenvalues. For example, if there are multiple eigenvalues on the line joining μ_a and μ_1 (Eq. (12.50) is not true), then $g_p = 0$ and $\theta^c = \arg(\mu_a) - \pi/2$ will give the desired result. This control law will be similar to the one presented in [797]. If there are multiple eigenvalues on the line joining μ_a and μ_b, then condition (12.26) will no longer remain as the necessary condition.

Theorem 12.1 *At steady state, the trajectories of the agents with kinematics* (12.4), *control law* (12.5) *and control parameters satisfying* (12.23), (12.25) *and* (12.26) *are given by*

$$x_{\infty,i} = r^*_{3i-2}(t), \quad y_{\infty,i} = r^*_{3i-1}(t), \quad z_{\infty,i} = r^*_{3i}(t), \qquad (12.27)$$

where for, $i = 1, \ldots, n$, $l = 1, 2, 3$,

$$r^*_{3(i-1)+l}(t) = 2|\zeta_{3(l)} v_{3a(i)} p_{3a}^\top r(0)| \cos(|\lambda_{3a}|t + \arg(v_{3a(i)} p_{3a}^\top r(0)) + \arg(\zeta_{3(l)})) +$$
$$2|\zeta_{3(l)} v_{3b(i)} p_{3b}^\top r(0)| \cos(|\lambda_{3b}|t + \arg(v_{3b(i)} p_{3b}^\top r(0)) + \arg(\zeta_{3(l)})). \qquad (12.28)$$

Proof Refer to [595].

We observe from (12.28) that at steady state, $r_{3(i-1)+l}(t)$ can be represented as the sum of two cos functions of the form $\alpha_1 \cos(\beta_1 t + \phi_1) + \alpha_2 \cos(\beta_2 t + \phi_2)$ where $\alpha_1, \alpha_2, \beta_1, \beta_2, \phi_1$ and ϕ_2 are real constants. Whenever (12.26) is not satisfied, (12.28) becomes simplified, further and we obtain only one term in cos similar to (12.17). Thus, the trajectory of the agents resembles a circle. Next, we analyse the trajectory arising from (12.28).

Remark 12.3 It can be observed from (12.28) that the centre of the patterns is the origin. It can also be verified that the Euler axis a is perpendicular to $[x_i, y_i, z_i]^\top$. Therefore, if we can appropriately select the origin of the coordinate axis and the Euler angle, we can obtain the pattern at a desired point on a desired plane. The patterns will be formed after the initial transient is over.

If the agents are restricted to move on a 2-D plane, the results obtained in this study will still remain valid. The results are summarised in the corollary below.

Corollary 12.2 *Let the control algorithm for kinematics* (12.4) *be given by* (12.5), *where* $r_i = [x_i, y_i]^\top$ *and* $R(\theta)$ *is a* 2×2 *rotation matrix given by*
$$R(\theta) = \begin{bmatrix} \cos(\theta) & \sin(\theta) \\ -\sin(\theta) & \cos(\theta) \end{bmatrix}$$

1. *If* (12.13) *is satisfied, the agents rendezvous at the origin.*
2. *If* (12.15) *and* (12.16) *are satisfied, then the agents undergo circular motion with period* $\frac{2\pi}{|\tilde{\lambda}|}$ *and the radius of the ith agent is given by* $2|\tilde{v}_{(i)} \tilde{p}^\top r(0)|$. *The steady-state motion of the ith agent is*

12 Trochoids

$$\begin{pmatrix} x_{\infty,i} \\ y_{\infty,i} \end{pmatrix} = \begin{pmatrix} \eta \cos(|\tilde{\lambda}|(t) + \arg(\tilde{v}_{(i)} \tilde{p}^\top r(0))) \\ \eta \sin(|\tilde{\lambda}|(t) + \arg(\tilde{v}_{(i)} \tilde{p}^\top r(0))) \end{pmatrix}$$

where $\eta = 2|\tilde{v}_{(i)} \tilde{p}^\top r(0)|$

3. If the control parameters satisfy (12.23) and (12.25) and (12.26), at steady state, the trajectories of the agents are given by

$$\begin{pmatrix} x_{\infty,i} \\ y_{\infty,i} \end{pmatrix} = \begin{pmatrix} a'_i \cos A'_i + b'_i \cos B'_i \\ a'_i \sin A'_i + b'_i \sin B'_i \end{pmatrix}$$

where $A'_i = |\lambda_{2a}|t + \arg(v_{2a(i)} p_{2a}^\top r(0))$, $B'_i = |\lambda_{2b}|t + \arg(v_{2b(i)} p_{2b}^\top r(0))$, $a'_i = 2|v_{2a(i)} p_{2a}^\top r(0)|$ and $b'_i = 2|v_{2b(i)} p_{2b}^\top r(0)|$.

Proof The eigenvalues of $R(\theta)$ are given by $e^{i\theta}$ and $e^{-i\theta}$ and the associated right eigenvectors as $[1, i]^\top$, $[1, -i]^\top$ and the left eigenvectors as $[1, -i]^\top$, $[1, i]^\top$, respectively. The rest of the proof follows similar arguments to those presented for the 3-D case and are omitted here.

Remark 12.4 The results presented in [797] and [441] are special cases of Corollary 12.2. Assuming $\theta^c = \pi/2$, $g_p = 0$ and undirected graph topology, Corollary 12.2 produces the results of [797]. Similarly, assuming cyclic topology, results in [441] can be obtained.

Remark 12.5 In a classical consensus problem, each agent evolves according to the relative information of its neighbours. However, the control law (12.5) requires both the relative information of the neighbours and the absolute information of itself. If relative information is available, then the absolute information can be estimated using a scheme as proposed in [441, Section II]. If absolute information is available, then this information can be exchanged to obtain the relative information.

Remark 12.6 The control law (12.5) has poles on the imaginary axis making the system marginally stable. In general, the Laplacian of a classical consensus law has an eigenvalue at zero. Many of the proposed consensus-based formation control laws for circle or patterns have eigenvalues on the imaginary axis [441, 649, 674, 798]. This, in general, raises the question of robustness of the control schemes. These design schemes can be considered a distributed reference trajectories generator as demonstrated in Sect. 12.5.2. The robustness analysis of the proposed control law remains as the future goal of this work. Building on the generalized consensus strategy developed for agents with single-integrator kinematics tracing trochoidal trajectories in a 3-D space, we now address a more complex scenario. In this next section, we extend the framework to agents governed by double-integrator dynamics. In this setting, we do not assume access to relative velocity information, adding to the challenge. Despite these limitations, the agents still interact through a general network topology. The upcoming section presents a distributed control approach that ensures trochoidal pattern generation under these conditions.

12.4 Distributed Control Law for Double Integrators

Consider n agents with double-integrator dynamics, given by

$$\dot{r}_i = v_i, \quad \dot{v}_i = u_i, \quad i = 1, \ldots, n, \tag{12.29}$$

where $r_i \in \mathbb{R}^m$ is the position, $v_i \in \mathbb{R}^m$ is the velocity and $u_i \in \mathbb{R}^m$ is the control input corresponding to the ith agent, respectively.

12.4.1 Existing Control Law

A distributed consensus algorithm with dynamics (12.29) is studied in [430] as for each $i = 1, \ldots, n$:

$$u_i = -g_p r_i - \sum_{j \in \mathcal{N}(i)} a_{ij} k_p (r_i - r_j) - \sum_{j \in \mathcal{N}(i)} a_{ij} k_v (v_i - v_j), \tag{12.30}$$

where $\mathcal{N}(i)$ are the neighbours of agent i and k_p and k_v are real positive constants. The agents exchange information through a fixed topology represented by a weighted digraph. The control law gives rise to trochoid-like trajectories of the agents. We extend the control law (12.30) to achieve the objective of this work as detailed below.

12.4.2 Proposed Control Law

In this section, we consider Cartesian coupling in the consensus algorithm. We modify the control law (12.30) as follows. In (12.30) there is a relative velocity information exchange, which can be stringent in many applications. Thus, we replace it with the absolute velocity term. Further, (12.30) is coordinate independent, which leads to the final motion of the agents being trochoid like. Thus, to overcome these demerits, we introduce Cartesian coupling along with the relative position term in the consensus law. We propose the control law as

$$u_i = -g_p r_i - g_v v_i - \sum_{j \in \mathcal{N}(i)} a_{ij} k_p R(\theta)(r_i - r_j), \quad i = 1, \ldots, n, \tag{12.31}$$

where $r_i = [x_i, y_i, z_i]^\top$ is the position; $v_i = [v_{x_i}, v_{y_i}, v_{z_i}]^\top$ is the velocity; $\mathcal{N}(i)$ are the neighbours of the ith agent, g_p, k_p and g_v are real positive constants; θ is a real constant; and $R(\theta)$ is a rotation matrix, as defined in Lemma 12.3. The control law (12.31) can be written in compact form as

12 Trochoids

$$\begin{bmatrix} \dot{r} \\ \dot{v} \end{bmatrix} = \Lambda \begin{bmatrix} r \\ v \end{bmatrix}, \tag{12.32}$$

where

$$\Lambda = \begin{bmatrix} 0_{3n \times 3n} & I_{3n} \\ -g_p I_{3n} - k_p \mathscr{L} \otimes R(\theta) & -g_v I_{3n} \end{bmatrix}, \tag{12.33}$$

$r = [r_1^\top, \ldots, r_n^\top]^\top$ and $v = [v_1^\top, \ldots, v_n^\top]^\top$ are the stacked position and velocity vectors, respectively. Let $\mathscr{L}$ denote the associated Laplacian matrix of the digraph. We assume that the weighted digraph has a directed spanning tree, which implies that $\mathscr{L}$ will have one simple zero eigenvalue. Let μ_i, $i = 1, \ldots, n$, be the ith eigenvalue of $-\mathscr{L}$ with the associated right eigenvector t_i and the left eigenvector w_i. Without loss of generality, we assume that

$$\arg(\mu_1) < \arg(\mu_2) \leq \cdots < \arg(\mu_n). \tag{12.34}$$

Since the graph has a directed spanning tree, $\mu_1 = 0$ and $\arg(\mu_i) \in (\pi/2, 3\pi/2)$, $i = 2, \ldots, n$. Let the eigenvalues of $-\mathscr{L} \otimes R(\theta)$ be ρ_i. Let the eigenvalues of Λ be denoted as λ_i. Then, from Lemma 12.1, for each eigenvalue of $-\mathscr{L}$, μ_i, there are three eigenvalues of $-\mathscr{L} \otimes R(\theta)$ given by

$$\rho_{3i-2} = \mu_i, \quad \rho_{3i-1} = \mu_i e^{i\theta}, \quad \rho_{3i} = \mu_i e^{-i\theta}. \tag{12.35}$$

Let $A = -g_p I_{3n} - k_p \mathscr{L} \otimes R(\theta)$. Let κ_i denote the eigenvalues of A. The eigenvalues of A are given as

$$\kappa_i = k_p \rho_i - g_p. \tag{12.36}$$

From Lemma 12.4, for each κ_j, $j = 1, \ldots, 3n$, the eigenvalues of the system matrix Λ are given as

$$\lambda_{2j-1} = \frac{-g_v + \sqrt{g_v^2 + 4\kappa_j}}{2}, \tag{12.37}$$

$$\lambda_{2j} = \frac{-g_v - \sqrt{g_v^2 + 4\kappa_j}}{2}. \tag{12.38}$$

For every complex eigenvalue in conv(ρ_{3i}), its conjugate eigenvalue is present in conv(ρ_{3i-1}) since, for every complex conjugate eigenvalue pair, μ_i, μ_j with $\mu_i = \overline{\mu_j}$, we can write using (12.35)

$$\overline{\rho_{3i}} = \overline{\mu_i e^{-i\theta}} = \overline{\mu_i} e^{i\theta} = \mu_j e^{i\theta} = \rho_{3j-1}. \tag{12.39}$$

The same property holds true for the eigenvalues κ_i, since from (12.36) and (12.39), we can write:

$$\overline{\kappa_{3i}} = \kappa_{3j-1}. \tag{12.40}$$

In our analysis of the control law (12.31), we consider either rotation in a clockwise direction or an anticlockwise direction. Since for any complex number χ, $\overline{\chi^k} = (\overline{\chi})^k$ for all $k \in \mathbb{Z}$, which implies that

$$\overline{\sqrt{g_v^2 + \kappa_{3i}}} = \sqrt{g_v^2 + \overline{\kappa_{3i}}} = \sqrt{g_v^2 + \kappa_{3j-1}}. \tag{12.41}$$

Let $h_{\rho_i} = \lambda_i^2 - g_v \lambda_i + (-g_p + k_p \rho_i)$. This is in accordance with equation Lemma 12.4. Then, the characteristic polynomial of matrix Λ can be written as $\mathcal{H} = \prod_{i=1}^{n} h_{\rho_i}$, We now study the stability of the following characteristic polynomial.

Lemma 12.8 *Denoting $\mathfrak{Re}(\rho_i)$ and $\mathfrak{Im}(\rho_i)$ as $\mathfrak{p}_i$ and $\mathfrak{q}_i$ respectively. The two roots of the polynomial*

$$h_{\rho_i} = 0, \tag{12.42}$$

lie on the left half of the s-plane if

$$\frac{k_p^2 \mathfrak{q}_i^2}{g_v^2} + \mathfrak{p}_i k_p < g_p, \tag{12.43}$$

Proof Refer to [596].

To analyse the stability of the system in (12.32), we need to study the eigenvalues of the matrix Λ. The following lemma examines the location of the eigenvalues of Λ in the complex plane.

Lemma 12.9 *Given a θ, g_v, k_p and $g_p \in \mathbb{R}^+$, the eigenvalues of Λ in (12.33) have negative real parts if*

$$-g_p + k_p |\mu_i| \cos(\arg(\mu_i) - \theta) + \frac{k_p^2}{g_v^2} |\mu_i|^2 \sin^2(\arg(\mu_i) - \theta) < 0$$

$$\forall i = 1, \ldots, n. \tag{12.44}$$

$$-g_p + k_p |\mu_i| \cos(\arg(\mu_i)) + \frac{k_p^2}{g_v^2} |\mu_i|^2 \sin^2(\arg(\mu_i)) < 0 \quad \forall i = 1, \ldots, n. \tag{12.45}$$

Proof Refer to [596].

Let the right and the left eigenvector of $-\mathcal{L}$ associated with μ_i be t_i and w_i respectively. Then, from Lemma 12.1 and Lemma 12.4, the associated right eigenvectors of Λ are given by

$$m_{3(i-1)+l} = \begin{pmatrix} t_i \otimes \zeta_l \\ \lambda_i t_i \otimes \zeta_l \end{pmatrix} \tag{12.46}$$

$i = 1, \ldots, n$, $l = 1, 2, 3$. The associated left eigenvectors of Λ are given by

$$p_{3(i-1)+l} = \frac{1}{(2\lambda_i + g_v) w_i^\top t_i v_l^\top \zeta_l} \begin{pmatrix} (\lambda_i + g_v)(w_i \otimes v_l) \\ (w_i \otimes v_l) \end{pmatrix} \tag{12.47}$$

$i = 1, \ldots, n$, $l = 1, 2, 3$. The solution of the system given in (12.32) can then be written as

$$\begin{pmatrix} r(t) \\ v(t) \end{pmatrix} = e^{\Lambda t} \begin{pmatrix} r(0) \\ v(0) \end{pmatrix} = \sum_{k=1}^{6n} e^{\lambda_k t} m_k p_k^\top \begin{pmatrix} r(0) \\ v(0) \end{pmatrix}, \tag{12.48}$$

where $\begin{pmatrix} r(0) \\ v(0) \end{pmatrix}$ is the initial condition. Also note that, when (12.44), and (12.45) are satisfied, $e^{\Lambda t}$ goes to 0, since all the eigenvalues corresponding to Λ have negative real part. It was shown in [673] that when all the eigenvalues are on the left side except one at the origin, it will lead to consensus. Similarly, when there is only a pair on the imaginary axis and the rest on the left side, it will lead to a circle formation. Likewise, logarithmic spirals are formed when there is a pair of eigenvalues on the right side. We can derive similar conditions on the gains θ, g_v, k_p and g_p to achieve circles and spirals. This will follow from (12.44) and (12.45).

Rather than one pair of eigenvalues on the imaginary axis, if there are multiple distinct pairs of eigenvalues on the imaginary axis [441, 595, 797] generates trochoidal patterns. In the next section, we consider such a pattern formation.

12.4.3 Pattern Formation

The trajectories of the agents will generate patterns if the system matrix Λ in (12.32) has two or more pairs of distinct eigenvalues on the imaginary axis and the rest on the left half of the complex plane [442, 595, 798]. In our case, we can place the eigenvalues of Λ by appropriately selecting the parameters θ, g_v, k_p and g_p. In this section, we derive the conditions on the gain parameters to ensure that Λ has two distinct pairs of eigenvalues on the imaginary axis while the rest are on the left side of the complex plane.

Let us define:

$$a := \arg\min_i (\mu_i - \mu_1). \tag{12.49}$$

Fig. 12.4 Representation of the eigenvalues of $-\mathscr{L}$ and the choice of μ_a and μ_b

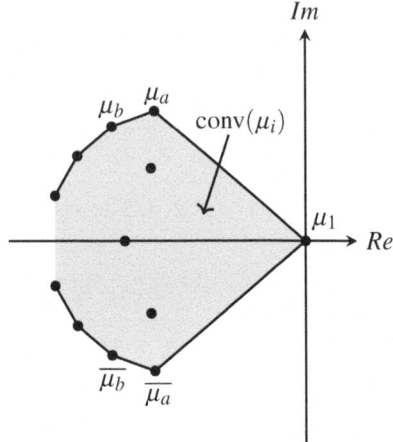

We assume that the index a satisfies

$$\arg(\mu_a - \mu_1) < \arg(\mu_i - \mu_1) < \pi, \qquad (12.50)$$

for all $i \neq a, 1$. From (12.34), it follows $\mu_a = \mu_2$. Let us also define:

$$b := \arg\min_i (\mu_i - \mu_a). \qquad (12.51)$$

We again assume that the index b satisfies

$$\arg(\mu_b - \mu_a) < \arg(\mu_i - \mu_a) \qquad (12.52)$$

for all $i \neq a, b, 1$. The assumptions in (12.50) and (12.52) imply that there exist no other eigenvalues on the line passing through (μ_1, μ_a) and (μ_a, μ_b). To develop the theory, we make these assumptions which will be relaxed further. Let

$$\mu_{\overline{a}} = \overline{\mu_a}, \quad \mu_{\overline{b}} = \overline{\mu_b}, \qquad (12.53)$$

then, for the convex hull $\mathrm{conv}(\mu_i)$, μ_1, μ_a and μ_b will be the three consecutive vertices in counterclockwise direction and μ_1, $\mu_{\overline{a}}$ and $\mu_{\overline{b}}$ will be the three consecutive vertices in clockwise direction. See Fig. 12.4 for a pictorial depiction. Also observing that all $g_v^2 + 4\kappa_i$ have nonnegative real parts, it follows that all $\lambda_{2i}, i = 2, \ldots, n$, have negative real parts if $g_v > 0$. Using (12.35), (12.36) and (12.37), we have $\overline{\lambda_{2a-1}} = \lambda_{\overline{2a-1}}$ and $\overline{\lambda_{2b-1}} = \lambda_{\overline{2b-1}}$ corresponding to $\mu_{\overline{a}}$ and $\mu_{\overline{b}}$. We aim to put the pairs of eigenvalues $\lambda_{2a-1}, \lambda_{\overline{2a-1}}$ and $\lambda_{2b-1}, \lambda_{\overline{2b-1}}$ on the imaginary axis.

Let us move one pair of eigenvalues to the imaginary axis. Let η_i be some value such that $\lambda_{2i-1} = i\eta_i$, where $\eta_i \in \mathbb{R}$. Solving for $\lambda_{2i-1} = i\eta_i$ using (12.37), we obtain:

$$g_p = k_p |\mu_i| \cos((\mu_i) - \theta) + \frac{k_p^2 |\mu_i|^2 \sin^2((\mu_i) - \theta)}{g_v^2} \tag{12.54}$$

and the value of η_i is given as

$$\eta_i = \frac{|\mu_i| \sin((\mu_i) - \theta)}{g_v} \tag{12.55}$$

However, we need to place two distinct pairs of eigenvalues on the imaginary axis to generate a trochoidal motion. We compute g_p and k_p and fix g_v and θ for ease of calculation. We solve for g_p and k_p because it turns out to be a linear equation to evaluate. If we solve for other pairs involving θ or g_v, it becomes either a messy trigonometric equation or a radical equation to evaluate. Thus, we choose to solve for g_p and k_p. Since we need to place two pairs of eigenvalues on the imaginary axis, we solve (12.37) simultaneously with $\Re = 0$ for μ_a and μ_b. Fixing θ and g_v, we solve for k_p and g_p using (12.37) for μ_a and μ_b. The eigenvalues corresponding to μ_a and μ_b for the block matrix (12.33) are denoted as λ_{2a-1} and λ_{2b-1}. Let

$$\lambda_{2a-1} = i\eta_1 \text{ and } \lambda_{2b-1} = i\eta_2. \tag{12.56}$$

Here, we assume that η_1 and η_2 are different, but later we show that with the choice of θ and g_v, both η_1 and η_2 can never be equal. By solving (12.56) simultaneously for k_p and g_p using (12.37), we obtain two values for k_p. Since k_p is a nonzero scalar, we ignore the case of $k_p = 0$. We obtain the following expression for k_p and g_p on solving (12.56):

$$k_p = -g_v^2 \frac{|\mu_a| \cos((\mu_a) - \theta) - |\mu_b| \cos((\mu_b) - \theta)}{|\mu_a|^2 \sin^2((\mu_a) - \theta) - |\mu_b|^2 \sin^2((\mu_b) - \theta)} \tag{12.57}$$

$$g_p = k_p |\mu_a| \cos((\mu_a) - \theta) + \frac{k_p^2 |\mu_a|^2 \sin^2((\mu_a) - \theta)}{g_v^2} \tag{12.58}$$

Given a θ and g_v, we can solve for (12.57) and (12.58). Here, the natural question arises if a given θ and g_v ensures that the remaining $3n - 4$ poles will be in the left half of the plane. Since the complex square root is not analytic on $\mathbb{C}$, the transformation by the gains θ, g_p, g_v and k_p on conv(μ_i) does not preserve the convexity after taking the complex square root as in (12.37). Therefore, to ensure that the gains θ and g_v make (12.33) stable, the gains must satisfy (12.44) and (12.45). We back substitute (12.57) and (12.58) in (12.44). On simplification, we obtain:

$$g_v^2 \cdot \frac{|\mu_a|\cos((\mu_a)-\theta) - |\mu_b|\cos((\mu_b)-\theta)}{(|\mu_a|^2 \sin^2((\mu_a)-\theta) - |\mu_b|^2 \sin^2((\mu_b)-\theta))^2}$$

$$\cdot \{(|\mu_a|\cos((\mu_a)-\theta) - |\mu_i|\cos((\mu_i)-\theta))$$

$$(|\mu_a|^2 \sin^2((\mu_a)-\theta) - |\mu_b|^2 \sin^2((\mu_b)-\theta))$$

$$+ (|\mu_a|\cos((\mu_a)-\theta) - |\mu_b|\cos((\mu_b)-\theta))$$

$$(|\mu_a|^2 \sin^2((\mu_a)-\theta) - |\mu_i|^2 \sin^2((\mu_i)-\theta))\} < 0 \quad i = 1, \ldots, n, \; i \neq a, b. \tag{12.59}$$

From (12.59), we can see that the stability condition of $2n-4$ poles is represented as a function of g_v, θ and the choice of the vertex μ_a and μ_b. Since g_v is chosen as a positive real number, and μ_a and μ_b are fixed, the stability of the remaining $2n-4$ poles can be viewed as a function of theta. The notion is to do a line search in $(0, \pi)$ satisfying (12.59). Therefore, a simple algorithm such as the one presented as Algorithm 1 can be used to find the range of θ and g_v to ensure stability and to place the two pairs of eigenvalues on the imaginary axis. The following algorithm gives the possible range of θ to choose from. Let $\mathbb{I}_s$ denote the union of disjoint intervals of θ that satisfies the stability condition (12.59). We can choose any vertex from the conv(μ_i) according to our convenience. Moreover, the gains k_p, g_p and g_v must satisfy (12.45) to ensure the stability of the rest of the n poles. Back substituting (12.57) and (12.58) in (12.45) and on further simplification, we obtain a similar expression as in (12.59):

$$g_v^2 \cdot \frac{|\mu_a|\cos((\mu_a)) - |\mu_b|\cos((\mu_b))}{(|\mu_a|^2 \sin^2((\mu_a)) - |\mu_b|^2 \sin^2((\mu_b)))^2} \cdot \{(|\mu_a|\cos((\mu_a)) - |\mu_i|\cos((\mu_i)))$$

$$(|\mu_a|^2 \sin^2((\mu_a)) - |\mu_b|^2 \sin^2((\mu_b))) + (|\mu_a|\cos((\mu_a)) - |\mu_b|\cos((\mu_b)))$$

$$(|\mu_a|^2 \sin^2((\mu_a)) - |\mu_i|^2 \sin^2((\mu_i)))\} < 0 \quad i = 1, \ldots, n, \; i \neq a, b. \tag{12.60}$$

Algorithm 1: Range of θ for stability

Data: $\mu_1, \mu_2, \ldots, \mu_n$.
Result: Interval of θ for stability
Fix μ_a and μ_b //initialisation;
while $0 < \theta < \pi$ **do**
    line search $\forall \mu_i, i = 1, \ldots, n, i \neq a, b.$;
    if (12.59) **then**
        | desired $\theta \in \mathbb{I}_s$
    else
        | θ not found;

12.4.3.1 Geometrical Notion Behind Stability and Pattern Formation

In the previous section, we discussed the suitable choice of choosing the gain parameters to achieve trochoidal formation. In this section, we consider the geometrical interpretation of placing the two pairs of eigenvalues on the imaginary axis and the rest on the left side of the s-plane. Since there is a square root operation involving complex numbers in the computation of the eigenvalues (12.37), unlike the case of a single integrator [595] in which the convexity of conv(μ_i) is preserved after various operations. In contrast, convexity is not preserved in the case of double integrators. The complex square root is not continuous on $\mathbb{C}$. Since convexity is not preserved, the geometrical insights provided in this section give further understanding of the control law (12.31).

Consider the eigenvalue given by (12.37). Assume θ and g_v are given, and we compute k_p and g_p given by (12.57) and (12.58), respectively. For the eigenvalue (12.37) to be stable, (12.44) and (12.45) must be satisfied. When equality holds for (12.44), the equation becomes:

$$-g_p + k_p|\mu_i|\cos(\arg(\mu_i) - \theta) + \frac{k_p^2}{g_v^2}|\mu_i|^2 \sin^2(\arg(\mu_i) - \theta) = 0. \quad (12.61)$$

Let $\mathbb{P}_s$ denote the loci of all such points that satisfy (12.61). Let ρ_{ix} and ρ_{iy} denote the abscissa and ordinate of all such points. Then (12.61) can be written as

$$-g_p + k_p\rho_{ix} + \frac{k_p^2}{g_v^2}\rho_{iy}^2 = 0. \quad (12.62)$$

Then $\mathbb{P}_s$ given as (12.62) represents a parabola with the vertex at $(g_p, 0)$. Given μ_a, μ_b, θ and g_v, the meaning of choosing the gain parameters geometrically is that we fit a parabola passing through μ_a and μ_b. The gains k_p and g_p represent the scaling and the shift of the parabola, respectively. As we know, the complex square root function $\mathbb{Z}^2 \to \mathbb{Z}$ maps parabolas into vertical lines (except the real axis). The vertices corresponding to μ_a and μ_b under rotation, scaling and shifting lie on the parabola before the complex square root operation. After the square root operation, they fall on the vertical line $x = g_v$ and are subsequently placed on the imaginary axis. If the eigenvalues are lying inside the parabola, then they are on the left of the vertical line $x = g_v$ and thereafter lie on the left side of the complex plane. This is geometrically explained in Fig. 12.5.

Instead of choosing any eigenvalues from μ_i, we prefer to choose from the consecutive vertices because this follows from the geometrical notion that, for stability, λ_{2i-1} must lie within $\mathbb{P}_s$. Therefore, we would prefer to choose any two consecutive vertices of conv(μ_i) to push the system eigenvalues (12.33) to the imaginary axis.

It is observed from the geometrical notion that if the poles lie within $\mathbb{P}_s$, then the system is stable. The choice of gains k_p and g_p represents the scaling and the

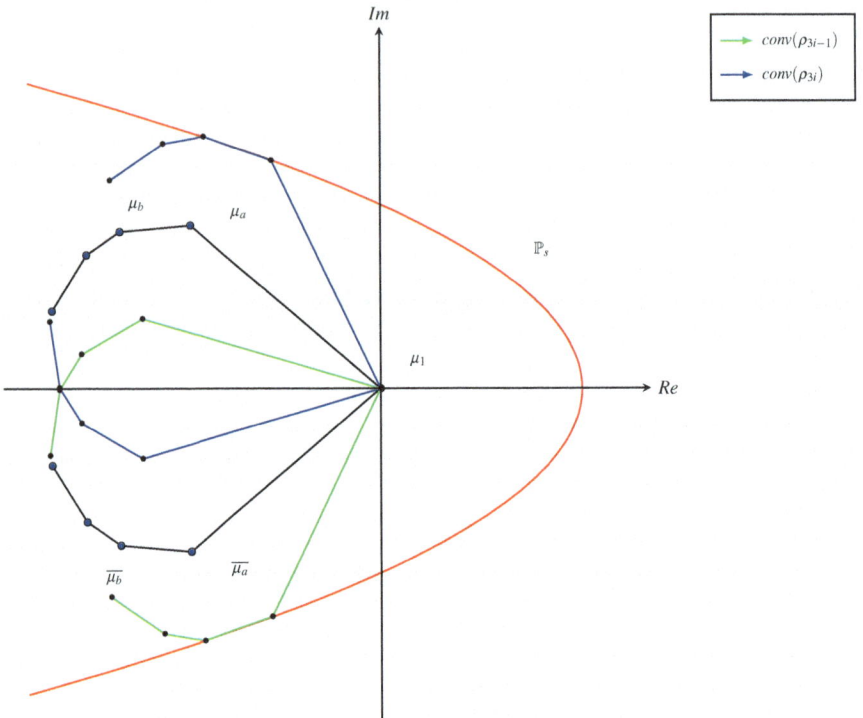

Fig. 12.5 Geometrical interpretation of placing two eigenvalues

shifting of the parabola. Depending on the choice of the vertex from $\text{conv}(\mu_i)$, the $\mathbb{P}_s$ can be made either open to the left or the right. Therefore, we can allow negative gain values for k_p.

If the assumptions (12.50) and (12.52) are not satisfied, Λ can still have multiple pairs of distinct imaginary eigenvalues. For example, if there are multiple eigenvalues on the line joining μ_a and μ_b, then we can achieve trochoidal motion. This is because the line joining μ_a and μ_b will fall within the $\mathbb{P}_s$ because of its convexity.

If the eigenvalues that are pushed to the imaginary axis are repeated, that is, μ_a and μ_b have an algebraic multiplicity of more than one, then we will not be able to obtain trochoidal motions at steady state.

We generally have positive gains for the gain parameters in the consensus law. A conservative bound on θ is obtained in [594] and reproduced in Lemma (12.10). The following lemma addresses when the gains (12.57) and (12.58) are positive.

12.4.4 Characterising the Pattern

Lemma 12.10 *Given a positive g_v, the values of the critical gains k_p and g_p as defined in (12.57) and (12.58) are positive when*

$$\max\{\arg(\mu_a) - \pi/2, \arg(\mu_b) - \pi, \arg(\mu_b - \mu_a) - \pi\}$$
$$< \theta < \min\{\arg(\mu_a), \arg(\mu_b) - \pi/2\} \quad (12.63)$$

Proof Refer to [596].

Given the set $\mathbb{I}_s$, we can find the intersection with the above range of θ as given in Lemma 12.10. Let the intersection be denoted as $\mathbb{I}_{ss}$. Picking up any θ from the set $\mathbb{I}_{ss}$ and any positive g_v would suffice to ensure that we are placing two pairs of eigenvalues on the imaginary axis. Since we need two distinct pairs of eigenvalues on the imaginary axis to achieve trochoidal motion, the following lemma guarantees the existence of two distinct pairs of eigenvalues on the imaginary axis. There is no designated method for choosing control parameters g_v and θ, although θ must lie within the bound given by $\mathbb{I}_{ss}$.

Lemma 12.11 *For the matrix Λ in (12.33), there exist two distinct pairs of eigenvalues on the imaginary axis and remaining on the left half of the complex plane if θ is chosen from $\mathbb{I}_{ss}$ and $g_v \in \mathbb{R}^+$ and the gains g_p and k_p are chosen as given in (12.57) and (12.58) and satisfying (12.60), respectively.*

Proof Refer to.

Theorem 12.2 *At steady state, the trajectories of the agents with dynamics (12.29), control law (12.31) and control parameters satisfying (12.57), (12.58), (12.60), $\theta \in \mathbb{I}_{ss}$ and $g_v \in \mathbb{R}^+$ are given by*

$$x_{\infty,i} = \gamma^*_{3i-2}(t), \quad y_{\infty,i} = \gamma^*_{3i-1}(t), \quad z_{\infty,i} = \gamma^*_{3i}(t), \quad (12.64)$$

where for $i = 1, \ldots, n$, and $l = 1, 2, 3$,

$$\gamma^*_{3(i-1)+l}(t) = 2|\zeta_{3(l)} t_{2a-1(i)} p^\top_{2a-1} \gamma(0)| \cos(|\lambda_{2a-1}|t$$
$$+ \arg(t_{2a-1(i)} p^\top_{2a-1} \gamma(0)) + \arg(\zeta_{3(l)}))$$
$$+ 2|\zeta_{3(l)} t_{2b-1(i)} p^\top_{2b-1} \gamma(0)| \cos(|\lambda_{2b-1}|t$$
$$+ \arg(t_{2b-1(i)} p^\top_{2b-1} \gamma(0)) + \arg(\zeta_{3(l)})). \quad (12.65)$$

Proof Refer to [596].

Trochoids are primarily 2-D curves, whereas the responses in (12.64) and (12.65) are in 3-D. In the following, we use Rodrigues' rotation formula to show that the response is indeed trochoid. We rotate the entire coordinate axis such that the Z axis intersects with the Euler axis. There is no need to translate the coordinate axis since the trochoids have the origin as their centre. The rotation angle ϕ between $a = [a_1, a_2, a_3]^\top$ and $\hat{z} = [0, 0, 1]^\top$ (unit vector along Z direction) is given by

$$\cos(\phi) = \frac{a.\hat{z}}{|a|} = a_3 \qquad (12.66)$$

The axis of rotation has to be orthogonal to a and $\hat{z}$. Therefore, the axis of rotation is given by

$$\alpha = \frac{a \times k}{|a|} = [a_2, -a_1, 0]^\top \qquad (12.67)$$

The unit vector along the new axis is given by normalising as (12.67)

$$\hat{\alpha} = \frac{\alpha}{|\alpha|} = \left[\frac{a_2}{\sqrt{(a_2^2 + a_1^2)}}, \frac{-a_1}{\sqrt{(a_2^2 + a_1^2)}}, 0\right]^\top \qquad (12.68)$$

We use Rodrigues' rotation formula with the rotation axis given by (12.68) and the rotation angle given by (12.66). The transformation matrix is given by

$$\begin{bmatrix} \cos\phi + \alpha_x^2(1-\cos\phi) & \alpha_x\alpha_y(1-\cos\phi) - \alpha_z\sin\phi & \alpha_x\alpha_z(1-\cos\phi) + \alpha_z\sin\phi \\ \alpha_y\alpha_x(1-\cos\phi) + \alpha_z\sin\phi & \cos\phi + \alpha_y^2(1-\cos\phi) & \alpha_y\alpha_z(1-\cos\phi) - \alpha_x\sin\phi \\ \alpha_z\alpha_x(1-\cos\phi) - \alpha_y\sin\phi & \alpha_z\alpha_y(1-\cos\phi) + \alpha_x\sin\phi & \cos\phi + \alpha_z^2(1-\cos\phi) \end{bmatrix}, \qquad (12.69)$$

where $[\alpha_x, \alpha_y, \alpha_z]$ represents the respective component along the α axis. In the new coordinate system, the trajectories (12.65) and (12.65) are computed as $[x_i', y_i', z_i']^\top = [a_i' \cos A_i' + b_i' \cos B_i', a_i' \sin A_i' + b_i' \sin B_i', 0]^\top$, where $A_i' = |\lambda_{2a-1}|t + \arg(t_{2a-1(i)} p_{2a-1}^\top \gamma(0))$, $B_i' = |\lambda_{2b-1}|t + \arg(t_{2b-1(i)} p_{2b-1}^\top \gamma(0))$, $a_i' = -2\sqrt{a_2^2 + a_3^2} \sin^2(\theta/2)|t_{2a-1(i)} p_{2a-1}^\top \gamma(0)|$ and $b_i' = 2\sqrt{a_2^2 + a_3^2} \sin^2(\theta/2)|t_{2b-1(i)} p_{2b-1}^\top \gamma(0)|$. We observe that the trajectories in the x-y plane of the new coordinate system fall in the family of trochoidal curves.

From the properties of trochoids [504, Chapter 6], the annular radius of the pattern for each agent is determined as $|a_i'| + |b_i'|$ and $||a_i'| - |b_i'||$. The eigenvalues on the imaginary axis determine the frequency of the solution. The relative phase of each agent is determined by A_i' and B_i'. Patterns are formed after the initial transient is completed. The pattern can either be a closed orbit or a transcendental curve. If the ratio $\frac{\lambda_{2b-1}}{\lambda_{2a-1}}$ is rational, then the trajectories of agents are closed curves, and if the ratio is irrational, then agents trace dense trajectories in the space.

Remark 12.7 It can be observed from (12.65) that the centre of the patterns is the origin. It can also be verified that the Euler axis a is perpendicular to $[x_i, y_i, z_i]^\top$. Therefore, if we can appropriately select the origin of the coordinate axis and the Euler angle, we can obtain the pattern at a desired point on the desired plane.

If the agents are restricted to move on a 2-D plane, the results obtained in this study will still remain valid. The results are summarised in the corollary below.

Corollary 12.3 *Let the control algorithm for dynamics (12.29) is given by (12.31), where $r_i = [x_i, y_i]^\top$ and $R(\theta)$ is a 2×2 rotation matrix given by*
$$R(\theta) = \begin{bmatrix} \cos(\theta) & \sin(\theta) \\ -\sin(\theta) & \cos(\theta) \end{bmatrix}$$

1. *If (12.44) is satisfied, the agent's rendezvous at the origin.*
2. *If the control parameters satisfies (12.57), (12.58), $\theta \in \mathbb{I}_{ss}$ and $g_v \in \mathbb{R}^+$, at steady state, the trajectories of the agents are given by*

$$\begin{pmatrix} x_{\infty,i} \\ y_{\infty,i} \end{pmatrix} = \begin{pmatrix} a'_i \cos A'_i + b'_i \cos B'_i \\ a'_i \sin A'_i + b'_i \sin B'_i \end{pmatrix} \tag{12.70}$$

where $A'_i = |\lambda_{2a-1}|t + \arg(t_{2a-1(i)} p_{2a-1}^\top r(0))$, $B'_i = |\lambda_{2b-1}|t + \arg(t_{2b-1(i)} p_{2b-1}^\top r(0))$, $a'_i = 2|t_{2a-1(i)} p_{2a-1}^\top r(0)|$ and $b'_i = 2|t_{2b-1(i)} p_{2b-1}^\top r(0)|$.

Proof The eigenvalues of $R(\theta)$ are given by $e^{i\theta}$ and $e^{-i\theta}$ with the associated right eigenvectors as $[1, i]^\top$, $[1, -i]^\top$ and the left eigenvectors as $[1, -i]^\top$, $[1, i]^\top$, respectively. The rest of the proof follows similar arguments to those presented for the 3-D case and is omitted here.

Remark 12.8 Suppose the graph $\mathcal{G}$ follows a cyclic topology. When the conditions for pattern formation are satisfied, the patterns fall into the class of epicycloids. This is a particular case of this section, with the graph being a unidirectional ring. The results can be derived by utilising the properties of the eigenvalues and eigenvectors of a circulant matrix.

12.5 Simulation and Experimental Results

12.5.1 Simulation Results

To illustrate, consider four agents with network topology given by Fig. 12.6. The Laplacian $\mathcal{L}$ of the graph $\mathcal{G}$ is given by

$$\begin{bmatrix} 5 & 0 & -5 & 0 \\ -3 & 4 & -1 & 0 \\ -1 & 0 & 3 & -2 \\ -1 & -5 & 0 & 6 \end{bmatrix}$$

Fig. 12.6 Interaction topology for four agents. The arrow in the edge indicates the direction of information flow

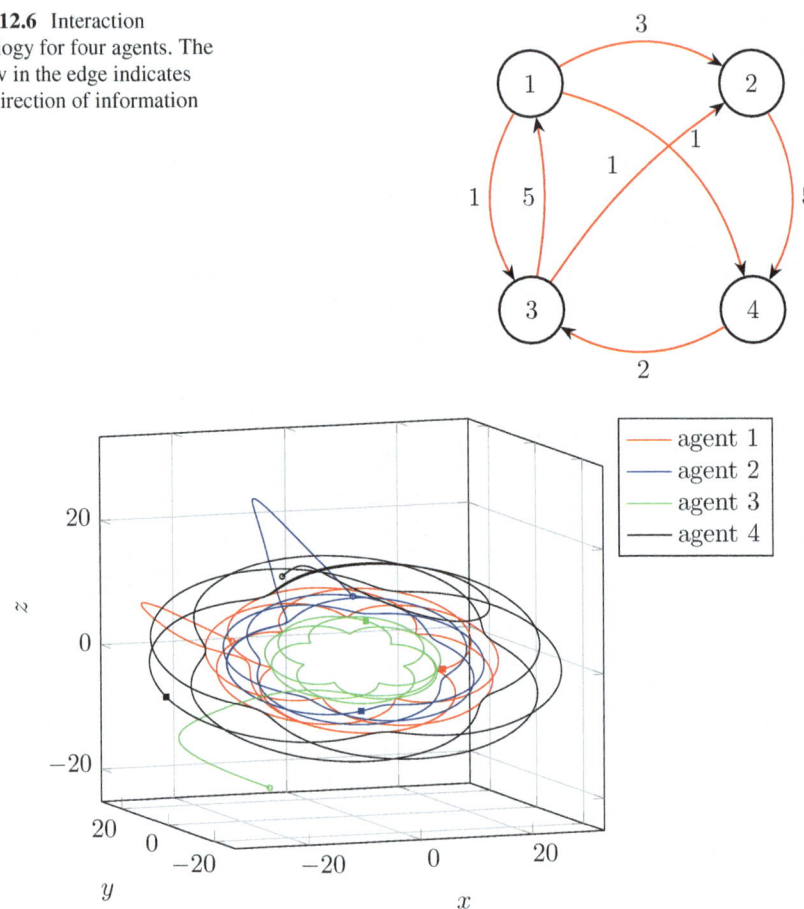

Fig. 12.7 Trajectories of four agents. The snapshot shows the agent trajectories at $t = 2{,}500$ seconds with random starting positions and velocities. Circles denote the starting positions of the agents, and the squares denote the snapshots of the vehicles at 2,500 seconds

The gain g_v is chosen to be 1. The parameter θ is chosen to be $2.28488\ rad$. The gain parameters g_p and k_p are calculated using (12.58) and (12.57) as 0.05063 and 0.00908. The interval $\mathbb{I}_{ss}$ is given as $(0.7848, 2.35)$. Figure 12.7 shows the trajectories of four agents having an interaction topology given by $\mathscr{L}$ at steady state. Figure 12.8 shows the location of the eigenvalues of $-\mathscr{L}$, $-g_p I - k_p \mathscr{L} \otimes R(\theta)$ and Λ. Note that agents trace an annular pattern on a plane.

Next, we present various graph topologies for four agents and the patterns formed by the agents. We highlight the cases in which the patterns are closed (see Fig. 12.9a–d) and not closed (see Fig. 12.9e–f). Table 12.2 shows the graph Laplacians and the imaginary eigenvalues of the corresponding system matrix Λ for each case. The ratio of the imaginary eigenvalues is also presented with the values

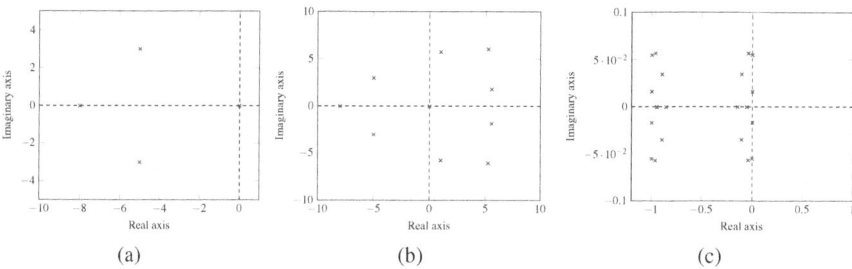

Fig. 12.8 (a) Eigenvalues of $-\mathscr{L}$. (b) Eigenvalues of $-g_p I - k_p \mathscr{L} \otimes R(\theta)$. The angle of rotation is chosen to be 2.28488 rad (c) Eigenvalues of Λ. The gains g_p and k_p are chosen such that there are two pairs of eigenvalues on the imaginary axis

rounded to four places after the decimal. For all the simulation, the Euler axis was chosen to be $a = \frac{[1,2,3]^\top}{\sqrt{14}}$ and random initial conditions are considered. The initial transients are omitted in the plots.

12.5.2 Experimental Validation

The proposed control law (12.31) has been implemented on ground robots. For the experiments, we use Fire Bird V differential drive robots[2] with linear and angular velocity inputs. The control law (12.31) provides the reference acceleration along the global x and y coordinates which we integrate using RK4 method to obtain the reference positions, $r_i^l(t) = (x_i^l(t), y_i^l(t))$. The robots track these reference positions using a tracking controller given in [68]. Robots track the positions since the proposed control law (12.31) is for double-integrator kinematics, and the robot is modelled as unicycle kinematics. The current position of the robots in the global coordinate frame is obtained from the VICON[3] motion caption system. Since Fire Bird V does not have the computing and communication capabilities, the computation for each robot is done remotely in a host PC running ROS. For each robot, a ROS node runs in parallel to imitate distributed structure with an individual bot id. These nodes communicate between them using the message structure available in ROS. These ROS nodes communicate the reference position to the neighbours. Essentially the implementation is typically done on a two-tier system. The upper tier is responsible for the control objective, the interaction of agents, receiving the actual position and orientation of robots, and generating linear

---

[2] http://www.nex-robotics.com/products/fire-bird-v-robots/fire-bird-v-atmega2560-robotic-research-platform.html
[3] https://www.vicon.com/products/camera-systems/vantage

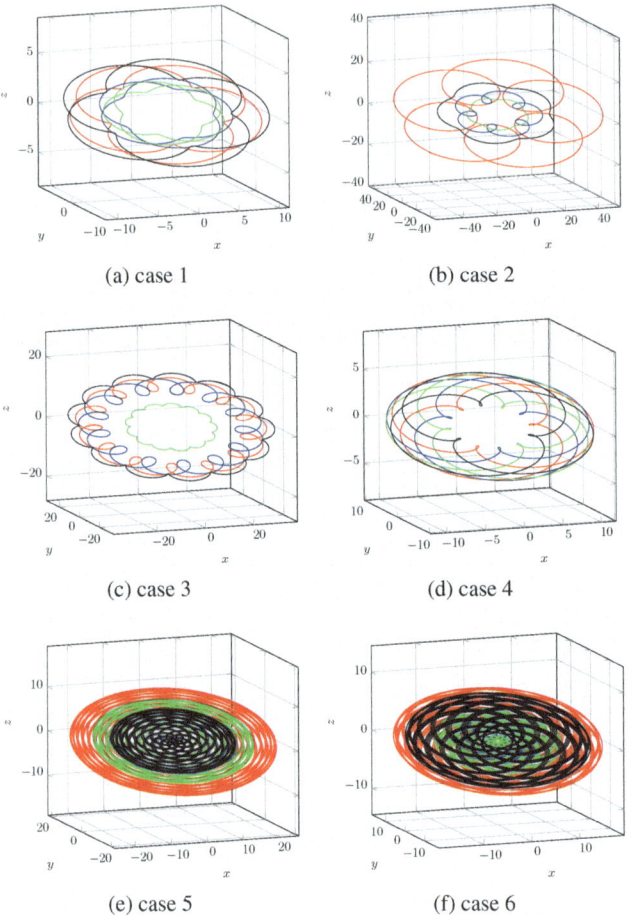

Fig. 12.9 Patterns when $\frac{|\lambda_{2b}-1|}{|\lambda_{2a}-1|}$ is rational ((**a**)–(**d**)) and irrational ((**e**)–(**f**))

and angular velocities. The lower tier is responsible for sensor data acquisition and wheel velocity control.

The communication between the host computer and the robots is through a Zigbee module.[4] The kinematic equations of the robot are given by

$$\dot{x}_i^f = V_i^f \cos(\phi_i^f(t)), \quad \dot{y}_i^f = V_i^f \sin(\phi_i^f(t)), \quad \dot{\phi}_i^f = \omega_i^f(t),$$

where V_i^f and ω_i^f are the linear and angular velocity of the ith robot. A heading command is calculated for the reference trajectory, given by $\phi_i^c =$

---

[4] https://www.digi.com/products/models/xbp24-awi-001j

12 Trochoids

Table 12.2 Simulation results

	Laplacian $\mathscr{L}$	$\|\lambda_{2a-1}\|$	$\|\lambda_{2b-1}\|$	$\frac{\|\lambda_{2b-1}\|}{\|\lambda_{2a-1}\|}$
Case 1	[4 0 − 4 0 ; − 2 3 − 1 0 ; 0 0 2 − 2 ; − 1 − 4 0 5]	0.04760	0.1665	3.5
Case 2	[6 0 − 6 0 ; − 1 3 − 2 0 ; 0 0 2 − 2 ; − 1 − 4 0 5]	0.0106	0.0640	6
Case 3	[5.5 0 − 5.5 0 ; − 3 4 − 1 0 ; 0 0 2 − 2 ; − 3 − 3 0 6]	0.0103	0.1454	14
Case 4	[1 − 1 0 0 ; 0 1 − 1 0 ; 0 0 1 − 1 ; − 1 0 0 1]	0.1904	0.4762	2.5
Case 5	[2.5 − 1.5 − 0 − 1 ; − 1 2 − 1 0 ; − 1 0 2 − 1 ; − 1 − 1 0 2]	0.2683	0.3933	1.4659
Case 6	[2 − 2 0 0 ; 0 1 − 1 0 ; 0 0 1 − 1 ; − 1 − 0.6 0 1.6]	0.2066	0.4432	2.1466

$\arctan\left(\frac{y_i^l - y_i^f}{x_i^l - x_i^f}\right)$. The angular speed of the ith reference point is computed as $\dot{\phi}_i^l(t) = \frac{d}{dt}\left(\arctan\left(\frac{y_i^l}{x_i^l}\right)\right)$. The linear and angular velocities of the ith bot are commanded as follows:

$$V_i^f = K_{V_i} D_i^l \cos(e_{\phi_i}), \quad \omega_i^f = K_{\phi_i} e_{\phi_i} + \dot{\phi}_i^l(t)$$

where $e_{\phi_i} = \phi_i^c - \phi_i^f$ is the heading error and D_i^l is the Euclidean distance between the ith robot and its leader position. K_{V_i} and K_{ϕ_i} are tracking law gains. The block diagram for the experimental implementation is shown in Fig. 12.10.

Two cases have been validated for the case of double integrators. In the first case, we used the Laplacian as in Case 6 of Table 12.2. The video link for the experiment is https://youtu.be/2z0futm9UMo. The left frame shows the robots, and the right frame shows the trajectories of the robot as recorded in the VICON data. The white cross mark on the left denotes the origin. We observe that all four agents trace a trochoid and cover an annulus. These trajectories can be useful in coverage applications such area monitoring or surveillance of different annular regions simultaneously. By changing the controller gains or initial conditions, the coverage region can be altered. For the second experiment, we used the Laplacian as given in Case 4 of Table 12.2. The video is available at https://youtu.be/pWxelWVdEuk. Since this is a case of a cyclic graph, all four agents traced the same curve and covered the same annulus. The simulation results for both cases are shown in Fig. 12.11. Figure 12.12 shows the annular trajectories traced by the team of mobile robots. The trajectories of the physical robot in the experimental part presented are as recorded in the VICON data.

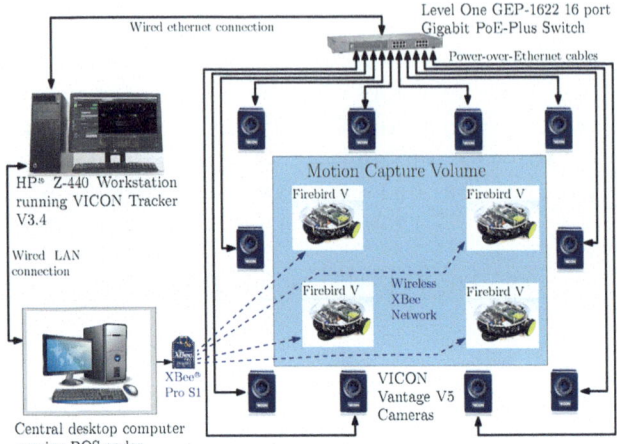

Fig. 12.10 Experimental set-up

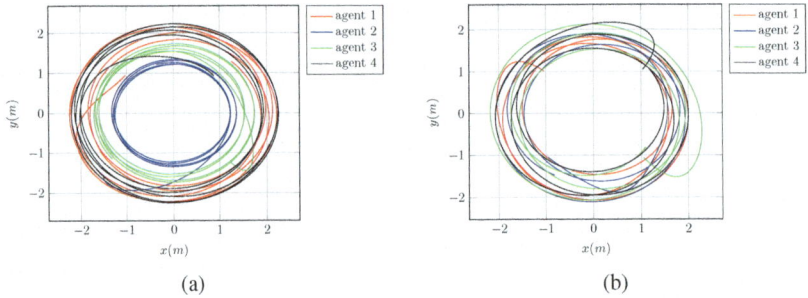

Fig. 12.11 Simulation results for double integrators: (**a**) Case 6 and (**b**) Case 4

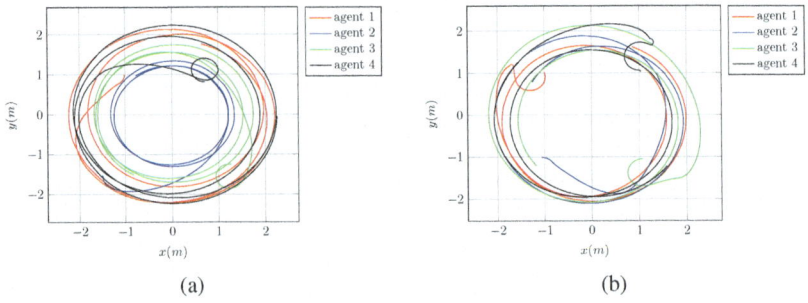

Fig. 12.12 Experimental results for double integrators: (**a**) Case 6 and (**b**) Case 4

12.6 Conclusion and Future Directions

12.6.1 Future Directions

In this section, we present some avenues for further research. Since the proposed control laws are linear in nature, the advantage is that we can characterise the solution exactly. For generating the trochoidal patterns, we are forcing the system to behave marginally stable. This will lead to the patterns being formed being sensitive to disturbances and parameter variations. To overcome these difficulties, we can introduce nonlinearity in the loop and force the system to exhibit a limit cycle so that the linearised system has two poles on the imaginary axis. In this way, the overall system can handle disturbances and will not be sensitive to parameter variations. There has been recent work on limit cycle-based controllers for circle formation [821]. Limit cycle-based controllers for trochoidal formation can be investigated to overcome the sensitivity issues.

There has been a most recent work to sustain trochoidal motions for the control law developed in Chap. 12.3. A disturbance-based observer controller is developed to ensure that the trochoidal motions of the agents are sustained. The other aspect to explore is to develop distributed tracking laws for the formation of trochoidal patterns. Tracking laws for consensus [463, 672] are predominant in the literature. Developing distributed tracking control laws for sustaining trochoidal motion can also overcome the sensitivity issues.

One possible extension from double-integrator dynamics is to consider agents modelled as a general linear system and derive the conditions for pattern formation. Another aspect to examine is the collision avoidance issue with finite-size agents. The agents are assumed to be point masses in this chapter. Either we can design initial conditions to avoid the collision or use some collision-avoidance techniques like the potential field method incorporated with the control law.

The design of the trochoidal pattern is another potential direction of future research. In this chapter, the control laws are used as a "distributed" trajectory planner. Since we can characterise the steady-state solution, we have the flexibility to generate patterns of our choice. The proposed control laws did not consider any constraints in the parameters. We can consider a few physical constraints, namely, turn radius and boundary constraints, to generate more physically admissible trajectories. The design aspect can be extended to consider more constraints, namely, velocity and turn rate, among others. This direction will be fascinating because we have more gain parameters to play with the control law to synthesise physically admissible trajectories that overcome physical constraints.

12.6.2 Conclusion

This chapter discussed distributed linear consensus strategies for pattern formation, namely, trochoidal patterns. First, we considered the problem of trochoidal pattern formation for multiple agents that follow single-integrator kinematics. Next, we considered the trochoidal pattern formation problem for multiple agents that follow double-integrator dynamics. For both cases, conditions were derived on gains for the formation of patterns. The theory was validated using various simulations and experiments.

Notes

The results in this chapter are based mainly on [[595] and [596]]. Acknowledgement is given to

© 2019 Elsevier. Reprinted, with permission, from Monsingh, J.M. and Sinha, A., 2019. Trochoidal patterns generation using generalized consensus strategy for single-integrator kinematic agents. *European Journal of Control*, 47, pp.84–92.

© 2024 Elsevier. Reprinted, with permission, from Monsingh, J.M., Sinha, A. and Chung, H,. Trochoidal patterns generation using generalized consensus strategy for double-integrator dynamic agents. *European Journal of Control*, 76, 100928.

Acknowledgments The authors acknowledge Elsevier for granting us the permission to reuse materials from our publications copyrighted by this publisher in this book.

Open Access This chapter is licensed under the terms of the Creative Commons Attribution 4.0 International License (http://creativecommons.org/licenses/by/4.0/), which permits use, sharing, adaptation, distribution and reproduction in any medium or format, as long as you give appropriate credit to the original author(s) and the source, provide a link to the Creative Commons license and indicate if changes were made.

The images or other third party material in this chapter are included in the chapter's Creative Commons license, unless indicated otherwise in a credit line to the material. If material is not included in the chapter's Creative Commons license and your intended use is not permitted by statutory regulation or exceeds the permitted use, you will need to obtain permission directly from the copyright holder.

Chapter 13
The Barriers and Opportunities of Effective Underwater Autonomous Swarms

Damien Guihen

This chapter discusses the concept of swarms of autonomous robotic platforms in the underwater environment. The submarine domain is made particularly complex to operate in by the high level of absorption of radio waves by water. There are thus large effects on navigation and communication, two things that are essential for effective autonomous swarms. Autonomous underwater vehicles (AUVs) and their origins are introduced, along with some definitions of the descriptions of various swarming configurations. The challenges of operations in the marine environment are explored for their impact on swarm dynamics and function. The systems that show potential in the land, air and space domains are not necessarily applicable underwater. Although AUVs are already seeing deployment for surveying tasks in oceans, lakes and rivers around the world, collaborative autonomy is a modality that suffers from low-bandwidth and high-latency communication, degraded positional information and extremely limited situational awareness. Not all of the challenges in the development of autonomous swarms are the consequence of the physical environment. Swarm developers and owners have different priorities and perspectives. The chapter addresses the need for stakeholders to understand each other and proposes a mechanism for communicating mission needs and design considerations.

13.1 Introduction

If one underwater robot is a hassle, a swarm must surely be a debacle. The operation of an individual AUV requires training, an understanding of the complex environmental forces to which they are subject and an appreciation of the risks inherent in their deployment. Hundreds, if not thousands, of individual AUV

D. Guihen (✉)
AMC Search, Australian Maritime College, University of Tasmania, Launceston, TAS, Australia
e-mail: damien@guihen.com

missions are undertaken successfully in Australia every year to the point that they can start to appear routine. This familiarity can set a trap for us as we calibrate our ambitions and seek to apply robots to increasingly bigger challenges underwater. The horizon for the development of effective underwater swarms, similar to in other domains, depends greatly on applications and expectations:

> If a reasonable launch schedule is to be maintained, engineering often cannot be done fast enough to keep up with the expectations of originally conservative certification criteria designed to guarantee a very safe vehicle. In these situations, subtly, and often with apparently logical arguments, the criteria are altered so that flights may still be certified in time. They therefore fly in a relatively unsafe condition, with a chance of failure of the order of a percent (it is difficult to be more accurate) ... For a successful technology, reality must take precedence over public relations, for nature cannot be fooled. [283]

A swarm of autonomous vehicles working in the underwater domain to achieve a common goal relies on, among other things, the design of a command-and-control architecture that is sensitive to the particular challenges of operating in this space. To set appropriate expectations, it is important that operational concepts and limitations are clearly communicated and risks mitigated to the greatest degree possible. On the latter point, the risk profile of an autonomous platform is unlike traditional, crewed vehicles. Autonomous and uncrewed systems present an asymmetric risk, in which the loss of a vehicle has a financial and operational impact, though the mechanism of its loss may entail a human cost to other users of the water space. An autonomous vehicle that is "at fault" is by its nature unhuman, with actions that are difficult for third parties to rationalise. For these reasons, there is a natural tension in the development of such systems. That is, in development it may be tempting to prioritise safety and reliability below operational capability and performance. Adopting this mindset can drive rapid advancement but runs a greater risk of stalling in the event of a significant accident or change of circumstances than would a community-minded, incrementalist approach.

A community-minded approach does not necessarily mean open source (although it can; see Ardupilot [1], MOOS-IvP [2, 111] and ROS [3, 661]) and is not incompatible with commercial and defence incentives. Instead, it means understanding the various perspectives of stakeholders. When done correctly this works to improve market awareness, set reality-based goals, highlight research and development opportunities, assess real-world operational readiness and crucially ease the path from concept to platform to capability. Building a community-minded approach entails an awareness of where AUVs and swarms come from: their design motivations and the limitations of their subsystems, the needs of the end users, their swarming dynamics and the challenges of operating in the specific environment.

On the low end, one might imagine a fleet of homogenous underwater vehicles that is controlled centrally, with some limited collaboration mediated by a shore-based mission planner that re-tasks vehicles according to mission completion rates. Such an approach would optimise for the filling of holes in surveys and aim for efficiency in node use. On the upper end, a swarm of underwater vehicles might be heterogenous and composed of nodes with specialisations. The swarm might form a distributed underwater network with decentralised control of task assignment and replanning. For instance, one node that is built for wide area surveying might identify a feature with medium confidence and make its details known to the network. Specialist nodes in the vicinity, and with mission capacity, might take on the job of surveying the feature in greater detail. Similarly to multinode systems in the surface, land, air and space domains, subsurface systems have the potential to extend the range of effect, provide wider coverage and increase operational duration, and even the dynamic formation of complex solutions to circumstances as they evolve.

The building of AUVs and the design of swarms means engaging skills that include sensor and, in particular, sonar engineering, navigation technologists, material scientists, hydrodynamicists, control engineers, software developers, data and computer scientists, lawyers, actuaries, naval architects, machinists and fabricators, oceanographers, marine surveyors, mariners, communications engineers and so on. These skills are found clustered in universities, research and development centres, large corporations, small companies and government civilian and defence institutes.

These actors do not usually coordinate efficiently or rationally, because often their incentives are different, unclear or compartmentalised. A commercial market or structured funding incentive is therefore frequently needed to marshal their efforts in a common direction. These approaches have a centralised structure of innovation, in which a market participant with enough resources draws in lower-level technologies, packages a solution to a problem of sufficient commercial importance and markets a product. One way of understanding this process is through the crossing of the Valley of Death between the lower/concepts and upper/products technology readiness levels where many ideas die on the vine because of a lack of resources to develop, test and make these ideas reliable [359]. When participants in the market can describe their needs and ideas in an accessible way, development cycles can become more targeted, and ideally more efficient.

The intentions of this chapter are threefold:

- to discuss the nature of autonomous swarms in the underwater domain
- to describe the challenges of the underwater environment and both the difficulty of operating and the potential benefit of swarming vehicles, which should serve as a primer on operation in the underwater domain
- to present an approach to communication and collaboration between diverse stakeholders.

The operation of AUVs, either individually or as part of a swarm, also has legal, philosophical and ethical implications that are not addressed here but are discussed in other chapters.

13.2 Autonomous Underwater Swarms

In April 1989, the oceanographer Henry Stommel published a feature in the form of a work of science fiction [762]. He wrote a fictional reflection, set in the year 2021, of his arrival 25 years earlier as a postdoctoral researcher at a field station in Cape

Cod, Massachusetts, USA, which was host to the imagined World Ocean Observing System. Staff at the facility monitored and controlled the missions of up to 1,000 low-speed diving vehicles that Stommel called Slocum gliders, in honour of Joshua Slocum, the first person to sail single-handedly around the world. The name stuck and is used today by an underwater platform produced by Teledyne Webb. (Spray, the name of Joshua Slocum's sloop, has been adopted by a competitor). The vision for the vehicles was that they could bring a new level of resolution to oceanographic measurement to help society meet the growing challenges of pollution and climate change. At any time, 480 vehicles were deployed to provide regular measurements of ocean structure:

> We have found, over the years, that the payoff in increase of knowledge often is greatest the more unconventional the idea, especially when it conflicts with collective wisdom. This policy has not always been easy to justify to our government sponsors, but they have become accustomed to allowing us to utilize a twenty - percent fraction of the observational resources they pay for in these imaginative risky speculative ways. So we have our fun, and they have learned that it pays off. [762]

Although the scale of Stommel's vision has not yet been realised with gliders, there are analogues in the Argo profiling float programme [157], and gliders have seen broad international adoption for smaller-scale oceanographic work. The article itself, along with some technical demonstrations, spurred the funding and development of technical programmes that led within 10 years to such vehicles being used in the field [690]. Stommel described his network as a fleet and given that the vehicles were individually controlled from the shore, it might fit the definition of a swarm. Scaling human effort to maintain such a fleet would itself be a massive task. Stommel's contemporaries in engineering did, however, consider vehicles as nodes within a greater system. James Albus, in his expansive 1988 technical note [425], discussed task decomposition and planning, control hierarchy, cooperative action and a roadmap for the developments needed to make such swarms real:

> There are still many issues of multivehicle command, control, and communication that still need to be addressed. Methods for transmitting commands with limited bandwidth, and with risks associated with communication emissions, need to be explored. The question of how to maintain cooperative group behavior when communications are lost needs much more study. [425]

Curtin et al. described the Autonomous Oceanographic Sampling Network [220] in 1993, with specific attention given to support infrastructure for power and data transfer, data processing and sampling performance by energy. They noted, with reference to other work:

> Intelligent control of multiple vehicle operations involves communication protocols, decision priorities, adaptive responses, distribution of control authority, relative navigation and sensor fusion. [220]

Thinking about swarms of underwater vehicles is thus a decades-old pursuit, with many of the original challenges persisting today. Designing and building swarms takes, and will take, the coordinated interaction of multiple layers of development.

Marshalling these efforts means ensuring that the goal is clear and expectations are managed. Central to this is the identification of the function and form of the swarm, in that order. Ideally, the following questions would be asked in sequence in advance of development:

- What measurements or effects are required?
- What temporal and spatial scales and resolutions are required?
- Is a swarm of AUVs the best way to achieve this?
- What exists in the market that can address this concern?
- What is the simplest swarm architecture that can achieve this outcome?
- Are the AUV nodes available, and must they be integrated?
- What are the communications and autonomous capacity of these nodes?
- Are you certain that a swarm of AUVs is the best way to achieve this?

13.2.1 Why Swarm?

Similarly to other domains, the use of swarms underwater has the potential to increase both the geographic range of robotic operation and the range of operations that can be undertaken. A swarm implies that there is an executive function that binds individual, collaborating robots to the meeting of a shared objective, rather than a critical mass of robots operating independently in a space. A swarm is coordinated, integrated and focused and might possess many of the characteristics of a unitary system, in which the robots within resemble subsystems.

It is conceivable that human operators might sit at the controls or planning interface of each robot and achieve coordination in a human way. The ambition of swarms, however, is to have a fast response to evolving environmental and operational circumstances, to be adaptive to unexpected changes in the swarm state or configuration and to dynamically adjust trajectories and functions to fill mission gaps and seize unexpected opportunities. The coordination of multiple humans at the scales and complexity envisaged of swarms is an enormous logistical and communication challenge, particular for robots acting remotely. Their future is therefore algorithmic. The design and purpose of these swarms to meet scientific, commercial and defence objectives is, however, decidedly human.

We might imagine a swarm of underwater vehicles that is deployed from an oil rig that stays in close formation but spreads out over the infrastructure to survey for any physical changes, such as deformation, shifting and cracking. A lead vehicle type ("surveyor") produces a wide-area high-resolution 3D LiDAR model of the infrastructure that is segmented, hashed and compared with previous surveys as the swarm progresses. Anomalies are quickly identified and a second, slower moving but more fully actuated vehicle type ("inspector") is directed to investigate. The inspector is then assigned to provide a further assessment of specific features using cameras and finer scale measurements at closer range. A third vehicle type ("manager") is responsible for the maintenance of communication and situational awareness between the surveyors and inspectors, replanning and re-tasking individuals and reporting to the surface. With only one manager in the fleet, it must travel through the swarm repeatedly, building a constantly updating picture of the swarm state. The manager estimates task completion rates and predicts delays caused by environmental factors. At the surface a small group of human operators watch a screen as reports from the manager appear as a sliding Gantt chart that show the evolving phases of the operation. It receives revised completion time estimates and prepares to recover the swarm. On a rare occasion, the manager might relay an urgent message that requires a human decision: perhaps a vehicle is performing poorly, and the swarm must reconfigure. The manager flags an inspector for maintenance, which will be conducted on recovery and requests that the human team confirm its updated plan because some lower priority tasks will not be fully completed. The mission planner notes the gaps and recommends that they receive a higher task priority on next deployment.

This example is, for now, fiction, but the concepts underpinning it are in active development and, in some cases, have been demonstrated in the field. The Monterey Bay Aquarium Research Institute has trialled quite a few vehicles working in collaboration. A particularly pertinent example is the use of multiple AUVs to travel through and with high-productivity features in the ocean and to observe their evolution in time and space [167]. The AUVs were used to identify features, with their activity coordinated through a Waveglider at the surface that provided localisation support and a communication gateway back to the research team. Features of interest, such as chlorophyll maxima, were detected and used to trigger the physical sampling of the water by a vehicle with a specialised collection tool.

The possible applications of underwater swarms are as diverse as the uses of the sea. These include resource mapping and exploitation, environmental assessment and damage mitigation, infrastructure inspection and intervention, border patrol and hydrographic survey. The applications of swarms entail tailored solutions to meet a set of common challenges. The marine environment presents technical barriers that are unique to the underwater domain, and these are discussed in a later section. A more easily forgotten challenge is that of the coordination of the interdisciplinary expertise needed to specify, design, build, test, refine, deploy, maintain and extract value from underwater swarms. This too is discussed.

13.2.2 Anatomy of a Swarm

The design of a swarm that is intended as a solution to a particular scientific, commercial or defence application will be guided by the parameters of the objective, not the least of which is cost. A swarm designed as a platform for the development of autonomous and communication technology might look very different, both in capability and architecture. They will have a few things in common, and a simplified view is presented in Fig. 13.1.

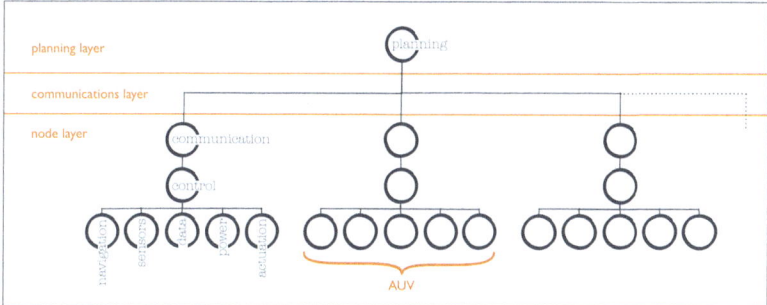

Fig. 13.1 A simplified view of the layers of design in an AUV swarm, in which three vehicles are shown, and the dashed line indicates the potential for more

It is difficult to know where to label the foundational technology layer of a swarm. Here it is drawn as the individual vehicle, with key subsystems connected to the node control unit. One might argue that the foundational layer exists right down to the development of individual components used in, for example, navigation or propulsion. These components speak directly to the suitability of the vehicle to perform within the swarm architecture. In addition, they are not stand-alone elements. Advances in one area can enable new potential in others. For example, higher situational awareness driven by a more sophisticated sonar would improve a vehicle's opportunity to improvise, thus enabling a push at other frontiers of autonomy. Even subtle improvements in the autonomous function of the vehicles can reduce the communications burden, thus shifting the architectural constraints of the swarm. The node layer is thus treated here as the foundational layer, because to go any deeper lacking specific context risks muddying the waters. Above this layer, the communication layer is implemented according to the operational scenario and the limits of the technology.

13.2.3 Node Heterogeneity

The function of the swarm indicates the node requirements and, specifically, their degree of heterogeneity. A swarm of vehicles that are performing the same task, such as bathymetric data collection, covering a wide area and each with no knowledge of other nodes, might be functionally homogeneous. Such a swarm thus has no ability to collaborate except through surface-side intervention in mission flow on an individual basis. At the other end of the spectrum is a swarm of autonomously planning vehicles, capable of internode communication and improvisation to achieve a common outcome. Such a swarm might be heterogenous, with different nodes offering individual capabilities that must be advertised or known within the network. To this end, there are multiple additional layers of complexity to address, including the distributed estimate of swarm state, task replanning and allocation.

13.2.4 Coverage and Density

The spatial density and extent requirements of nodes determine the modes of subsurface communication that are feasible for use. Underwater communication is predominantly acoustic, and given the frequency-specific absorption rates and relationship between bitrate and frequency, communications' bandwidth can be considered inversely proportional to range. Densely arranged vehicles have a better likelihood of maintaining higher throughput of bidirectional communication, but specific environmental factors, to be discussed in the next section, can have a significant impact on performance. Conflicting and interfering communications must also be considered, because the acoustic medium can quickly become saturated. Lacking an extensive data transmission infrastructure, widely distributed nodes must rely on longer-range, lower-bandwidth communication, and potentially be robust to periods of no interaction with the network. Figure 13.2 summarises these design choices in swarming architecture against sophistication, specialisation and coverage.

13.2.5 Locus of Control

The operation of multiple AUVs from a centralised location is not the only control architecture, but it is the most straightforward, from the perspective of task assignment. It does, however, necessitate central contact with each node. A distributed mode of swarm control might have more flexibility, where nodes are not necessarily aware of all the others all the time. Here one might imagine dynamically forming mesh networks of vehicles that communication when within range. There might even be specialised vehicles that roam through the operational area and serve as something like a router, with enormous latency. The decentralised

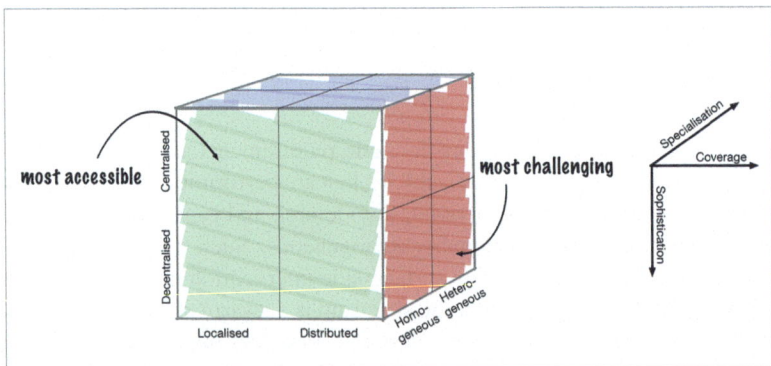

Fig. 13.2 A 3D perspective of underwater swarm considerations for describing the design of swarming architecture

case requires more than task decomposition, assignment and monitoring. It also requires a brokerage system that is tolerant of communication gaps and ensures that multiple vehicles do not end up assigned to the same task. A key consideration when contemplating swarm architecture is the design intention of the nodes. Were they built with swarming in mind, or are they being retrofitted for it? AUV development, bills of material and testing can become very expense; thus, if a new AUV type is needed, it is a significant undertaking to build it.

The question of control is more nuanced than a binary presentation of centralisation or decentralisation might suggest. Clusters of vehicles may form discrete units under decentralised control, with overall command centralised and higher level instructions issued to groups that relay information back through a gateway. Three distinct modes of communication that might be used to enable various control structures are shown in Fig. 13.3. Subsurface vehicles might, for instance, be paired with surface vehicles that maintain proximity to the underwater vehicle but can provide satellite or terrestrial radio to maintain high-bandwidth communication with a command centre. This mixed vehicle/communication control architecture is employed by Ocean Infinity Ltd. [4] and was deployed in search of the lost Malaysian Airlines aircraft of flight MH370 [101]. The arrangement has the additional potential to aid in the localisation of the underwater vehicle, and the deconfliction of the space with other water users. Such a surface-underwater pairing allows for the development of distributed networks of underwater vehicles but may limit operational concepts and increases the logistical effort involved in moving assets into the field. A longer-term objective is to develop planning and communication systems that are optimised for the unique challenges of underwater operation.

The development of swarming algorithms and use cases is continuing to develop in parallel to AUV control and communication systems. Vehicles that can use the MOOS-IVP middleware can make use of the growing library of toolboxes to

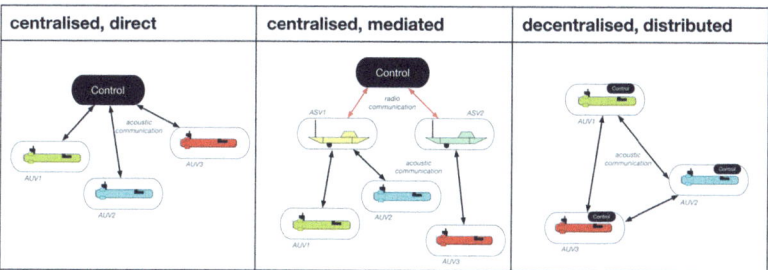

Fig. 13.3 Modes of communication and control of multiple vehicles, in order: direct acoustic communication between underwater vehicles and a mission manager (control), communication with underwater vehicles mediated by surface vessels that may use radio links to the control, and a decentralised control of underwater vehicles capable of internode communication and distributed decision-making

support advanced capabilities. One such toolbox is Swarm Autonomy [5]. This toolbox provides for the decentralisation and brokerage of tasks using an auction behaviour. The development of community-supported toolboxes over time speaks to the benefit of planning AUV control architecture with expandability in mind. The Swarm Autonomy toolbox is currently most suited for surface applications because it relies on internode communication and the reliable transfer of state across the network. The challenges of communication mean that it is likely to remain the case for some time that remote underwater nodes will need to digest larger chunks of tasks and have flexibility in achieving them with more intermittent communication with the network than a surface architecture might.

13.2.5.1 Planning

Communication is foundational to both centralised and decentralised control. In the case of centrally controlled, homogeneous fleets, there is little or no communication between vehicles during deployment, with their effort coordinated centrally. The communication overhead is that of mission monitoring and the sending of relatively basic messages, such as activation, termination, navigation correction and, for more advanced systems, the movement between different objectives. Where control is decentralised, a planning system must monitor the performance of individual nodes against assigned tasks and replan as necessary. Replanning considers task completion, remaining endurance and mission priorities but also must be flexible to adapt to changing connectivity between nodes. A summary of planning systems for autonomous vehicles is given in Thompson and Guihen [783]. The TREX (Teleo-Reactive EXecutive) planner is one particularly notable example of a goal-orientated remote agent as it has seen deployment on an AUV [569], and related concepts used for space exploration robots [608]. "Teleo" refers to the Greek *téleios* ($\tau\epsilon\lambda\epsilon\iota\varsigma$), which means "perfect" or "complete". The term "teleo-Reactive" was introduced by Nils Nelson in 1994 [626] to describe a robotic control system that is responsive to changing environmental conditions with a feedback loop referencing a prescribed goal. The system makes use of the paradigm of sense-deliberate-act, in which goals are expressed in the form of set points in time. Multiple concurrent control loops of feedback and goal comparison and command output are coordinated by function—specific teleo-reactive agents that are tightly integrated with a time-aware messaging protocol. Similarly to the MOOS-IvP Swarm Autonomy toolbox, ROS is also extended by community developers and includes a TREX implementation [6, 569], along with other planners such as ROSPlan [7, 179]. A comparison of TREX and ROSPlan deployments on an AUV is provided by Xue and Lekkas [845].

Adaptive replanning is central to the vision of autonomous underwater capability in the Royal Australian Navy's RAS-AI 2040 strategy [689]. Other models of interaction exist, such as the Blueswarm platform [112], which was inspired by the shape and collective behaviour of schooling fish. Blueswarm vehicles use optical cues of colour, flashing and motion to coordinate their motion. This is an interesting capability but in the context of RAS-AI 2040 is limited to optical ranges and focuses on formations for preservation of the school, rather than mission.

The logic of planning, the power density and battery performance and the platform computing performance develop at a pace equal to, or slightly lagging behind, other domains of operation. The principles of underwater navigation, propulsion and hydrodynamics are also mature concepts of naval architecture. It is the challenges and development of communication systems underwater that determine the pace of progress towards highly complex, capable, self-organising underwater swarms. Although systems communicating in the air or in space can make use of the electromagnetic (EM) spectrum for communication, this is severely curtailed underwater. Systems exist that allow for the high-speed, low-latency transfer of data between underwater nodes, but they are not in common usage, are unidirectional and have range limitations that are dictated by the optical clarity of the water and power budget. The application of lasers around 450 nm has been demonstrated to deliver a bandwidth of over 100 Mbps in laboratory conditions [449] but is not yet mature for use in the field. For this reason, most underwater systems currently rely on the transmission of data using acoustic modems. Depending on water conditions, the speed of sound in water is approximately $1,500$ ms^{-1}, implying latency and bitrate orders of magnitude inferior to EM transmission. Higher-frequency modems can transfer data at a rate of kilobits per second, whereas lower frequencies may be limited to hundreds of bits. The absorption of acoustic energy by water, measured in dB m^{-1}, increases with frequency. Builders of underwater platforms are thus required to choose between shorter range and higher bandwidth or longer range and lower bandwidth.

A future planning system should be flexible to incorporate uncertainty into mission state. If nodes are uncontactable, a planning system could use a probabilistic approach to estimating network status. A decentralisation of planning allows for the ad hoc formation of task-orientated squads and the propagation of messages over a wider area. An additional benefit is that this decentralisation allows a swarm to travel independently of communications and computing infrastructure. Nodes may also benefit from the experience of others, such as the sharing of environmental conditions to optimise trajectories [464]. Decentralisation is dramatically more complex than a centralised controller, and not necessarily a descendent of technological development. Decentralised messages must be efficient for transmission in underwater conditions. They must also allow for periods of noncontact, which may also mean nodes taking on larger "chunks" of a task than moment-to-moment collaboration. It is likely that a hybrid model will emerge, in which underwater platforms are multimodal in their communication, chunking large tasks when distributed and negotiating in more detail when in proximity.

13.2.5.2 Where Do Platforms Come From?

Robot design begins with an intention, which may be intrinsic, extrinsic and sometimes assumed (more on that point later). That is, people make them to meet their own ends, or those of some other group. The motivating factors may be commercial, academic and so on. There is thus not one pathway for the development of AUVs but many, the design intent of which may change over time as a market develops, stakeholders gain or lose interest and circumstances evolve. Many AUVs in use today are provided by (often large) commercial companies but have their origins in academia, for example, REMUS [761], Bluefin and Slocum [741]. With notable exceptions, for example, Autosub [207], there is a natural progression through the technology readiness levels [8] from concept and early prototype in a university setting to commercial applicability in companies primed for manufacture and distribution. Crossing the technology readiness levels, Valley of Death is a daunting and expensive proposition [360] and is the reason that many promising concepts do not progress to commercial reality.

The emergence of groups that have developed highly capable components of autonomous systems, for example, Ardupilot [1] and Mavlink [9, 483], both of which were initially developed to support the hobbyist model aircraft community, has lowered the barrier to entry by providing one of the more challenging innovations. Operating underwater means no continuous GPS feed, which means that it is inherently more expensive than operating at the surface, because the components needed to provide a reasonable quality of positioning are beyond the reach of most hobbyists. Simulations of AUVs and swarm control in community-supported projects such as MOOS-IvP (Mission Oriented Operating Suite—Interval Programming) allow for the development of swarm control algorithms to test and refine concepts without the capital drain of node acquisition and the risk of field trials on unproven systems.

A perspective of platform development, whether individual node or integrated swarm, is shown in Fig. 13.4. The process of the development of a solution to fit a problem statement implies a relationship that must be managed, and objectives that must be clearly articulated. Some platforms are built by individuals, or small, well-integrated groups. This approach is suited to problems that have a niche application and might be the result of an enthusiast in a group. Sustaining such projects in the long run is difficult if they rely on the skills of an individual that leaves.

Collaborative groups, for instance between university departments, can bridge some skill shortages but must navigate the incentive structures that may not be aligned between groups. To illustrate this point, consider a collaboration between environmental science and mechatronics departments. The solution to an environmental science question may be the application of a robotic system that the mechatronics team can deliver. In principle, this seems to be a mutually beneficial relationship. The reality of funding and academic promotion, however, is that from the outset, both groups are pressured to publish and to show tangible results. The project timing is likely such that both groups are starting at the same time. Thus, when the robotic system might reasonably be expected to be reaching field applicability, the environmental scientists are out of options, potentially with little immediate scientific benefit from the robot. Rather than wait, they would need to undertake other work to build a record of success to apply for further funding. This is principally a scheduling and funding structure issue, and one that can be mitigated. University departments are not, however, generally equipped with the means to easily change funding body rules, and it can be less risky to stay in their lane and buy in services or products.

	Description	Challenges	Benefits	
	Builder/User (solo)	availability of skills and resources	simplicity of purpose and communication	
	Builder-User (as peers)	differing professional incentives; inter-team communication	skill specialisation and development time	collaborative
	Builder-User (as a service)	communicating objectives; managing expectations and long-term support	clear incentives; selection of providers and expertise	
	Builder-Users (as a product)	achieving scale; lead times	established concept/product; long-term support (hopefully...)	
	Builders-Integrator-Users (integrator mediated)	communication bottlenecks, cost, and solution suitability; flexibility and 3rd party involvement	adherence to standards and documentation; long-term support; managment of relationships	off-the-shelf
	Builders-Integrators-Users (ecosystem)	oversight and coordination; data formats and management	flexibility; long-term viability; 3rd party involvement	standards-based

Fig. 13.4 A perspective on where platforms come from, with associated challenges and benefits

With market maturity, opportunities develop for integration specialists, specifically those who acquire or license technology to package with support, training and ancillary services to one or more customers. This is a model common in the provision of a platform at scale to defence forces and the oil and gas industry. Products in this domain are generally stable, changing little with updates. The trade-off is the dynamism and flexibility that can be characteristic of smaller-scale development. This is often seen as a lower-risk approach to the acquisition and sustainment of capability but can come with significant upfront and ongoing costs.

Communication from end user to fundamental developer must also be mediated by the integrator; thus, it can take a long time for off-the-shelf products to meet emerging needs. Once the market opens somewhat, through the introduction of innovations such as Ardupilot, something of an ecosystem of builders and users can flourish. There is still space for integrators, but innovation is not regulated to the same degree. When the remotely operated vehicle company Blue Robotics [10] started selling kits and spare parts, it didn't take long for the marine robotics community to start using them in many [13] unrelated platforms. The company solved a core component issue with a Kickstarter campaign, backed by 360 supporters, to develop the T100 thruster to propel vehicles through the water at a small fraction of the cost of previous products [11].

13.2.6 Scaling from One to Many

Commercial autonomous underwater platforms currently in use require the attention of trained human operators to structure, execute and monitor missions. Further human effort is required to recover the platform and interpret the collected data

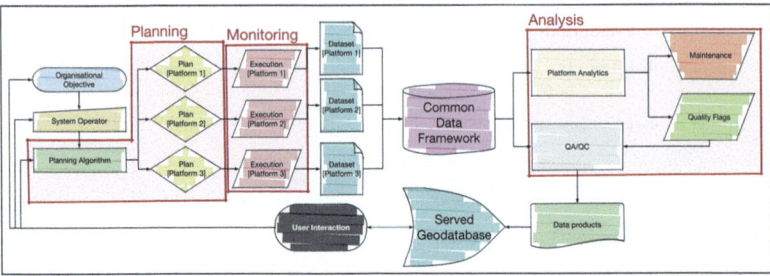

Fig. 13.5 A flow diagram of an idealised workflow from objective development, through planning, monitoring and analysis. Product sharing is related back to organisational objective and planning resources. The workflow highlights the distinction between individual platform planning and data modes and the utility of a common data framework. The role of platform performance analysis to inform both maintenance and payload data quality is illustrated

for assessment of mission success, maintenance warnings and vehicle faults. These are time-consuming tasks and impose an often steep learning curve on operators. Scaling from single to multiple vehicle operations requires an almost linear increase in human resources. Scaling from infrequent to regular operation imposes a time pressure on the analysis of vehicle performance. Tools for such analysis vary by manufacturer, with common spreadsheet software the lowest common denominator. Mistakes are easy to make, and early signs of performance degradation are easy to miss. This is particularly the case when the analysis of payload data takes precedence. Progress in this model might be described as improving the robot to human ratio, in which a small team can effectively run numerous vehicles. This requires the automation of large swathes of the current workflow and represents the immediate opportunities for achieving scale with existing platforms. A model of such a workflow is shown in Fig. 13.5. The workflow presented hinges on the development of a common standard for data storage and analysis, which is easier said than done. Those who would attempt to create such a standard should heed the warning of Randall Munroe [12].

13.2.7 Surely

It is a truth universally acknowledged that a scientist in possession of a good fortune must be in want of robots. However little known the feelings or views of such a scientist may be on their first entering a department, this truth is so well fixed in the minds of the surrounding engineers that they are considered the rightful property of some one or other of their projects (with apologies to Jane Austen).

The misattribution of intent by Mrs Bennet in the opening chapter of *Pride and Prejudice* [69], rephrased above, is analogous to many conversations involving collaboration or requirement communication. The hasty generalisation fallacy of logic can lead one to assume that the designer (or user) of a system shares their priorities and expectations. As an exercise, count how often the word "surely" is used in conversation, with each use marking an expectation that may, or may not, be reasonable. If, when discussing AUVs and swarms thereof, one finds oneself using the word "surely", it is a sign that there is, perhaps, an unaddressed assumption that may be a bigger issue later. In terms of design and problem statements, at least, there is no such thing as too much information. A great number of assumptions of operation and development in the underwater domain can be assuaged with a consideration of how the environment affects functions that we take for granted at the surface.

13.3 The Marine Environment

The unique conditions of the underwater domain mean that the design of multiple systems and concepts differ from that of their counterparts on the surface, on land and in the air. The design of a single node must account for the difficulty in communication, navigation, situational awareness, power and prediction of environmental forces that are caused by the presence of water. By turns, a swarm of underwater vehicles can exacerbate or mitigate these challenges.

A defining characteristic of the subsurface domain is the encapsulation of objects in water, which is thus the primary medium for the transmission of signals. The absorption of EM radiation by water molecules, and other components of the water column, is efficient, as described in the previous section. Although the absorption of specific wavelengths is a complex, nonlinear function, it is reasonable to generalise that we cannot rely on the mechanisms routinely used on the surface for communication and navigation. The laser-based communication and sensing systems in development and trial promise greater bandwidth and precision, though the ranges will be measured, at best, over tens of metres. When considering the stakeholders and applications of underwater robotics, such capabilities at these ranges may make all the difference or become insignificant.

The limiting factors of AUV swarms are thus variable by scales, both temporal and spatial. Over long periods of time, power becomes a predominant concern. The energy density of batteries will continue to define operational time limits and will determine the infrastructure requirements to push through these barriers. Over long distances, the limiting factor becomes communication. Figure 13.6 illustrates how distinct design limitations exist at various intersections of temporal and spatial scales. This form of time and space diagram with log-log axes is, incidentally, often called a Stommel diagram, after the same Henry Stommel who proposed the Slocum mission described earlier. Stommel used the diagram to describe the variability of processes in the ocean and to plan and assess sampling regimes accordingly [810]. Beyond power and communication, navigation is a crucial concern that affects every mission and at midscale can be a limiting factor. At large extremes, sustainment and maintenance are an issue, with questions of wear and replacement. At the other end

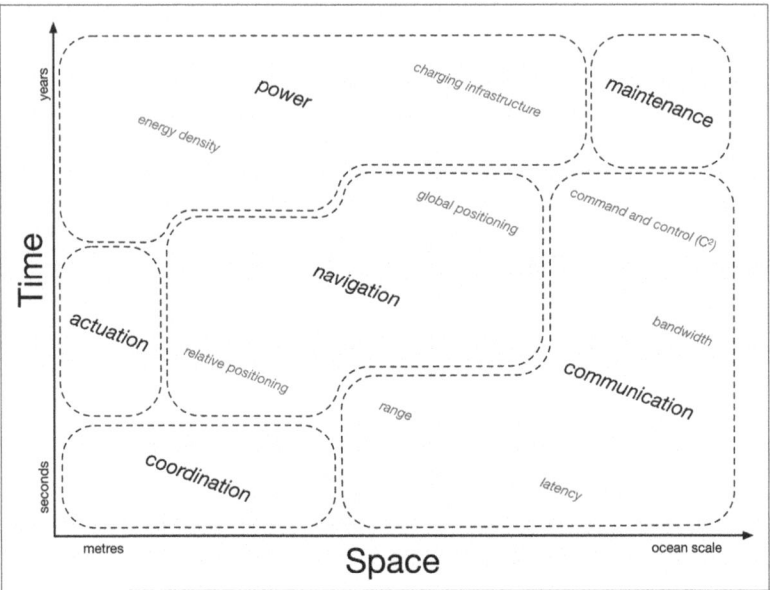

Fig. 13.6 Typical limiting factors for swarm function over different temporal and spatial scales

of the scales, the operation of equipment in close confines over short periods means careful coordination of positioning and function to avoid kinetic or communications effects.

13.3.1 Communication

The three principal considerations of communications systems are bandwidth, latency and range. That is, how much data can one send in a given time, and how long does it take to be received? Certainly, there are factors, such as security and error control, but the former three constrain their solutions. Data are generally transmitted by the modulation of a carrier wave, or variable field, transmitted through a medium. The higher the frequency of these waves, the more potential they have for modulation and thus the transmission of data. The faster they travel, the lower the latency of the signal. With an EM wave travelling at the speed of light, a complex message can be sent to the moon and back in a few seconds. Underwater, we must typically rely on acoustic waves, which to travel a kilometre or two effectively are usually in the 10–30 kHz range. That low frequency, relative to the MHz and GHz signals used at the surface, already means that the transmission rates are many orders of magnitude lower underwater than radio systems at the surface. The speed of sound underwater averages about $1{,}500\,\text{ms}^{-1}$, with significant variability in the water column driven by differences in temperature and salinity. Together, this means that the transmission of a basic several kilobytes message underwater can take multiple seconds. By the time it travels to a receiver 2 km away, it might be received 4 or 5 seconds after it began transmitting. This low-bandwidth and high-latency regime eliminates any human-in-the-loop, low-level remote control and illustrates the importance of autonomy in the underwater domain. One interesting approach is the communication of information such as complex trajectories in data-efficient forms. For example, a series of coefficients and shared algorithms can be used to share intricate detail. See Chap. 12 for a treatment of trochoidal formations and their efficient description as an example.

The underwater communications regime also has a profound influence on the architecture of potential swarming networks. Systems must anticipate periods of no, or poor, communication given environmental factors. They must not immediately assume that a node is dead if it is not responding. They must plan for low data volumes in their protocols, and they must be robust to errors of transmission.

13.3.2 Network Design

A future for network state forecasting by individual nodes is tantalising. An AUV forming part of a wider swarm will be expected to lose connection with radio-based communication once it goes beneath the surface. It will retain the potential for acoustic communication with other nodes to a range proportional to the square of its modem power and inverse to its bandwidth, a function of its operating frequency. One envisages a robot drifting from the herd hearing less and less from the others until it is alone. That robot, were it capable of estimating the trajectories of other nodes within the swarm might, however, be able to schedule its goals with an adaptive planning system, such as TREX, such that it would come within the expected range of other nodes for a duration sufficient to transfer information that could ultimately propagate across the swarm.

The Opal distributed agent-based system [401], developed by Australia's Defence Science and Technology Group, has been demonstrated to provide an analogous resilience in uncrewed aerial vehicles, providing ground-based stations with network access [400]. Individual elements calculate the probable locations of other nodes and might adjust their trajectories accordingly to re-establish communication. The design of Opal focused on its use to solve problems associated with the Radio Frequency domain, but the concept of planned re-encounters may work underwater. Several factors complicate the analogy but also point towards potential domain-specific solutions.

Underwater operations, particularly those conducted over longer durations, are subject to drift. The environment exerts forces that induce accelerations that may equal or exceed those produced by an AUV. The pattern of such environmental forces is complex, and by their nature chaotic. Confidence in the estimation of swarm configuration calculations will therefore fall once communication is lost.

Nodes that base their path planning on projected communication opportunities under such uncertainty will then be subject to error and inefficiency and with no recourse to correction short of returning to the surface. The consequence of this is that mission design will need to consider the maximum geographic extent of the swarm as a function of the planning algorithm's tolerance for positional and state uncertainty, the range of the modems and the probability of stochastic reencounter.

A network that includes nodes dedicated to mitigating the problem of time away from communication could significantly extend the planned geographic extent of the swarm. For example, a communications-focused node type might weave between the wider swarm, updating its understanding of both swarm configuration and relevant environmental conditions at a much higher rate than the task-orientated nodes. Limiting the searching behaviour to one side of any peer-to-peer communication reduces the likelihood of missed connections because of a locked-in state of binary searching, for example, a pair orbiting a point but out of phase and failing to establish contact. A roving communicator would also allow for information propagation across the swarm by physical transfer as opposed to successive opportunistic handoffs. The effect of this implementation on planning would be to reduce the density of "worker" nodes, increasing the swarm footprint but maintaining the situational awareness of a much denser swarm, and at a potentially lower modem power. A hierarchy of communication strategies might even be employed, in which low-frequency acoustic signals are used to acquire, locate and identify nodes. Higher-frequency acoustics could be used to negotiate a rendezvous at which high-bandwidth optical communication can occur at closer range. The implied heterogeneity adds complexity but pays off in a more robust communication system that has consequent benefits to adaptive planning and the likelihood of mission completion in the face of operational uncertainty.

Network design must also treat silence as a resource. Multiple payload and navigation systems rely on the use of sound. Active acoustics have a large potential to interfere with each other and to significantly compromise the performance of passive systems, in addition to the increased detectability of noisy platforms, which can be a substantial drawback in certain circumstances. Therefore, subsystem designers need to be community-minded in their use of sound and their scheduling of pings. This is as true for navigation as it is for communication.

13.3.3 Navigation

The comforting blue dot drifting over a tan and green map on a phone screen has massaged the direction-finding anxiety of millions of drivers around the world since at least 2008. Satellite navigation was a feature in cars before this, but its mainstream explosion was signalled by the release of the Apple iPhone 3G, which included a GPS receiver, Google Maps and a data connection sufficient to query directions and download map tiles. GPS and other global navigation satellite systems (GNSS) have existed for decades, and new features, resolution and coverage are added regularly [32]. The hardware required by navigation clients to make use of the satellite signals is small, requires low power and can be provided in even the lowest-end smartphone. The same is true of navigation systems used in autonomous vessels used at the surface, which significantly reduces the barriers to entry when building nodes and swarming systems.

Although most, if not all, AUVs include a GNSS receiver to establish an initial fix at the surface, once underwater they must rely on other means to establish their location. The main technique used is dead reckoning, in which the speed in a direction is calculated and integrated to provide position updates. This can be further augmented with Doppler velocity logger (DVL) bottom-tracking system, which measures the impact of motion on the frequency of a returned echo to establish velocity over a surface. An inertial motion unit (IMU), the most sophisticated of which include solid-state fibre-optic gyroscopes (FOG), provides an even greater precision in navigation by detecting changes in rotation rates of pitch, roll and heading, along with acceleration in three dimensions. A navigation system keeps track of the IMU and DVL inputs, along with other data, such as ambient pressure, to calculate a position that can be accurate to the order of 0.1% of the distance travelled. Travelling underwater, one might thus expect the positional uncertainty to grow by approximately 1 m for every kilometre travelled when conditions are optimal. Over short durations, such a dilution of precision may not be overly concerning. For longer missions the precision of localisation, and thus the success in navigation, can be severely compromised. AUVs are generally be programmed to return to the surface at regular intervals to re-establish a GNSS fix, but this is not possible, or desirable, in all circumstances. Vehicles operating in areas of heavy surface traffic or under structures or those seeking to avoid detection may be unable to return to the surface. Vehicles operating at depth may exceed the maximum range of their DVL when climbing and diving; thus, the impact of navigational drift may be worsened by coming to the surface rather than reduced. The inclusion of each supporting technology to assure navigational accuracy can mean that the cost of each node in a swarm increases quickly.

Whereas a GNSS receiver can cost A$50, the cheapest DVL is approximately A$5,000 and IMUs of sufficient quality may be 10–100 times that. The navigation system on an AUV is therefore an enormous proportion of the cost of its materials. Determining the scale achievable with an AUV swarm depends to a great extent on the number of nodes required and the quality of navigation necessary to meet the design objectives. Building a swarm does provide some economies of scale. External reference systems, including acoustic beacons and phase-detecting sonar arrays, can be used by multiple AUVs, assuming the acoustic noise can be managed. The use of navigation infrastructure, to include beacons and data transfer stations, can be a part of a solution in solving the large-scale, long-term challenges of managing communication, navigation and power. Curtin et al. provided an early description of what this might look like.

13.3.4 Power

The gliders described by Henry Stommel made use of thermal energy in the environment for power [156]. This is a line of engineering research [158], but the power requirements of modern systems and payloads mean that for underwater operation, lithium or alkaline batteries are the only viable options. Nuclear-powered AUVs are certainly possible, and the capability is claimed by some states [243], but the shore-based infrastructure and supply chains to host one are enormous. It is difficult to conceive of an efficient use of reactors at swarm scale. The energy-density of batteries, and their limits, are parameters under scrutiny by many industries, from the aforementioned smartphones to cars. Electric cars, for example, need a reliable, distributed charging infrastructure before they are relied on for long-distance journeys. The charging infrastructure of AUVs is, thus, a key enabler for delivering swarms that persist over longer temporal scales. The development challenge then is docking in an environment with forces that can be difficult to predict.

13.3.5 Prediction of the Environment

Newton's second law of motion, approximately that force is equal to mass times acceleration, is as true underwater as above it. Maintaining stable and predictable trajectories means the balancing of these forces through the delivery and vectoring of thrust and drag through a field of environmental forces that include gravity, buoyancy, waves (both surface and internal), eddies, jets and turbulence. Accurate prediction of these environmental forces means that a vehicle can optimise its propulsion. The general unpredictability of small-scale ocean currents, and the chaotic nature of turbulence, means that driving precise trajectories underwater, where AUVs are typically travelling at velocities of 2–5 knots, can be a hit-or-miss affair. The building of infrastructure to support the delivery of power to nodes in an AUV swarm entails a robust docking mechanism that necessitates the vehicle's navigation back to, and mating with, a receiver. The challenge is a complex interaction of vehicle control, environmental state estimation and docking design. Concepts such as travelling upstream towards a vaned dock with optical positioning used for final alignment are in development [855].

AUV swarm design needs a careful consideration of the motivating objective and the temporal and spatial scales required. The limiting factors, and thus the opportunities for progress, are specific to the swarm's coordinates within these dimensions. In the world of small and fast operation, the emphasis should be on planning and coordination, while the broader scales require investment in infrastructure and technologies to make it interface correctly with the nodes. Figure 13.7 provides a summary of enabling concepts for swarming scales.

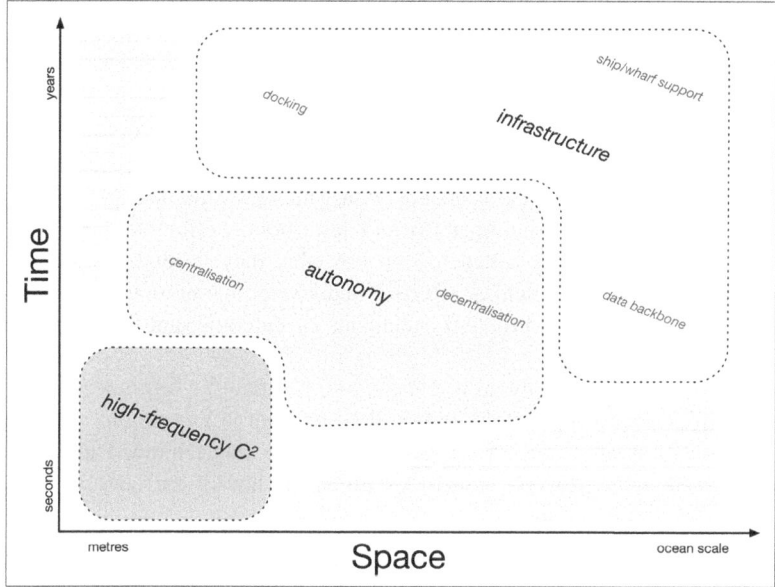

Fig. 13.7 The enabling developments that will allow swarms to scale over time and space

13.4 On Communication

The development of functional and effective AUVs is, as discussed, broadly multidisciplinary. The intricacies of multivehicle task allocation in adverse communication conditions must be ironed out. Standards that might support wide-area infrastructure and docking must be agreed upon to enable broad multiparty adoption. Workflows and supporting technologies must be developed to smooth human interaction with swarms, from mission concepts to planning to execution, analysis, interpretation and assimilation. Operational concepts must be carefully constructed with consideration of the capabilities and development trajectories of the technologies on which they are dependent. At each point, the problem, or environmental challenge, that is to be addressed should remain a guiding principle. This problem should then be articulated unambiguously.

13.4.1 The XY Problem

A common problem with stated problems is the so-called XY problem, in which the true problem is obscured with a problematic solution. The term XY problem was coined by Eric S Raymond in a list of questions not to ask software developers [669]:

Q: How can I use X to do Y? A: If what you want is to do Y, you should ask that question without pre-supposing the use of a method that may not be appropriate. Questions of this form often indicate a person who is not merely ignorant about X, but confused about what problem Y they are solving and too fixated on the details of their particular situation. It is generally best to ignore such people until they define their problem better. [669]

The terse nature of the advice, that it is best to ignore people who have an ill-defined question, perhaps speaks to the frequency with which the situation arises. The XY problem is older than the quotation above and describes the failure that occurs when someone receives what they asked for but not what they needed. At the root of commissioning the development of effective underwater autonomous swarms is the formulation of unvarnished problem statements of outcome and objective, rather than platform and workflow.

The mirror of the XY problem is *Einstellung*, German for "attitude", described by Abraham S Luchins in 1942 [543]. Whereas the XY problem refers to a miscommunication of a need, *Einstellung* implies a single-minded approach to solving a problem. A proverb, sometimes called the law of the instrument [153] or Maslow's [562], puts it thus:

> When all you have is a hammer, everything looks like a nail. [562, Traditional]

Robotic platforms, or their potential components, developed in a vacuum can suffer for not having natural paths to progress further. Upstream and downstream stakeholder engagement reduces this likelihood, although research on the cultural differences of collaboration between distinct groups is not without its challenges.

13.4.2 Talking to Stakeholders

By avoiding "surely", discussions between builders and users can proceed on solid ground if both groups, and intermediaries, are methodical in their specification of objectives and performance expectations. Reciprocally, the readiness of available technology to meet these expectations should be a bilateral exchange. Atomisation of requirements into the simplest parts can help to illustrate the complexity profile of the whole. It is unlikely that all aspects will be equally complex; thus, decomposing the challenge and weighing component readiness can highlight areas in need of attention. The most suitable axes of complexity measurement will vary by application, but in the first instance, the perspective in Fig. 13.2 could be used as an aid to discussion. Figure 13.8 shows a reworking of this figure, in which area coverage can be remapped as needed. This mapping could be used in development to identify the niche complexity profile of a particular application. Existing concepts that resemble these profiles may be found more quickly or projected against the application map to highlight needed development.

The axes of the map will need to be further refined in consultation with manufacturers and end users. Some axes may be binary states or require the development of indices, such as the degree of communication decentralisation. Axes

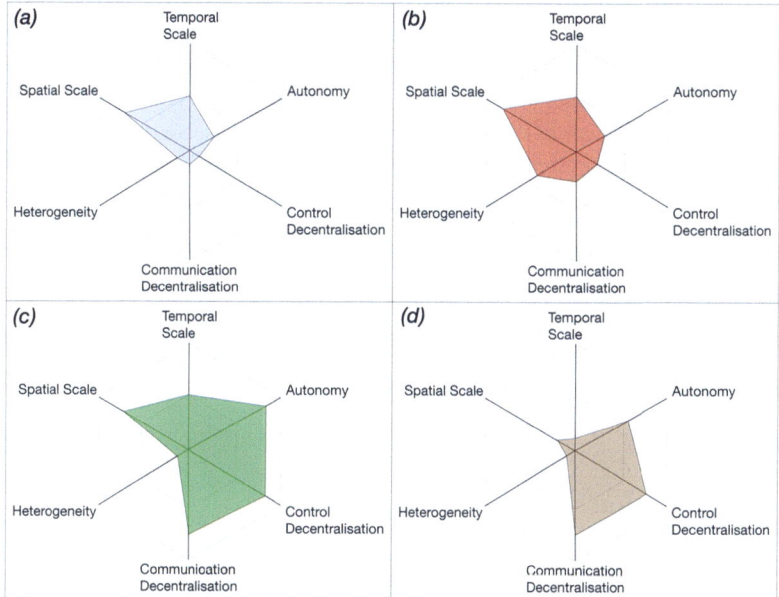

Fig. 13.8 A suggested description of autonomous underwater swarm complexity with (**a**), (**b**) and (**c**) an approximation of complexity profiles for scenarios in Fig. 13.3; (**d**) is an approximation of the profile of the Blueswarm platform

such as "autonomy" may require clarification as to whether the metric refers to the individual node or the system. A communicated understanding of swarming objectives will make the progress of each underpinning technology more cohesive to the overall vision of autonomous underwater swarms.

13.4.3 Of Hassles and Debacles

If making autonomous underwater swarms were easy, one suspects they would be common. The need for resolution of measurement in the ocean, and the variables to measure to help us understand our world, have only increased. It is clear that we cannot adequately scale human effort or time at sea accordingly. AUVs are helping scientists to map the unexplored ocean, with the cryosphere, or polar regions, a particularly harsh proving ground of technology [766]. The use of AUVs working in close partnership with other observing technologies has been acknowledged at the highest levels of policymaking as essential for meeting ocean governance objectives [712]:

> Critical Ocean S&T infrastructure includes ships, data buoys and mooring systems, seafloor cables, borehole observatories, satellites, fixed and airborne remote sensing platforms,

scientific drilling platforms, floats, gliders, Autonomous Underwater Vehicles (AUVs) and other unmanned systems, sensors (including optical, acoustic, and animal borne), submersibles, and modelling and computational infrastructure (including cloud computing) all working in concert with one another. [712]

The design of AUV swarms, along with the infrastructure needed to support them, can optimise sampling systems' effectiveness at resolving features at the appropriate temporal and spatial scales.

Attention to detail in problem descriptions, awareness of the differences in stakeholder language and incentives and, where useful, the use of tools such as the complexity map outlined here can help prevent swarm development from becoming a debacle.

It is likely that node development, testing and integration into networks will remain a hassle. However, is that not the fun of it?

13.5 Conclusion

The vision of fleets or swarms of underwater robots deployed to address uncertainty in oceanography and climate interactions, seabed mapping and infrastructure support and defence applications long predates the development of viable underwater robotic platforms. The potential applications have only increased over time as computing power, energy systems, sensor size and capability and planning algorithms have advanced. The underwater domain has unique environmental challenges that mean that solutions that might migrate seamlessly between land, surface, air and even space domains need considerable re-imagining before application below the surface. Navigation and communication, both affected by the impact of the opacity of water to EM signals, relay to a much greater extent beneath the waves. The consequences for localisation accuracy and communication range and bandwidth are profound. Underwater swarming networks, which rely heavily on communication and navigation in one form or another, must be tolerant to information latency, loss of communication and localisation accuracy, as well as the complexity of moving through a viscous, dynamic fluid and around poorly mapped structures.

The future development of underwater swarms therefore needs a range of innovations in parallel and drawing from multiple disciplines. High levels of individual vehicle autonomy, efficient communication protocols, swarm control architectures and task arbitrage in an uncertain messaging environment, environmental modelling, edge computing, swarm state estimation and power storage are only some of the lines of relevant development. Research in this space, if it is to be relevant and coherent, needs to keep in mind the requirements from the component and subcomponent level through to node and network design paradigms and protocols. Associated data architecture and infrastructure elements for launch and recovery, charging, communication and navigation are equally important. Research is also needed to develop accurate and reliable simulations for the assurance of behaviour in uncertain and complex and dynamic environments. System integrators need a keen awareness of the maturity and limitations of these technologies but also insight into the questions that ocean-based end users are trying to address.

The building of effective underwater autonomous swarms is thus a very human endeavour. Developers of swarm concepts need to communicate their needs and ambitions to a wide range of stakeholders who may not all speak the same technical language. One lesson is central: surely assumptions are to be avoided. There are debates for human society to be had on the legal, ethical and philosophical implications of autonomous swarms operating in the world's oceans. These debates will flounder without a shared human understanding of our robotic vision. Expressing our swarm design and application ambitions clearly will facilitate clear-eyed human engagement.

Open Access This chapter is licensed under the terms of the Creative Commons Attribution 4.0 International License (http://creativecommons.org/licenses/by/4.0/), which permits use, sharing, adaptation, distribution and reproduction in any medium or format, as long as you give appropriate credit to the original author(s) and the source, provide a link to the Creative Commons license and indicate if changes were made.

The images or other third party material in this chapter are included in the chapter's Creative Commons license, unless indicated otherwise in a credit line to the material. If material is not included in the chapter's Creative Commons license and your intended use is not permitted by statutory regulation or exceeds the permitted use, you will need to obtain permission directly from the copyright holder.

Part IV
Thinking Swarms: Cognition

Chapter 14
Introducing Lifelong Learning in Swarm Robotics

Phillip Smith, Aldeida Aleti, Asad I. Khan, Vincent C. S. Lee, and Robert Hunjet

This chapter presents the first known attempt to enable a lifelong machine learning system for adaptive emergent behaviours in swarming agents. The proposed system builds on our decentralised swarm behaviour selector, R-HGN, and uses post-operation simulations to improve behaviour selection and expand the swarm's behaviour repertoire. Via simulation, we demonstrate that this lifelong learning swarm is able to learn from past experiences, emulating hindsight, and use the collective findings of all swarm members to improve the swarm as a whole, emulating an adaptive culture. The learning swarm is evaluated in a nontrivial task and found to outperform a once-trained swarm by up to 23% of the fitness range.

14.1 Introduction

As humans we are always learning and improving ourselves, so should our attempts at artificial intelligence not do the same? This question is asked in the relatively new field of lifelong machine learning (LML) [193, 279, 643], a field that aims to have machines improve their execution of tasks as they perform them. Similarly to the learning of an individual, communities often adapt and improve via the circulation

---

P. Smith (✉) · R. Hunjet
Defence Science and Technology Group, Edinburgh, SA, Australia
e-mail: phillip.smith14@defence.gov.au; robert.hunjet@defence.gov.au

A. Aleti · V. C. S. Lee
Monash University, Clayton, VIC, Australia
e-mail: Aldeida.Aleti@monash.edu; vincent.cs.lee@monash.edu

A. I. Khan
Sensor Analytics, Clayton, VIC, Australia
e-mail: asad.khan@sensoranalytics.com.au

© The Author(s) 2025
S. Ng et al. (eds.), *Thinking Swarms*, https://doi.org/10.1007/978-3-031-82790-7_14

of ideas and behaviours. In this work, we attempt to emulate this social learning in a robotic swarm with LML.

In most robotic swarm implementations, individual members of the swarm are controlled via sensor-input-actuator-output rules, which we call "behaviours". Via the interaction of these individual behaviours, a collective swarm behaviour emerges. In our prior studies [749], we explored a decentralised selector, called robotic hierarchical graph neurons (R-HGNs), which in situ switched the active individual behaviours of each swarm member, allowing task-appropriate group behaviour. However, these selections were limited by the behaviours provided to the selector and training conditions. In contrast, our other prior studies [752, 753] of swarm behaviour creation found that environment-specific behaviours could be created, but it requires time and significant environment knowledge a priori.

The proposed LML swarm aims to mitigate these limitations by using both algorithms and is the first known instance of lifelong learning in a robotic swarm. This LML improves the swarm performance over time by expanding and refining the R-HGN condition awareness and by adding to the swarm's behaviour repertoire when it is found to be ill-equipped for a previously unseen environmental condition. Unlike single-agent LML, this swarming approach uses the combined experiences of all swarm members to collectively improve the shared behaviour selection and thus emulates social learning [52]. In addition, the proposed LML uses simulations of past experiences to have the swarm learn from its mistakes, thus emulating the human trait of hindsight.

To validate the proposed system, the performances of several LML swarms were compared with a statically trained R-HGN swarm in a simulated nontrivial task. This evaluation found that continuously retraining the behaviour selector produces the greatest improvement in performance whereas adding new behaviours to a swarm's repertoire can improve performance, though to a lesser extent and requires some moderation.

The remainder of this chapter proceeds as follows. We present a background of related works in Sect. 14.2. In Sect. 14.3 we define the task used to evaluate the swarm. The proposed LML swarm is presented in Sect. 14.4, and the experiment conditions are outlined in Sect. 14.5. Finally, Sects. 14.6 and 14.7 present the results and conclusion of this study.

14.2 Related Work

To understand the proposed LML swarm, we first discuss LML in general, then in relation to robotic agents. Following this, we discuss swarm behaviour creation and selection, because these are the foundations on which we build the LML swarm. However, we keep these discussions brief because other chapters, specifically Chap. 15, give a greater exploration of these ideas.

14.2.1 Lifelong Learning

In [193], LML is defined as a learning system that grows in knowledge and ability after each experience. Formally, Chen et al. [193] defined an LML system as one that learns from the experiences of task 1 to task $n-1$ such that it can solve task n with greater effectiveness. Therefore, the standard form of LML requires full retraining after each experience to prevent biasing the system towards the $(n-1)^{th}$ data set.

In contrast to this full (batch) re-training, *cumulative learning* [279] is a subset of LML that extends the existing model with only the $(n-1)^{th}$ data, similar to the recursive process of reinforcement learning. Fei et al. [279] compared this approach to a standard (full retraining) LML in a data classification task. This comparison found cumulative learning achieved higher accuracy with significantly reduced retraining time. To achieve this time reduction, the system separated the classification results of task $n-1$ into new and existing classes. The former were appended to the classifier, and the latter refined the classifier such that any classes not re-experienced in task $n-1$ were unaffected.

Traditionally, LML is implemented in nonagent domains; however, Parisi et al. [643] found agent-based LML to be achievable, though with a greater potential for catastrophic forgetting. That is, an agent unlearns prior knowledge in favour of new knowledge. Specifically, in relation to robotic-agent LML, several works have explored learning schemas similar to, but distinctly different from, LML. One such approach is *transfer learning* [309, 779, 784], which has a "master policy" learnt during a training stage and used in all subsequent operations via periodic switching [779] or as the foundation of environment-specific learning [309, 784]. That is, the agent's experiences during a set of training tasks improve the agent's abilities during actual operations. Although this transfer learning explores similar principles, because learning stops after the training tasks, it is not seen as true LML. Therefore, no literature (to our knowledge) has implemented lifelong learning in robotic agents (simulated or physical), and thus this research is seen as novel both in the swarming and general robotic aspects.

14.2.2 Swarm Behaviour Creation

The field of autonomous creation of robotic behaviours is well established, dating back several decades [846]. Given the (relatively) recent influx of swarming robot research, many works have explored this autonomous creation process for swarm behaviours. However, as discussed in Chap. 15, such behaviour generation is challenging because of the non-linear (and potentially nonconvex) relation between individual and swarm emergent behaviours [835]. To counter this challenge, the genetic algorithm (GA) meta-heuristic has become arguably the most prominent method of autonomous swarm behaviour creation [329, 357, 363, 651, 752, 791, 827] because of its ability to find pseudo-optimal solutions in a nonconvex space [413]. In these genetic algorithm works, two noticeable limitations are the timeliness of creation and the accuracy of the evaluation conditions.

In [363, 651, 739] and [827], "online" creation had the swarm generate and trial behaviours in situ. This reduced the inaccuracy of evaluation because there was no "reality gap" [480]. However, this came at the cost of evolutions taking hours or days [739]. Further, this training time was required at the start of each deployment, making the swarm unfit for operation in real-world applications with time restrictions.

In contrast, "offline" evolution, as seen in [329, 357, 752, 791], could be conducted in advance of the operation, allowing for timely deployment of the swarm. However, a behaviour generated in this approach was specialised to the conditions known a priori. Should such a behaviour be deployed in a distinctly different environment or task instance, it could become unstable and thus cause reduced swarm performance [739, 752].

Given that the proposed LML swarm aims to avoid re-creating behaviours for each task instance and may use the experiences of past tasks, the latter approach to behaviour creation, offline evolution, was used in this study. We used the rule-set evolving algorithm, known as swarm learning classifier system of [750, 752]. This algorithm generates a human-readable collection of low-level condition-action rules via a series of simulated evaluations with a Lamarckian-style replacement process at the micro/rule level and a novelty domination search [581] at the macro/rule-set level.

14.2.3 Swarm Behaviour Selection

In contrast to behaviour creation, behaviour selection has an agent hold in memory a set of pre-defined behaviour controllers. Via some selection policy, the agent implements one of these behaviours for a period, until a reselection is made. In the literature, this policy has been built via a uniform-trial approach [609], a probabilistic selection [280] or via the classification of the agent's surroundings and problem instance [160, 351, 367, 541, 749, 754, 781]. From this list, only the latter method used prior-collected data to prevent ineffective behaviour execution during an operation and is thus applicable to LML.

Further dissecting the latter approach, classifications of the environment or problem instance and relating these classifications to a behaviour have been made via feature vectors [160], neural networks [754], decision trees [351, 781], hard-coded rules [367, 541] and R-HGN [749]. Of these approaches, R-HGN is the most appropriate for cumulative learning because it is an effective one-shot learner [611].

As a brief introduction to R-HGNs, graph neurons [460] are an approach to discrete pattern matching via a collection of "neurons" that each responds to a unique value at a unique index of a pattern string. Via the collective recognition of the neurons, the full pattern may be identified. Nasution et al. extended this work

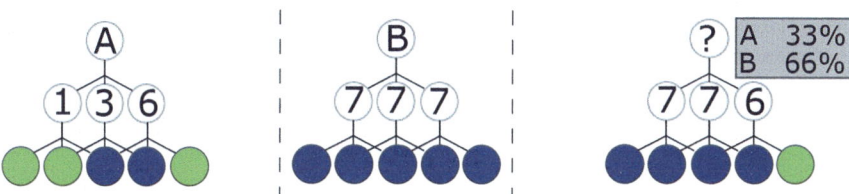

Fig. 14.1 Example of HGN training on two patterns (A and B), then presented with a B pattern with noise. This example-scale HGN is able to make a 66% confident prediction of the noisy B pattern

with hierarchical graph neurons [611], in which a triangular hierarchy was added to overcome the "crosstalk" issue, a problem for single-layer graph neurons that resulted in prior unseen patterns being falsely identified as known patterns during the sub-pattern recombination stage.

An example of an HGN is shown in Fig. 14.1. In this example, the HGN is trained on five-value binary (green or blue) patterns. In the first panel, a pattern is presented to the HGN for one-shot training. In the middle layer, the three-value combination from the node directly below and one adjacent in the bottom layer (pattern input layer) causes a unique neuron of the rows to activate. That is, the input "green-green-blue" triggers the first neuron of the first row, while "green-blue-blue" triggers the third neuron of the second row. The top layer then trains that a "1-3-6" input from the middle layer is pattern A. In the second panel, a new pattern is presented for training and the HGN trains that "7-7-7" is pattern B. Finally, in panel three, the HGN is in recall mode. A modified (noisy) version of pattern B is presented middle-layer nodes identifies the known subpatterns. The third node observes an unknown pattern (and does not make a new id because it is in recall mode rather than training mode), and thus the top-layer node is unable to make a direct match. However, using the closest match on "7-7-6", the middle-layer pattern "7-7-7" can be matched with 66% certainty.

Nasution et al. [611] evaluated the HGN algorithm via a comparison with a single-cycle backpropagation multilayer perceptron neural network. The HGN achieved pseudo-real-time recall speeds with notable accuracy even with high pattern distortion. The HGN also demonstrated significant superiority to the multilayer perceptron regarding pattern size and quantity scaling. Further, it was observed that these recalls could be made on computationally limited devices (such as the single-board computers used for swarm robots) because the complexity is $O(1)$. A comparative study between HGN and a recurrent neural network was also conducted in [372], finding that HGNs required smaller training data sets before effective pattern matching could be achieved.

In [749], the HGN was extended to R-HGN to match pattern strings with nonuniform and potentially large character ranges (as one might expect from a set of robot sensors) without unnecessary neuron creation. That is, a pattern might consist of a binary input (range = 2) and an 8-bit integer input (range = 255).

The R-HGN would create appropriately sized neuron rows for each data type, whereas a standard HGN assumes uniform patterns and thus requires each row to be sized to the maximum range. Further, the R-HGN dynamically grows each row as the pattern component is observed, further removing unused neurons from the graph. In addition, the R-HGN associates each pattern identification with a row of a statistics table, allowing a sequence of matches from one or more swarming robots to be combined into an environment estimate and thus a behaviour selection. In this prior study, the swarm agents were able to classify three key environment features observed during training and switch to behaviours designed explicitly for such features. In environments with multiple of these features, the swarm members were able to recognise the most relevant of the features in their current surroundings via a nonuniform selection consensus between members called "soft consensus". In doing so, the swarm was able to switch between behaviours as its surroundings changed. That being said, this initial study was limited to three manually designed behaviours, and the R-HGN was trained on only six distinct environments. When implemented in randomly generated environments that were noticeably different from training conditions, this R-HGN swarm saw limited success.

14.3 Problem Domain

In this study, we trialled the proposed system in a nontrivial networking assistance swarming task. In this task, multiple mobile communication devices, hereby known as "network nodes", were spatially separated such that direct wireless communication was prevented by obstacles (such as buildings), noise and distance. To facilitate data exchange between these network nodes, a robotic swarm was tasked with the transfer of p_{max} data packets from source(s) to destination(s). This process was simulated with discrete time-step agents and used the log-distance path loss model [667] with Gaussian noise for signal strength at the receiver to determine channel contention and the success or failure of transmissions:

$$P_r = P_t + 20 \log_{10}\left(\frac{\lambda}{4\pi d_0}\right) - 10n \cdot \log_{10}\left(\frac{d_{rt}}{d_0}\right) - \chi_\sigma \qquad (14.1)$$

where P_r is the received signal power in decibels (dB), P_t is the sent signal power (dB), λ is the signal wavelength (m), n is the unit-less signal exponent, d_{rt} is the distance between receiver and transmitter (m), d_0 is a reference distance (m) and χ_σ is a Gaussian random value (dB) with σ standard deviation and mean 0.

Signals from device i are acknowledged by the receiver if a power threshold is met, $P_{r,i} \geq P_{r,\text{req}}$ and if the signal to noise and interference ratio (SNIR) is above a threshold, SNIR_{th}:

$$\text{SNIR}_{th} < P_{r,i} - 10 \cdot \log_{10}\left(\sum_{m \in M} (10^{\frac{P_{r,m}}{10}}) + 10^{\frac{CN}{10}}\right) \qquad (14.2)$$

where $P_{r,m}$ is the interference caused by the communication device m, M is all communication devices also transmitting in the time step (including other swarm members, network nodes and jammers), noting $i \notin M$, and CN is the constant background noise (dB).

For each task instance, a randomly generated obstacle layout was produced, emulating an urban environment. The task was run until all data packets were transferred or a time limit, $T_{\max}$ was reached. To make the data-transfer task more challenging, each task was broken into stages. In each stage an immobile jamming device (initially unknown to the swarm) may be active or inactive, the network nodes may relocate, and the data traffic requirements between nodes may change. We assumed a stage had started when all transfers of the prior stage were complete.

The behaviour fitness metric for this task, executed in κ instances, is expressed as

$$\text{Fitness} = \frac{\sum_{k=0}^{\kappa} p_k}{p_{\max} \cdot \kappa} - \begin{cases} \frac{\sum_{k=0}^{\kappa} T_k}{T_{\max} \cdot \kappa}, & \nexists k \in \kappa : p_k < p_{\max} \\ 1, & \text{otherwise} \end{cases} \quad (14.3)$$

where p_k and T_k are the number of packets successfully transferred and total time steps of instance k, respectively. For all κ instances of a task, the data-transfer challenges defined above remained constant. However, a unique seed was used for the stochastic network in each k instance. Finally, it was also noted that the condition in Eq. (14.3) restricted the fitness to the range $(-1, 0]$ unless the behaviour was able to complete the data-transfer task in all κ instances, in which case the fitness was in the range $(0, 1)$.

14.4 System Design

In this section, we define the swarm agents of this study, including the sensing abilities and actions they can perform. Following this, the LML architecture and learning process of this study are presented. Finally, this section discusses the theoretical social learning aspect of this system.

14.4.1 Swarm Agents

In keeping with the philosophy of swarm robotics [147], each agent is relatively simple in terms of sensor and actuator complexity.

Each agent was equipped with a low-fidelity LiDAR and wireless connectivity module. Via the LiDAR, the agent was made aware of the nearest obstacle. Via the wireless module, it was assumed that an agent could detect the range and bearing [499, 522, 638] of other swarm members, jamming devices and the networking nodes within the agent's one-hop communication range. Using multihop information propagation, agents were also periodically updated on the location of other swarm members and networking nodes outside direct connectivity. The swarm agents are also able to use the wireless module to take a discretised measurement of the current noise and interference strength on the wireless channel. Finally, each swarm agent could buffer up to ten data packets at a time and was made aware of this buffer's state via a binary "empty/nonempty" signal. The agents could also read the destination id of the packet at the top of this buffer, which directs them towards one of the network nodes.

Using this information, agents could perform two types of action: communication and locomotion. For communication, agents could attempt to transfer the top packet in their buffer to another swarm member or a network node or could request a packet from a network node. For locomotion, the agents used the virtual forces of Boids flocking [677] with a simple unicycle model [604] to move towards, away-from or perpendicular to a known obstacle, swarm member, network node, jammer or some combination of the above.

In the behaviour-switching swarm, each member was equipped with a copy of the R-HGN pattern matcher and a repertoire of behaviours. These behaviours were a collection of condition-action rules, using the knowledge and abilities defined above. Initially, the R-HGN was trained to select between three manually designed behaviours (MBs) of [749]: a relaying behaviour based on [512, 539], a ferry behaviour based on [428, 863] tuned for obstacle circumvention and a second relay behaviour tuned to avoid areas of high network traffic.

The initial training of the R-HGN was as detailed in [749]: the swarm was run with each behaviour in the six training environments shown in Fig. 14.2, the feature patterns of the environment were learnt by the R-HGN and the most effective behaviour of each environment was associated with the learnt patterns.

14.4.2 Lifelong Machine Learning

The proposed LML consists of two learning processes: repertoire growth and selection retraining. The former uses a swarm learning classifier system to evolve new, task instance-specific behaviours after operations. This allows the swarm to be better prepared when confronted by similar tasks in the future. The latter adds new patterns to the R-HGN and updates the environment statistics table with the observations of the operation, improving the swarm's environment recognition range and thus the selection generality and quality.

This learning process is graphically presented in Fig. 14.3 and further discussed in the remainder of this section. It should be noted that only the first step of this process requires the swarming agents to embody physical platforms. After the task operation of Step 1 (when the swarm returns from deployment), the collected data can be transferred to a central machine in which a simulated (digital twin) swarm undergoes LML. A copy of the updated R-HGN and repertoire can then be distributed to the physical swarm platforms before the next deployment (end of Step 5).

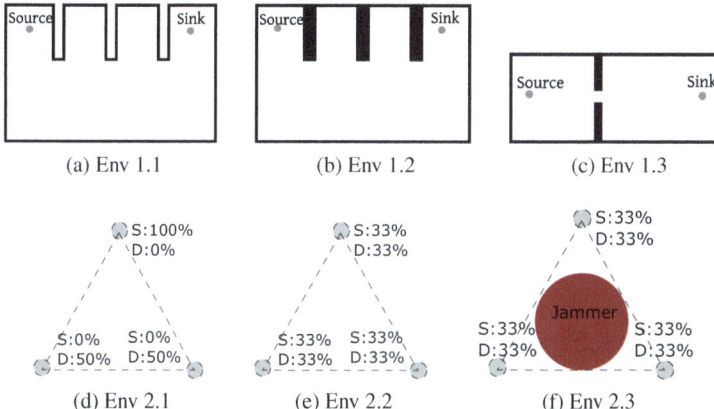

Fig. 14.2 Manually designed training environments for R-HGN. (**a**) inside a building with thin walls, low communication restriction, (**b**) inside a building with thick walls, high communication restriction, (**c**) building with narrow passage, (**d**) 3 network nodes in broadcast, (**e**) three network nodes in equal communication and (**f**) as prior, with a jammer in the centre

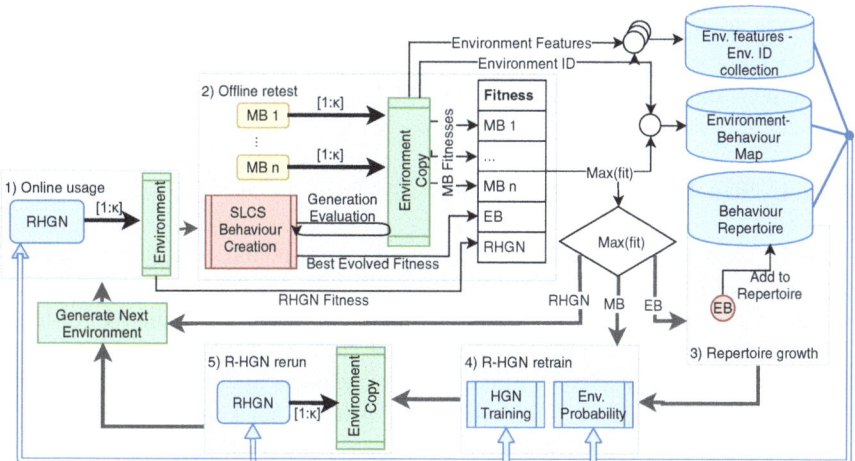

Fig. 14.3 Proposed lifelong learning cycle, training robotic hierarchical graph neurons to select between manually designed behaviours and evolved behaviours

14.4.2.1 Step 1: Online Usage

This step represents the deployed usage of the proposed swarm. The agents use their current R-HGN to switch between behaviours and solve the data-transfer task as quickly as possible. For this study, we executed the task in κ instances to prevent statistical anomalies in our later analysis. During this step, agents also recorded all pattern observations for the retraining in Step 3.

14.4.2.2 Step 2: Offline Retest

After deployment, problem instances are recreated for offline testing. In this study, we simplified this process by reusing the simulation environment of Step 1. However, future work will aim to use the collected data from the swarm during Step 1 to generate an environment digital twin.

First, a new evolved behaviour (EB) was created. Second, the existing manually designed behaviours (MBs) were implemented in isolation (not switched between by the selector).

The next step was determined by the performance results:

1. If the new EB has the greatest performance, the process moves to Step 3.
2. If one of the MBs has the greatest performance, the process skips to Step 4.
3. If the current swarm has the greatest performance, no further LML processing is required in this cycle and the system waits for the next task.

14.4.2.3 Step 3: Repertoire Growth

If this step is executed, the new EB of Step 2 is added to the repertoire, and the statistics table is expanded to accommodate the new behaviour for each R-HGN pattern, ρ, as shown in the example of Fig. 14.4.

14.4.2.4 Step 4: R-HGN Retraining

Using the observation patterns recorded by all swarm members in Step 1 and the best-found behaviour in Step 2, the statistics table of the R-HGN is updated to better associate the patterns with the pseudo-optimal behaviour in a cumulative learning approach. Along with the normalised values for each behaviour, the statistics table holds the total occurrence count of each pattern which is used for weighted matrix addition during this update. In addition, any patterns not known to the R-HGN are added via the one-shot learning HGN process, and a new statistics table row is generated.

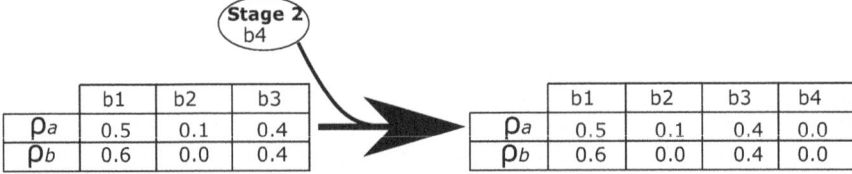

Fig. 14.4 Example of a behaviour being added to the statistics table. A new column is added for b4 with each cell being initialised to 0. This initial value is corrected in Step 4

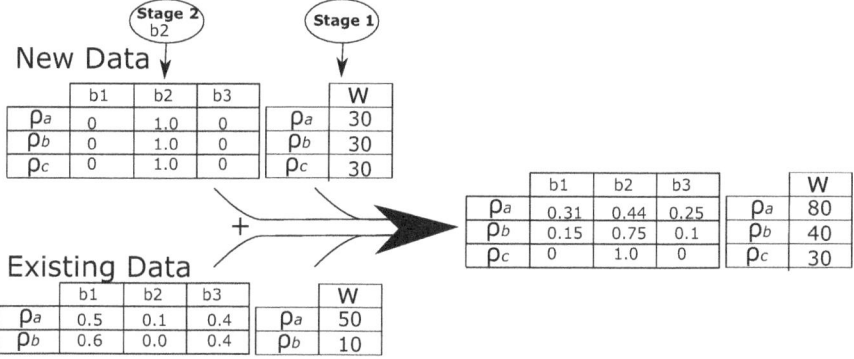

Fig. 14.5 Example of a statistics table update. For the two known patterns, ρ_a and ρ_b, this update moves the table toward advocating $b2$. However, the weights of the existing table cause this shift to be stronger in ρ_b, because it has fewer prior occurrences. This example also demonstrates a new pattern identification being added to the table

In Fig. 14.5 an example statistics table update is depicted. In this example, behaviour $b2$ was identified in Step 2 as pseudo-optimal, and the patterns ρ_a, ρ_b and ρ_c were observed 30 times each in Step 1. These data were converted to a matrix, allowing a simple matrix addition and re-normalisation of the rows to update the statistics table.[1] Further, ρ_c was prior unknown and thus a new row is added to the statistics table, which was associated with the new R-HGN output for ρ_c.

14.4.2.5 Step 5: R-HGN Rerun

The updated R-HGN swarm is tested in the recreated environment of Step 2. This final step is not required for learning but allows task-specific performance improvement of the swarm to be evaluated.

14.4.3 *Social Lifelong Learning via R-HGN*

In the above system design, it is noted that the patterns observed by all swarm members were used for a central R-HGN update. Thus, the unique experiences of one swarm member may influence the future selection of all members, and the shared experiences of the swarm will influence each individual, thus introducing a passive form of social learning [52] and memetic exchange [362].

---

[1] Example of a ρ_a update is old data $[0\ 1.0\ 0] \cdot 30$ plus new data $[0.5\ 0.1\ 0.4] \cdot 50$ becomes $[25\ 35\ 20] = [.31\ .44\ .25] \cdot 80$.

a)

	b1	b2
ρ_a	0.5	0.5
ρ_b	0.5	0.5

	W
ρ_a	20
ρ_b	20

b)

	b1	b2
ρ_a	0.5	0.5
ρ_b	0.86	0.14

	W
ρ_a	120
ρ_b	70

Fig. 14.6 Example of group learning in R-HGN statistics table. (**a**) the initial statistics table. (**b**) The table after two learning cycles

As an example, consider a simple ten-agent swarm that operated in Step 1 for ten time steps. This example swarm uses an R-HGN that can identify patterns ρ_a and ρ_b for behaviours $b1$ and $b2$ and has a statistics table shown in Fig. 14.6 a.

During the first operation, all ten agents observe patterns ρ_a and ρ_b equally (five occurrences each per agent). After this operation, $b1$ is found to be the most effective behaviour. In the second operation, five agents observe ρ_a in all time steps (ten occurrences per agent) and $b2$ is most effective. After the two learning cycles, the statistics table becomes that shown in Fig. 14.6 b. In this updated table, it can be seen that the normalised values for $b1$ and $b2$ remain the same for ρ_a. That is, the shared experiences of the swarm majority (for $b1$) and the focused experience of the swarm minority (for $b2$) influence the learning in equally valid and impactful ways.

As another example of social learning, consider in the second case of the above example that one agent (an agent that did not exclusively observe ρ_a) observed a new pattern, ρ_c. From this novel observation, a new pattern row would be added to the table, as seen in Fig. 14.5. That is, the experience of one agent would allow the full swarm to recognise this environment feature and behave accordingly.

We can further examine this emergent knowledge sharing by acknowledging that the initial statistic row of ρ_c is overly advocating $b2$ (at 100%) which is an immature statistic to share with the swarm. However, we also predicted two cases to emerge over the swarm's lifespan: ρ_c is observed multiple times by multiple swarm members, and thus the statistic row for ρ_c matures over the swarm's lifetime to better reflect the pseudo-optimal behaviour or ρ_c remains an infrequently observed pattern and thus the swarm's behaviour selections are minimally affected by this immature knowledge. This maturation of prominent patterns can be related back to our analogy of cultural idea circulation in Section 1 by noting that even well-established scientific facts are the result of hypotheses being refined, improved and built on by multiple generations of scholars.

14.5 Experiment Design

To validate the proposed LML, we implemented a swarm in a sequence of 50 tasks. The performance (Eq. (14.3)) over the learning swarm's lifetime was compared with a swarm that had undergone initial R-HGN training (defined in Sect. 14.4.1) but not further trained between the 50 evaluation tasks. This baseline swarm is referred to as the "static swarm".

Given that the proposed LML system has multiple learning aspects, we also isolated the behaviour addition and R-HGN retraining to better understand each component's contribution to the behaviour selection quality growth. Therefore, we implemented three forms of the LML swarm:

LML_{class} R-HGN retraining only updates the classification and selection of existing MBs (Option 1 in Step 2 is disabled)
$LML_{beh.}$ R-HGN retraining is only for behaviour additions (Option 2 in Step 2 is disabled)
LML_{full} the full LML with both R-HGN retraining and behaviour addition

For this study, it was hypothesised that all three LMLs would outperform the static swarm. Further, it was hypothesised that the combined learning in LML_{full} would produce greater growth than either LML_{class} or $LML_{beh.}$ individually.

14.5.1 Environment Settings

This study used the multi-agent simulator toolkit Mason [546] to create environments with continuous 2-D space and discrete time steps. For each task, 1,000 data packets were to be transferred within a time limit of $T_{max} = 50,000$, and the communication model uses the values: $Pt = 12\,\text{dBm}$, $\lambda = 0.125\,\text{m}$, $d_0 = 1.0\,\text{m}$, $P_{r,\text{req}} = -83\,\text{dBm}$, $\text{SNIR}_{th} = 10\,\text{dB}$, $CN = -95\,\text{dBm}$.

Each task was tested with $\kappa = 5$. For the evolution of behaviours, the generation limit was set to 1,000. For R-HGN, the time-step period between behaviour reselections was set to 500.

To create the environments of this study, 2–12 building-sized obstacle structures were placed in the environment at random coordinates. A single jammer was placed in the centre of the network nodes' initial positions. For the networking task in each environment, the set of network nodes was $\phi \in \Phi$, which was randomly sized between two and four (inclusive). The networking *stage* of the operation, $s \in S$, was randomly sized between one and three (inclusive). For each stage, the data packets were assigned a source and destination network node. This randomised assignment was limited such that $\forall \phi \in \Phi, \sum_{s \in S} (source_{\phi,s} + destination_{\phi,s}) > 0$; that is, no network node was idle for the entire operation. To add mobility to the network nodes, waypoints were created for each network node in the stages. Waypoints were traversed by the mobile nodes at half the maximum velocity of the swarming agents and via straight lines. On reaching the final waypoint of a stage, the node became stationary until the next stage. Waypoint coordinates were randomly generated such that the paths were feasible in the optimal network stage time, the network node did not collide with any obstacles, and the nodes did not enter the jammer's effective range.[2] In addition, in each stage, the jammer's state was randomised to be active or inactive.

14.5.2 Analysis Process

To analyse the overall performance growth and generality of the LML swarms, we report the fitness differences (in Stage 1) between the LML swarms and the static swarm:

$$\Delta \text{Fitness} = \text{Fititness}_{LML} - \text{Fititness}_{\text{static}} \tag{14.4}$$

We visualised this comparison via a moving window average (window size of ten tasks).

To analyse the ability of the LMLs to improve the swarm for each specific task, we defined "improvement" as the binary metric:

$$\text{improvemnt} = \begin{cases} 1, & \text{fitness}_{\text{stage1}} < \text{fitness}_{\text{stage5}} \\ 0, & \text{otherwise} \end{cases} \tag{14.5}$$

which we also visualised via a moving window average (window size of past ten tasks).

In addition to the performance examination, we reviewed the additional computation and memory required by the extended behaviour selector at the end of the swarm's lifespan. That is, if the swarm learns to solve all tasks effectively but requires unrealistic memory or computational power, the behaviour selector cannot be used in real-world implementations. For computation time, the mean time per time step per agent of the final task simulation is reported for the individual behaviours, the static swarm and the LML swarms. These simulations were run on a computer cluster with 2.1 GHz processors, with each simulation run on a single core. For memory usage, the peak RAM utilisation in these final task runs divided by the swarm size is reported. This study did not examine the computation time of the LML cycle (testing each MB and evolving new EBs) because this process can be conducted offline and thus time and memory are not restricted.

---

[2] We prevented network nodes entering the jammer because this can result in the task being unsolvable, given that data cannot be sent to a jammed destination node.

14.6 Results

14.6.1 Performance Improvement

Figure 14.7 presents the fitness differences between LML and static swarms (Eq. (14.4)) over these lifetimes. Figure 14.8 shows the average task-specific improvement of the swarm (Eq. (14.5)) over the three lifetimes.

First examining the two incomplete LML swarms, LML$_{beh.}$ had a high fitness difference in early life as novel EBs were added to the repertoire and the swarm was able to solve more tasks. However, the performance begun deteriorating at the 15th task and became worse than the static swarm by the end of the swarm's lifetime. That being said, Fig. 14.8 shows that LML$_{beh.}$ continued to improve the swarm for each specific task, as shown by the high average improvement throughout the lifetime. In contrast, LML$_{class}$ initially showed a negligible difference from the static swarm, but over the swarm's lifetime, there was a steady growth, resulting in a notable performance difference (up to 23%) in the late-life tasks.

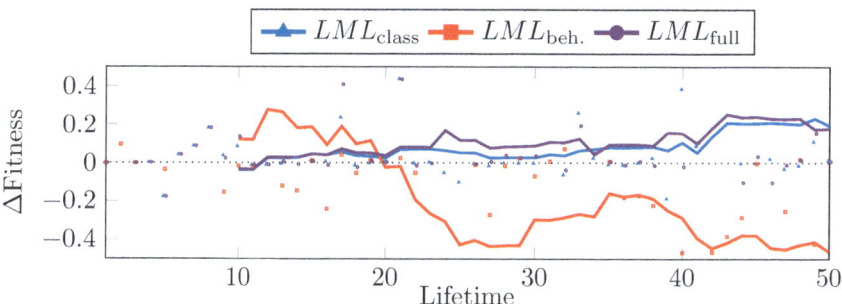

Fig. 14.7 Fitness difference between the three LML swarms and static swarm. Fitness for each environment (dots) and averaged over the past ten tasks (lines)

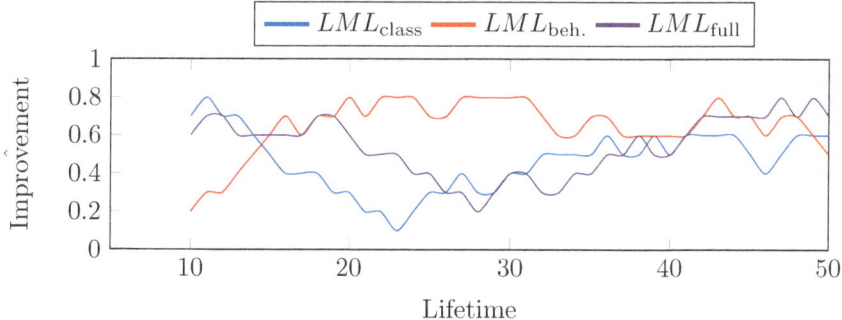

Fig. 14.8 Task-specific improvement (fitness in Step 5 is greater than Step 1) of the three LMLs using ten-task moving-window-average

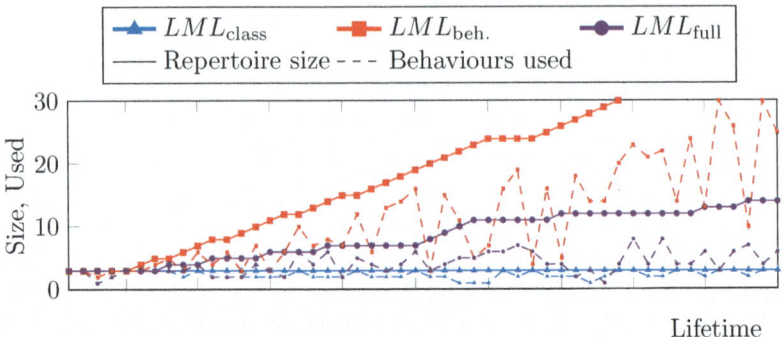

Fig. 14.9 Repertoire size and behaviour usage for LML repertoire in each task

This more gradual improvement by LML_{class}, compared with $LML_{beh.}$, is further demonstrated in Fig. 14.8 by the averaged improvement of the former being predominantly less than that of the latter, without converging to 0 (and thus demonstrating learning has not stopped) throughout the examined lifetimes.

Concerning the full LML swarm proposed, LML_{full}, the performance steadily improved similarly to LML_{class} and with the same maximum difference to the static swarm of 23%. However, the overall fitness difference of LML_{full} was slightly higher. LML_{class} had an average fitness difference of 6.3%, while LML_{full} has an average difference of 8.6%. In addition, LML_{full} showed higher task-specific improvement, including a smaller decline during midlife tasks and a higher improvement rate in late-life operations.

To further investigate the fitness decline of $LML_{beh.}$, Fig. 14.9 presents the repertoire size and behaviour use in the 50-task lifetimes. This figure shows that both LMLs that extended their repertoires continued to grow throughout their respective lifetimes with no evidence of convergence, though this growth was more significant in $LML_{beh.}$. It is also seen that the average number of behaviours used in an operation scales with the repertoire size. From this, it can be deduced that the repertoire of $LML_{beh.}$ became overpopulated and the statistics tables of each R-HGN identification became diluted. Therefore, the swarm agents repeatedly switched to different behaviours throughout the operation rather than identifying an appropriate behaviour and executing it long enough for the intended behaviour effect to come to fruition. Further, by examining this repertoire growth alongside the task-specific improvement seen in Fig. 14.8, it can be observed that this limited LML fell into a recursive loop. That is, the overpopulated behaviour selector performed poorly in a task during Stage 1. This caused even a suboptimal EB of Stage 2 to have higher fitness and thus be added to the repertoire. This new task-specific EB (and the resulting R-HGN retraining) allowed the swarm to perform correctly in Stage 5 (resulting in the high improvement in Fig. 14.8) at the cost of generality, thus further hindering the behaviour selector in future tasks. It should, however, be noted that this overpopulation was not due to behaviours being re-evolved and the repertoire being filled with similar EBs; all EBs added to the repertoire for both $LML_{beh.}$ and LML_{full} had an average difference from prior EBs of 99.8% (as derived from the rule set novelty metric in [751]).

In relation to the performance difference between LML_{class} and LML_{full}, the two produce equivalent results until the eighth task, in which an EB was added to the repertoire of LML_{full} and the performances diverged. Over the following 42 tasks, a further 10 EBs were added to the LML_{full} repertoire, which were noted to be successfully reused by the swarm in later operations. The most significant case of this reuse was task 24 in which the three MBs were inadequate and thus the static swarm and LML_{class} produced a fitness below 0 (indicating that all data could not be transmitted in time). In contrast, the majority of the LML_{full} swarm members were able to successfully select a priori evolved EB and use it to complete the data-transfer task. In doing so the swarm achieved a high fitness of 0.79.

To summarise this performance analysis, it was found that adding behaviours to the swarm's repertoire had some value but can reduce performance if overused. Meanwhile, retraining the R-HGN with the observations of each task generally improved the swarm's ability to solve future task instances, though at a slower rate and the swarm remained ill-equipped to solve problems outside the scope of the provided behaviours. Finally, the higher performance and wider task generality of LML_{full} justified the combination of retraining and adding behaviours to improve the swarm.

14.6.2 Computation Cost

Table 14.1 presents the mean time per time step and the peak memory usage per agent for the individual behaviours and the behaviour selector swarms.

As could be expected, the most significant computation cost was the addition of a behaviour selector to each swarm member. Comparing the simulations with individual behaviours (MBs or the EB), the static swarm showed a time increase of $\sim$400%. In contrast, comparing the static swarm and the LML swarms that did not develop overpopulated repertoires (LML_{class} and LML_{full}), there was no statistically significant change in simulation time. This was to be expected because HGN recall has complexity O(1) [611] and thus is not affected by the number of patterns stored or behaviours selected across the way. The only LML that had a significant computation time increase was $LML_{beh.}$. This increase of $\sim$21% was due

Table 14.1 Computation time and memory usage of swarms in the final task

	MB1	MB2	MB3	EB	R-HGN	LML_{class}	$LML_{beh.}$	LML_{full}
Mean time per time step (ms)	0.45	0.44	0.35	0.43	2.04	1.8	2.48	2.03
Peak memory (MB)	13			77	137	1,249	2,881	1,324

to the previously discussed frequent behaviour switching, which is computationally expensive when performed en masse because of this study's implementation of behaviour loading/unloading,[3] which could, in the future, be optimised.

In relation to memory consumption, all LMLs cause a noticeable memory increase, 908%, 20,955% and 963%, respectively. Such growth was expected because the R-HGN algorithm scales $O(|\rho| \cdot |b|)$ with the number of unique patterns identified, $|\rho|$ and the number of behaviours stored, $|b|$. This linear growth relative to patterns was considered reasonable because the number of novel patterns observed by the swarm (and thus the R-HGN size) grows sub-linearly relative to tasks executed. That is, in early life many new conditions will be observed by the swarm, but in later life only rare corner-case observations will be novel to the R-HGN and thus pattern adding will subside over the lifetime. In addition to this R-HGN growth, we note for completeness that the large memory of $LML_{beh.}$ was due to the aforementioned overpopulated repertoire.

In terms of hardware feasibility, modern single-board computers such as the well-known Raspberry Pi [288] feature ≥ 4 GB of RAM. Therefore, the behaviour selector is expected to be executable on swarming robot hardware at any point of the explored LML lifetime.

14.7 Conclusion and Further Work

This study presented a self-improving behaviour selector for robotic swarms. The proposed system introduces social learning and task-solving hindsight to have the swarm improve over its lifetime. This improvement was empirically validated via a simulated sequence of nontrivial task deployments. Via this investigation, it was found that both retraining the behaviour selector and adding behaviours were beneficial (in some way), but using both in combination produced the highest performing swarm.

In regard to adding behaviours to the swarm's repertoire, it was found that not restricting this process can reduce the swarm performance and cause the lifelong learning process to fall into a recursive loop of failure which, in turn, leads to high computation and memory costs. This issue was only observed in the limited LML (with only a behaviour addition). However, the full LML system showed no evidence of behaviour growth convergence, and thus it may be speculated that with a sufficiently long lifespan, the repertoire of LML_{full} may too become overpopulated. Therefore, further work will investigate techniques to mitigate this issue, such as the swarm selectively forgetting unused behaviours or behaviours identified to not be generalisable to other task instances, as seen in [561].

Returning to the problem domain of this study, the implemented swarm was tasked with facilitating network traffic between multiple mobile networking devices

---

[3] Using compressed behaviour files.

in an obstacle-filled (urban) environment. Via the proposed lifelong learning, the swarm improved its ability to solve variations of this problem, such as the number of networking nodes, urban density and jamming intensity. In doing so, the swarm became more generally applicable and thus prepared for real-world applications.

Acknowledgments This research was supported by an Australian Government Research Training Program scholarship. Additional funding for this research was provided by both Cyber and Electronic Warfare Division and Land Division, Defence Science and Technology Group, Commonwealth of Australia.

Open Access This chapter is licensed under the terms of the Creative Commons Attribution 4.0 International License (http://creativecommons.org/licenses/by/4.0/), which permits use, sharing, adaptation, distribution and reproduction in any medium or format, as long as you give appropriate credit to the original author(s) and the source, provide a link to the Creative Commons license and indicate if changes were made.

The images or other third party material in this chapter are included in the chapter's Creative Commons license, unless indicated otherwise in a credit line to the material. If material is not included in the chapter's Creative Commons license and your intended use is not permitted by statutory regulation or exceeds the permitted use, you will need to obtain permission directly from the copyright holder.

Chapter 15
Learner-Centred, Teacher-Centred and Blended Curriculum Design in Swarm Systems

Aya Hussein, Sondoss Elsawah, Eleni Petraki, and Hussein A. Abbass

Robot swarms have been used in various civilian and military applications, from entertainment to serious missions. Complex swarm tasks involve multiple interdependent skills that swarm members need to possess for successful operation. Swarm operation environments can be dynamic, noisy or with limited communication resources, which exacerbates task complexity. The non-linear interactions between swarm members make it challenging for experts to formulate agent-level rules that result in the emergence of a desirable swarm-level behaviour. Until now, there has been no general methodology that algorithm designers can follow to go from the individuals to the group behaviour. Automated swarm behaviour design has demonstrated its potential in enabling the learning of several swarm tasks. Nonetheless, incorporating human knowledge in these automated techniques is instrumental in accelerating the learning process and ensuring that the learning results in the intended behaviours. The inclusion of domain expertise in automated swarm design has mostly been conducted in an ad hoc manner without sufficiently studying how to maximise the benefit of domain expertise while not overwhelming the human expert. This chapter investigates the use of machine education as a holistic approach for teaching swarm members the required skills in complex swarm

---

A. Hussein (✉)
Faculty of Science & Technology, University of Canberra, Bruce, ACT, Australia
e-mail: aya.hussein@canberra.edu.au

S. Elsawah
Capability Systems Centre, University of New South Wales, Canberra, ACT, Australia
e-mail: s.elsawah@unsw.edu.au

E. Petraki
Faculty of Education, University of Canberra, Bruce, ACT, Australia
e-mail: eleni.petraki@canberra.edu.au

H. A. Abbass
School of Systems and Computing, University of New South Wales, Canberra, ACT, Australia
e-mail: h.abbass@unsw.edu.au

tasks. Specifically, we study various ways of designing a curriculum to systematise the process of learning and the incorporation of expert knowledge. We focus on three curriculum design approaches: learner centred, teacher centred and blended. We present case studies for each of these approaches and present the key lessons learnt and recommendations for future studies.

15.1 Introduction

Robot swarms consist of large groups of robots that use local interactions to coordinate their actions. Robots in a swarm are typically characterised by limited sensing, processing and communication capabilities [478]. The local interactions between swarm members can lead to the emergence of complex swarm-level behaviours, including flocking and shape formation [214].

Robot swarms have three main advantages over single-robot systems: scalability, robustness and flexibility [693]. Scalability means that swarm performance experiences low variations in response to the number of swarm members [405]. Swarms are scalable because each swarm member only perceives its vicinity and interacts with its neighbours regardless of the size of the swarm. Robustness refers to the ability of swarms to continue to operate after the loss or failure of some individual members. Swarms are robust because their inherent redundancy means that failed or lost individuals can be compensated for by other members. Flexibility refers to swarms' versatility and their ability to perform a wide range of tasks. Adjusting the local interactions among swarm members enables the emergence of different swarm-level behaviours, to meet different task requirements. Thanks to these desirable characteristics, swarm robotics can be used in different applications, including agriculture [37], humanitarian operations [177] and monitoring and surveillance [700].

One of the key challenging questions in swarm robotics is how to define individual agents' behaviours in a way that leads to the emergence of an intended swarm behaviour [251]. This is commonly referred to as the micro-macro problem. Users usually define the desired behaviour on a macroscopic level (swarm level); however, we can only program the behaviours of the individuals (micro-level) [270]. Deducing the agent-level rules that can achieve a given swarm-level behaviour is challenging because of the non-linearity of swarm interactions [694].

Existing approaches for designing swarm behaviours can be classified as manual and automatic approaches. Manual approaches require solid expertise in swarm systems and undergo laborious trial and error of agent-level behaviours before the intended swarm behaviour can be attained. These approaches have been heavily inspired by biological swarm systems to generate behaviours similar to those found in nature, for example, flocking and shepherding [406]. However, automatic approaches use artificial intelligence (AI) for the design of agent behaviours to deliver the intended system operation. Automatic approaches greatly reduce the burden on humans to experiment with wide ranges of possible agent behaviours [289].

This however comes at the cost of predictability of the resulting swarm behaviour and the degree to which it aligns with the required behaviour.

This chapter discusses the use of machine education (ME) as a methodology for systematically combining manual and automatic approaches to swarm behaviour design to obtain the best of the two worlds. ME is a theory-driven approach inspired by the well-established field of human education with the objective of teaching machines desirable skill sets and values [517]. Using ME, we can design a curriculum that enables the teaching of a set of skills to the swarm members while taking into account both learners' and teachers' abilities [405].

The objective of this chapter was to introduce the emerging field of ME and demonstrate its potential in supporting the design of complex swarm behaviours. Given that it allows the systematic incorporation of domain expertise, ME improves the alignment between user expectations and the learnt swarm behaviours. The use of machine learning (ML) approaches within ME enables reasonable degrees of autonomous learning by swarm members to alleviate the burdens on humans to fully specify agent behaviours [405]. In addition, the holistic approach of ME considers important education components, including the specification of the learning contexts, the creation and sequencing of the lessons and the design of assessment instruments and performance measures. All these components are necessary for the design, realisation and assurance of the learning process [20].

We first briefly discuss the main approaches to swarm behaviour design and highlight the strengths and limitations of each approach in Sect. 15.2. In Sect. 15.3, we introduce ME and describe using a case study how ME can be used for teaching swarm members the skills needed for a required swarm behaviour. We then dig deeper into the factors that need to be considered while designing a curriculum for swarm members, in Sect. 15.4, and describe various curriculum design approaches in Sects. 15.6–15.8. Finally, we finish with a summary and provide suggestions for future work in Sect. 15.10.

15.2 Swarm Behaviour Design

In their recent reflection on the future of robot swarms [251], Dorigo et al. identified the problem of how to design a desirable swarm behaviour given that we can only program the individual robots (i.e. the micro-macro problem) as one of the key challenges that require further research. We argue that ME can make a significant contribution to addressing this challenge. However, before we show why this is the case, we start by discussing the existing approaches to swarm behaviour design and describe the pros and cons of each approach. Two main swarm design approaches can be found in the literature: manual design and automatic or AI-enabled design.

15.2.1 Manual Design of Swarm Behaviours

The classical approach to swarm behaviour design is to manually craft agent-level rules and search for appropriate parameters for these rules to generate the intended behaviour. Manual design approaches have been heavily inspired by the behaviours of biological swarm systems. For example, Reynolds [676] formulated the bird flocking behaviours into a computational model to regenerate similar behaviours in computer games. Reynolds' model and its variants have been extensively used to generate flocking behaviours in robot swarms [281, 404, 799]. Another example is the modelling of honeybee interactions to generate cooperative swarm behaviours [508, 670]. Shepherding is a third example of bio-inspired behaviours. It has received considerable attention in the swarm robotics literature because it facilitates swarm guidance and control [266, 406].

Taking inspiration from nature has played a great role in advancing swarm robotics research. Nonetheless, many questions have to be answered before a bio-inspired algorithm is applied to a physical robot swarm, for example, how to select an appropriate parameter setting for a given robotic platform, how to determine context changes and how to adapt to such changes. In addition, as Dorigo et al. noted, the design of robot swarms needs to adopt a more engineering-minded approach to make them relevant for real-world applications [251]. That is, although natural swarms can provide novel principles to guide the swarm design process, biological inspiration should not be taken too literally, because more focus needs to be devoted to the application domain and its requirements.

There is a lack of generic methodologies for designing individual robot behaviours that lead to the emergence of an intended swarm-level behaviour. A number of swarm design methodologies have been proposed to facilitate the manual design process (e.g. [144, 670]). However, these methodologies are domain specific because they do not generalise to different swarm applications and they are developed within relatively simple problems but quickly show their limits in complex settings [251].

15.2.2 Automatic Design of Swarm Behaviours

Methods for the automatic or AI-enabled design of swarm behaviours have proved their potential in reducing the load on humans during the design process [289]. In particular, evolutionary algorithms (EAs) and reinforcement learning (RL) can offer great solutions in this domain [251]. In these methods, a human formulates the required swarm-level behaviour into an objective function that can be used to provide guidance to these AI methods in their search for an optimal agent behaviour.

15.2.2.1 Evolutionary Algorithms

An EA is by far the most commonly used approach to automatic swarm behaviour design [251]. EAs [88] are a class of heuristic-based optimisation techniques inspired by Darwinian evolution. Starting from a random set of solutions, EAs employ evolutionary operators including reproduction, crossover and mutation to evolve new generations of solutions that are likely to have better fitness for the problem. In EAs, the objective function is required to estimate the fitness level of individual solutions to enable individuals that have high fitness values to be selected more often for reproduction. This fitness-biased selection guides the search towards individuals that have increasingly higher fitness, hence maximising the objective function.

EAs have been applied to swarm behaviour design in which an EA's individual represents a possible robot controller. Research in this area has shown that it is very challenging to use EAs to evolve an appropriate robot controller from scratch in swarm tasks [259]. Alternatively, hierarchical evolution methods have proved feasible in a number of common swarm tasks, for example, aggregation [291] and foraging [328]. In hierarchical evolution methods, the search problem is staged in two steps. In the first step, a diverse set of primitive robot behaviours are evolved. In the second step, a high-level arbitrator is evolved by using the diverse primitive behaviours to find a solution to the problem in question.

The application of hierarchical EAs to realistic swarm tasks needs to consider a number of factors. First, it is important to note that the quality of the primitive behaviours and their level of diversity will affect the final solution. Therefore, human expertise and judgement are needed for evaluating the evolved primitive behaviours before proceeding with the high-level arbitrator. There is a need for a methodology for guiding such human evaluation processes to avoid overwhelming the evaluator with hundreds of primitive behaviours. Second, the well-documented issue of misspecified fitness can affect both stages of hierarchical EAs. Misspecified fitness occurs when the objective function has some imperfections that are exploited by the EA in a way that leads to evolving individuals that maximise the fitness function while exhibiting unexpected and undesirable behaviours [253].

15.2.2.2 Reinforcement Learning

Despite the astonishing results achieved by ML techniques across various domains, their application to swarm robotics has so far been very limited. ML techniques can facilitate handling the complex, unpredictable contingencies that characterise swarm behaviour [251].

Particularly, RL [772] can be useful for learning agent behaviours for a given task. RL uses agent interaction with the environment via trial and error to learn an action-selection policy that maximises the associated reward function. The objective of the task in question is formulated into a reward function that is used to provide an agent with feedback on the effectiveness of its behaviour.

The framework of multi-agent RL (MARL) extends RL to deal with multi-agent tasks in which the agents learn simultaneously [394]. MARL has been used mainly in multi-agent games, for example, poker and video games [498, 819]. The application of MARL to swarm robotics has been very limited, and only a handful of studies have investigated how MARL can be used to design swarm member's behaviours [93, 407, 600]. The main challenges faced by MARL are environment nonstationarity and the exponential increase in complexity as the number of agents increases [340]. The latter is amplified in swarm contexts because of the typically large number of swarm members. In addition, similar to the issue of misspecified fitness in EAs, reward hacking is common in RL domains [859] and is exacerbated by the complexity of formulating the reward function in swarm settings [340].

15.3 Machine Education for Swarm Behaviour Design

ML research has focused mostly on designing algorithms that can learn a desirable behaviour from training experiences. Recent ML algorithms have had tremendous success in a variety of tasks and have consistently increased their presence in our day-to-day life. Despite these remarkable achievements, ML is facing a multitude of technical and social challenges that limit its potential. A major technical challenge is the need for excessive amounts of data from which the ML algorithms can learn. These data are not always available for new domains or in scenarios in which access to data can be a costly endeavour [610].

In addition to the technical challenges, there is consistent evidence in the literature suggesting that fully autonomous learning can result in social integration issues [178]. For example, autonomously learnt behaviours can be poorly understood by humans and can consequently lead to automation rejection or ineffective human-machine teaming [16, 745].

Increasing the involvement of human users or domain experts in the learning process of ML agents can greatly mitigate the aforementioned technical and social challenges. For example, when compared with autonomous RL, learning from demonstrations provided by a domain expert leads to considerably quicker learning and better alignment between the intended and learnt machine behaviours [350]. User involvement in the learning has also been shown to boost trust in the resulting behaviour [348], which is a critical factor for determining user adoption of autonomous systems [862]. Methods for incorporating human knowledge in learning has existed for decades [701]; however, these methods mostly use humans as a data generation source rather than as a part of a more comprehensive learning framework. ME aims to provide a systematic approach to incorporating human requirements, knowledge and priorities into the learning process of machines.

ME is an emergent research direction that aims to offer a meaningful and systematic involvement of humans in the teaching of ML agents [22]. Being inspired by the long-standing field of human education, ME aims to adopt well-established education theories and investigate their applicability in teaching artificial agents [566]. ML research has focused mainly on the design of computational models and learning algorithms that enable a machine learner to maximise a given performance metric (e.g. accuracy).

ME brings to the equation its pedagogical, theory-driven and holistic approach to the learning and teaching process [20]. ME acknowledges the critical role of ML algorithms in learning, but it also takes into account other key considerations including the relationship between an overall learning goal and the different learning exercises, the appropriate form for transferring domain knowledge, the design and sequencing of lessons and the design of assessments and performance measures for evaluation [405]. In a prior work [405], we investigated the use of ME for a swarm collective decision-making problem. Below, we briefly present that work as a case study of applying ME to a multiskill swarm problem.

15.3.1 Case Study 1: Machine Education in Swarm Decision-Making

Swarm collective decision-making is a class of problems in which swarm members need to make a decision in a distributed manner according to the local knowledge and perceptions of individuals in the swarm. The case study presented in this section considers a large tiled environment in which each tile can be black or white. The objective was to enable swarm members to explore various parts of the environment and collectively decide whether black tiles were more common than white tiles. In a prior work [405], we used ME as a framework for teaching swarm members the required skills to autonomously perform this task. We borrowed the Dick and Carey model [240] from the human education literature and used it for our application of ME. The model gives step-by-step practical guidance for curriculum design, as shown in Fig. 15.1.

Applying the model to the swarm decision-making problem commences with the identification of the learning goal, which is to enable swarm members to autonomously perform the specified collective decision-making problem while achieving high effectiveness and efficiency. This goal acted as the driver for the following steps of the model, such that the goal was analysed to identify the supportive skills needed to achieve the goal, as shown in Fig. 15.2. Functional analysis, described in Chap. 1, can be used to support identifying the required skills in a structured way. The swarm members' sensing and acting abilities were also analysed to identify the gaps between the existing and required capabilities. These gaps represented the skills to be learnt. The learning context was characterised to identify any infrastructure required for learning. Chapter 1 gives a detailed discussion on infrastructure analysis including its use to identify physical requirements, policies and procedures and human-swarm interaction requirements.

After analysing the learners and learning contexts, the performance objectives were formulated for each skill, together with the specifications of the assessment

Fig. 15.1 Steps of the Dick and Carey model

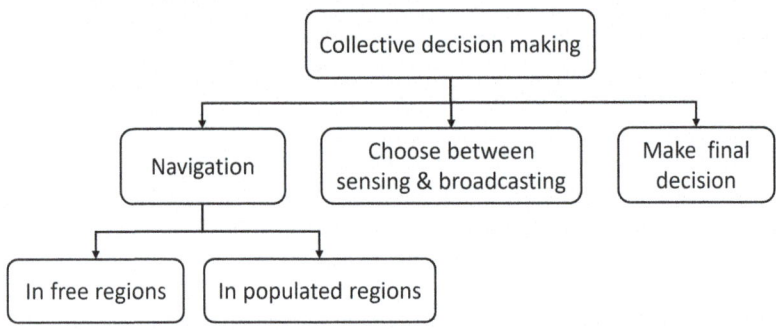

Fig. 15.2 Skill analysis in the swarm decision-making case study

instruments in terms of the evaluation processes and conditions used. Functional analysis can greatly facilitate this step because it enables generating a list of competencies required by each skill, which can be used as assessment criteria. The evaluation specifications were decided at such early stages to make them derived from the learning needs independent of the curriculum implementation.

The detailed design of the learning strategy was then selected. We identified RL as an appropriate approach for learning the required skill and designed the requirements of the lessons to ensure their effectiveness. This was followed by identifying the interdependencies between skills, the order of skill teaching and the order of lessons for teaching each skill. The design of the curriculum then underwent formative evaluation, in which the curriculum was enacted with a relatively small-sized swarm in a miniature environment to obtain quick initial results on the effectiveness of the curriculum. These results were then feedback to earlier steps of the model to review the curriculum to address some of its identified issues.

Our application of the Dick and Carey model to swarm decision-making gives a concrete example of one way of using ME to systematically design swarm behaviours. However, given that the Dick and Carey model was specifically designed for human education contexts, several human-machine factors beyond the model need to be further investigated. The next section discusses a framework that was specifically designed for ME to consider these human-machine factors.

15.4 Curriculum Design in Machine Education

Given the benefits of educational models for systematising the teaching-learning process, there is a requirement for models specifically designed for ME that builds on human education models and extends them to consider human-machine aspects. Such a requirement stems from the need for models that can educate both humans and machines; human education has been offering solutions for the first half of this puzzle. In our prior work [566], we proposed Human Education AI Teaming (HEAT), an educational framework for human-AI teaming. HEAT aims to facilitate the social integration of AI by enabling domain experts to directly communicate a body of knowledge to the machine without requiring them to possess considerable computational literacies.

The HEAT framework includes six types of agents performing complementary roles within the ME system: stakeholder agent, curriculum agent, TAISOL[1] agent, ME agent, machine teaching agent and ML agent. The stakeholder agent represents the problem owner and other interested parties that have interests and requirements to be satisfied through the ME system. The curriculum agent formalises the requirements of the stakeholders and designs the main learning modules to meet these requirements and the assessments and performance measures to be used for the verification of the requirements. The TAISOL agent decides on the skills required for a lay user or domain expert to be able to teach a machine. They formulate a curriculum for teaching these skills to the human teachers.

The ME agent designs a curriculum for delivering each teaching module identified by the curriculum agent to the machine learner. To successfully achieve this aim, the ME agents need to have a sufficient level of both ML and domain literacies to enable them to design a curriculum that meets its objectives while taking into account the realistic characteristics of the teachers and the machine learner. The machine teaching agent is the domain expert who knows how to perform some aspects of the tasks to be learnt or can guide the machine learner through its learning process. Finally, the ML agent is the one that has the computational skills

---

[1] TAISOL stands for Teaching AI as a Second or Other Language.

and expertise required to develop appropriate ML models for learning the desirable learner skills.

Only two HEAT agents have received considerable attention in the literature: the ML and machine teaching agents. The former has been investigated by studying various ML schemes, computational models and learning algorithms that improve the learning from a set of experiences. Meanwhile, the latter has been investigated by studying various techniques for humans to present their domain knowledge to the machine. The other HEAT agents have received only slight mentions, if any at all, from the AI community. In this section, we start unfolding the role of the ME agent in designing a curriculum for swarm tasks. The rest of this section discusses the factors that need to be considered when designing a curriculum for the machine.

Given a task that requires learning, the ME agent needs to design a curriculum that identifies the lessons needed for the machine to learn the task, lesson sequencing and evaluation. In doing this, the ME agent has to consider several aspects, including the characteristics of the learners and teachers, the task complexity and the available educational resources.

Learner Characteristics include the existing behaviours, abilities and constraints of the learner, for instance, built-in sensors and functionalities or actuator limits. The curriculum should use and build on existing abilities when learning a new skill while being realistic about the learner's constraints and whether it contains the foundational elements that allow its limits to be pushed.

Teacher Characteristics include their skills, level of domain expertise, physical and cognitive abilities and time constraints. Teachers' skills and expertise determine the quality of teaching. The cognitive and physical abilities should be used for selecting appropriate forms for knowledge delivery. Time constraints affect teacher availability through the learning process.

Task Complexity stems from factors including the size of state (input or feature) and action (output or decision) spaces, the neighbourhood topology connecting the state space, the complexity of the mapping required to move from a state to an action, the level of dynamism and nonstationarity of the environment and the proximity to the goal state. The size of the state space defines the number of situations that the learning agent may encounter, and the size of the action space defines the number of available choices/actions that agents need to compare and learn, which is better. The representation language for the state space is a crucial contributor to complexity because, together with algorithmic operators, it defines the proximity of one state to another. The mappings from a state to an action could be highly non-linear; thus, the complexity of the task could become complicated because of the mapping. Environment dynamism affects the uncertainty about the future state in response to the agents' actions. Goal proximity affects the temporal aspects of the task in terms of the number of sequential actions to be executed by the agents before the task can be completed. Goal proximity also affects the need for reasoning about the future while selecting the current action. Learning can become very challenging because these factors coexist and could interact in different manners in different parts of the state space.

Educational Resources include computational resources, existing data, platforms and time. Computational resources include memory and storage units and processing power. Existing data in the form of raw or processed data or relevant pre-trained models should be used to enable efficient teaching and learning. Supportive platforms can be any hardware or software components required for learning, for instance, the robotic hardware or available GUI. The required supportive platforms should be explicitly accounted for in the curriculum design. Time requirements come as constraints on the time limit for designing the curriculum, teaching the lessons and learning.

The above aspects are examples of the factors that need to be considered by the ME agent to enable designing a realistic curriculum that can be enacted to generate the intended effect. The rest of this chapter focuses on how learner and teacher characteristics can be incorporated in the curriculum design. The incorporation of the other aspects is left for future work.

15.5 Learner-Centred, Teacher-Centred and Blended Curriculum Design

In human education, curriculum design approaches can be classified along various dimensions. For example, they can be classified according to the sequence of designing the curriculum components (i.e. contents, teaching methodology and learning outcomes) [679], the underlying educational philosophy (e.g. rationalism and constructivism) or the roles of the learner and teacher (i.e. learner centred versus teacher centred) [149].

Of particular relevance to ME contexts is the classification of curriculum design according to whether it is a learner-centred, teacher-centred or blended approach. This classification is important because it touches on the design of the interaction space between the human teacher and the machine learner and their influence in designing a curriculum. Figure 15.3 shows the difference between these three approaches. In learner-centred approaches, the curriculum is mainly designed with the learner's needs in mind. The literature on curriculum learning [610] is an example of learner-centred approaches. Teacher-centred approaches focus on teacher characteristics, skills and knowledge when designing the curriculum. Teacher-centred curriculum design approaches are not sufficiently present in the ML literature. The closest to teacher-centred curriculum design are the techniques for learning from human preferences [198], although they lack key aspects of curriculum design (e.g. lesson selection and sequencing). Blended approaches are those that consider both learner and teacher characteristics when designing the curriculum. The next three sections describe case studies for these curriculum design approaches.

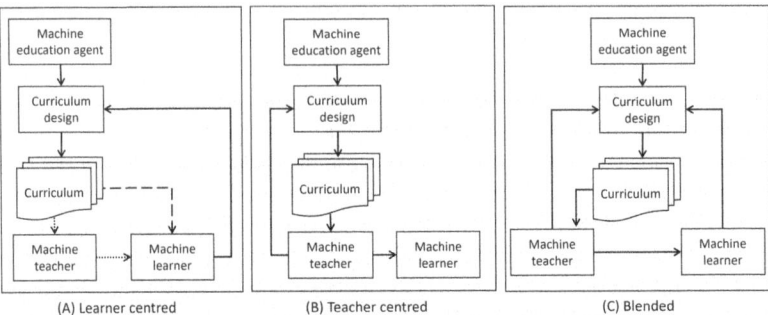

Fig. 15.3 Learner-centred, teacher-centred and blended curriculum design. The dotted/dashed arrows in subfigure (**a**) refer to the possibility of delivering the curriculum via a teacher or via autonomous learning of the curriculum lessons

15.6 Learner-Centred Curriculum Design

In learner-centred curriculum design, learner characteristics and needs are given most of the attention when designing the curriculum and deciding on its pace during learning. Curriculum learning approaches [610] in the ML literature have mainly adopted learner-centred curriculum design, because the key interest of this literature is in machine performance. It is worth mentioning that in learner-centred curriculum, the student can learn curriculum lessons autonomously or from a teacher.

Mount and Beesley, in their chapter on hermeneutics, offered opportunities for applying the theory of hermeneutics to ME. For example, a learner-centred curriculum could consider each student type as an interpretation of the curriculum, in which an ontology could be used to situate the curriculum in various interpretations. Such an approach is very powerful because a single curriculum forms a backbone that allows multiple interpretations to present it differently according to student's needs.

This section presents two case studies for learner-centred curriculum design and shows how this approach enables learning successful behaviours in complex tasks, as compared with learning without a curriculum. In the first case study, the curriculum modules are defined according to the learner's abilities. Further, autonomous learning is used such that learners progress on the curriculum according to their performance. The second case study demonstrates that autonomous learning in student-paced curriculum may not be sufficient for highly complex tasks, such that using a teacher's guidance can be necessary for learning.

15.6.1 Case Study 2: Autonomous Unmanned Aerial Vehicle Operation in Search and Rescue

Search and rescue (SAR) operations require navigating unknown, cluttered environments to locate missing victims in a timely manner. The use of robots and unmanned vehicles can bring significant advantages to SAR operations because of their ability to navigate to locations that are inaccessible or risky for human rescuers. Increased autonomy of these robotic systems is required to alleviate the workload on human rescuers in such high-pressure situations and to allow for continued operation in environments with no or limited communication. Scripting the complete robot behaviour in SAR operations is not feasible, because the situation cannot be fully anticipated beforehand. Instead, the novelty of a SAR environment requires making decisions on the spot using real-time observations.

In our prior work [776], we used RL to learn the behaviour of unmanned aerial vehicles (UAVs) in wilderness SAR operations. Learning this task is challenging because the UAV needs to use its sensorial inputs (RGB and depth images and position information) to decide on its actions in a vast, cluttered and partially observable environment. Our initial experimentation quickly suggested that the task is too complex for a UAV to learn in one go (i.e. using end-to-end RL). To deal with this learning complexity, we needed to use human expertise to design a curriculum that decomposed the task into a number of interconnected modules to make it easier for the UAV to learn.

The task was decomposed into five modules, each specialised in one aspect, as shown in Fig. 15.4. The obstacle detection module aimed to detect obstacles (e.g. trees) and terrain information and calculate their positions relative to the UAV. The human detection module used an off-the-shelf, pre-trained ML model for detecting humans in an image. Detected obstacles and humans, as well as UAV position information, were sent to the information map module, which organised all this information in a spatial map that started empty and was filled over time as the UAV

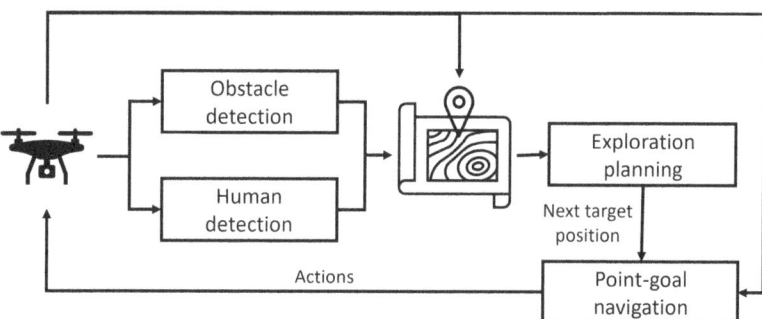

Fig. 15.4 Decomposition of the unmanned aerial vehicle behaviour in the search and rescue case study

explored more parts of the environment. The exploration planning module used the information map and the past trajectory of the UAV to decide which region should be explored next. Finally, the navigation planning module issued low-level action commands to the UAV to navigate it towards the next target area.

When the problem was decomposed into these modules, it became much more manageable to find a solution for each module. The obstacle detection and information map modules did not need learning, because efficient scripted programs could be easily developed for them. The human detection module needed to be performed via ML. To alleviate the learning requirements on the learner, an off-the-shelf ML model was selected to perform this function. Only the exploration planner and the navigation modules needed learning. We used RL for both modules to learn how to generate the module output given the relevant aspects of the task state.

The navigation module was particularly difficult to learn, because it required finding the shortest path from the UAV to a target position while avoiding obstacles and changing UAV altitude according to the terrain level, all in an unknown environment. Using traditional RL for this module, the UAV rarely reached a given target region because it would collide with an obstacle, crash or go astray. It was found that a curriculum is necessary for accelerating the learning of this module. The aim of the curriculum was to expedite learning by starting with easy lessons before progressing to more complex ones. We controlled goal proximity to affect the lesson complexity. In initial lessons, the target region was selected to be very close to the UAV, which made it relatively easy for the UAV to find a sequence of actions that led to the goal. As the UAV performance progressed over time, the next lessons in the curriculum with further target regions were used for learning. This resulted in a very good navigation performance that rarely encountered collisions. Once the navigation module completed its learning, the exploration planner module started using RL to learn how to use the information map to generate the next target point.

15.6.2 Case Study 3: Autonomous Air-to-Air Combat

The autonomous air-to-air combat domain has seen increasing interest recently. The high risk of pilot fatality motivates serious efforts to enable unmanned aircraft operations. The US Defense Advanced Research Projects Agency launched the Air Combat Evolution program to explore the potential of AI in the air combat domain. The program's AlphaDogfight Trials aimed to investigate the feasibility of the autonomous piloting of F-16 aircraft in simulated air-to-air combat. This competition showcased a number of autonomous agents defeating an expert human pilot, a remarkable success of AI in such a complex setting. Although there is not much available information on the design of the successful agents, our research group sought to develop an autonomous agent for aircraft piloting in a similar simulated environment [678].

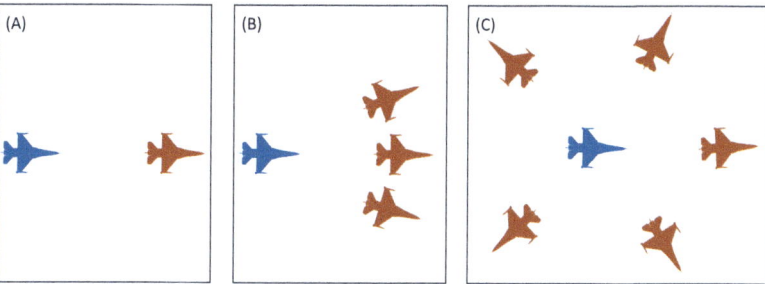

Fig. 15.5 Three lessons of increasing complexity in the curriculum for autonomous learning in the air combat task. The blue aircraft shows the initial state of the learning agent, while the red aircraft represents the possible opponent initialisation in a lesson

Our team formulated the problem as an RL problem in which the RL agent needed to learn a policy for selecting its action given its own and the opponent's state. To facilitate learning, we designed a learner-centred curriculum in which the lessons increased in complexity by controlling the opponent's strategy. Figure 15.5 shows three lessons of different complexity levels that were included in the curriculum. In early learning iterations, the RL agent competed against a deterministic move-forward opponent that was positioned along the heading direction of the learning agent. Curriculum progression was learner paced such that as the RL learnt to defeat its simplistic opponent, it moved to more complex lessons by increasing the angle between the heading of the agent and its opponent.

The curriculum enabled the learning agent to bootstrap strategies that successfully defeated the opponent when the opponent was initialised within a 10° cone from the agent's aircraft nose. The learning agent could not progress to more complex lessons because it failed to locate and target its opponent. It was apparent from these results that the curriculum for self-learning reached its limit in this domain and the agent could no longer handle more complex situations. This led us to redesign the curriculum by incorporating a teacher to guide learning in the complex lessons of the curriculum. We used learning from demonstration [701], in which an expert provides the learning agent with a set of demonstrations on how to defeat its opponent in the advanced curriculum lessons. The agent then learns to mimic the human behaviour before autonomously learning to improve the mimicked behaviour.

Although we did not have access to an expert pilot, a member of our research group was fairly familiar with aircraft piloting in the simulation environment; he acted as the expert and provided demonstrations to the learning agent. The demonstrations were clearly suboptimal; however, they enabled the learning agent to learn a reasonable targeting policy using the Conservative Q-Learning algorithm [489].

The key lessons learnt from this case study are as follows: (1) some tasks are too complex to be autonomously learnt even when the learning is staged via a curriculum, (2) even a fair level of teacher expertise in the application domain can make the difference between the success and failure of learning and (3) the selection of an appropriate learning algorithm based on the quality and distribution of the teacher data is critical. These insights should be considered when designing a curriculum to increase its effectiveness.

15.7 Teacher-Centred Curriculum Design

In addition to task complexity, there are other reasons that learning through a domain expert could be preferred over autonomous learning. Specifically, the need for building user trust in and acceptance of the learnt behaviour and ensuring the machine/swarm behaviour aligns with users' standards are among the key motives for incorporating humans in the learning process. User involvement in learning has been shown to boost trust [348], which is a critical factor for determining user adoption of autonomous systems [862]. Ensuring the harmony between the behaviours of team members in deployment environments is critical for effective human-swarm teams. If the learning process results in different tactics than those adopted by humans, this can detrimentally affect the team performance [745].

One important factor to consider when incorporating a human teacher in a curriculum is that humans come with their own skills, needs and constraints. The abilities and skills of human teachers play an important role in determining the format, type and quality of knowledge to be transferred between the teacher and the learner. Curriculum design should consider the teacher's level of domain expertise, mastery of the required skills and physical and cognitive capacity. For instance, although it is common for an RL agent to spend millions of time steps to learn a moderately complex task, a human teacher typically does not have the capacity to be actively engaged in supervising the agent throughout such extended time. Further, the teacher's skill level can be a constraint that needs to be acknowledged in the curriculum design process. Section 15.6.2 gives an example of situations with no access to professional teachers. Although a modestly skilled teacher had a considerable positive effect as compared with autonomous learning, the curriculum effectiveness could have been enhanced by considering teacher characteristics. The rest of this section presents a case study for teacher-centred curriculum design in which teacher skills and needs were used to inform the lesson design.

15.7.1 Case Study 4: Shepherding a Swarm of Physical Robots

Shepherding is a bio-inspired guidance technique where a sheepdog agent influences the movement of the sheep agents (i.e. swarm members) towards a designated goal area [531]. By monitoring the behaviours of biological sheep and sheepdogs in nature, prior studies have proposed computational models for shepherding that enable scripting behaviours for a robotic shepherding agent [592]. Nonetheless, existing shepherding models fall short in responding to emergent changes in the environment (e.g. environmental structure and noise) or in the swarm behaviour (e.g. getting familiar to the sheepdog) [274]. This gave rise to ML models because of their ability to learn adjusted shepherding behaviours to adapt to changes in the task [406]. Incorporating a domain expert in learning can be necessary to minimise potential damage or unrecoverable failures during learning and to improve the predictability of the learnt shepherding behaviour.

In our prior work [619], we developed a system for a human to teach a UAV how to shepherd a physical robot swarm to move the robots into a designated goal area. Dealing with physical robots is known to be greatly more challenging than simulated ones, because of many factors including sensing and positioning noise, execution errors and performance sensitivity to battery level. Therefore, including a human teacher can be critical for ensuring the learning of an effective shepherding behaviour within a reasonable amount of time. We used learning from demonstrations, in which the teacher teleoperates the UAV when delivering the lessons. These demonstrations are then presented to the learning agent to guide its learning.

In the first low-autonomy experiments, we explored the ability of a human teacher to use low-level commands to control the UAV through a number of shepherding scenarios. This low-level autonomy experiment allowed the human to exert a higher level of control on the low-level actions of the sheepdog. We designed a user interface that used a set of hand gestures for issuing commands. These were then mapped to the actions needed for low-level autonomy. The position of the right hand specified the longitudinal/lateral movement of the UAV (forward, backward, left, right or hover), and the position of the left hand specified the vertical movement (up, down, maintain height). A new low-level command was formed from the position of the two hands each time step and was sent to the UAV to execute during a demonstration. Our analyses revealed that this level of control had serious issues that hindered teaching. Even though the human teacher usually had a good plan in mind, plan execution was very hard. The translation of an intended command into the correct combination of hand gestures was often confusing. In addition, robot sensing and acting errors and performance variations between swarm robots led to divergence between the desirable and actual effect of a command. Further, the need for the teacher to quickly change the hand gestures to switch between commands generated considerable workload. All these factors suggested that low-level teaching may not be the best option given the cognitive and physical workload imposed on the teacher.

Based on these findings, in the second phase, we developed another teaching setting that increased autonomy to consider teachers' factors. The new design needed to include fewer commands to reduce confusion, require a lower rate of issuing commands to give the teacher sufficient time to plan for the next action based on the evolution of the situation and provide support with handling the frequent robot errors. To meet these requirements, the new design supported teaching using a higher level of autonomy, that is, replacing low-level actions with high-level UAV behaviours. The new behaviours were selected according to the position of the right hand (collect or drive) and the left hand (activate or deactivate collision avoidance). The collect, drive and collision avoidance behaviours were in one case scripted and in another based on models learnt from the low-level autonomy data, to enable the automatic selection of low-level UAV commands to deliver the intended behaviour. The role of the teacher in the new design was to teach the UAV when to collect sheep that has gone astray, when to drive clustered sheep to the goal and when to consider collision avoidance when planning its path.

The experiments with the new design showed that high-level teaching resulted in notably less workload on the human teacher, shorter task completion time and higher task success rate. Task completion time and success rate are important indicators of the quality of the demonstrations, which, in turn, affect the ability of the UAV to learn effective shepherding behaviours. The key message from this case study is that lesson design should consider teachers' skills and constraints, because this will not only affect their user experience but also teaching quality.

It is worth noting that to perform their role effectively, a human teacher needs more skills than a swarm operator. A teacher should not only focus on the success of swarm operation but also act as a diagnostic system to identify skills that the swarm is lacking, devise new lessons or modify existing ones to address the identified limitations, sequence and repeat the lessons as needed and design testing procedures to assess swarm capabilities. Although all of these tasks are part of the curriculum design, when delivering the lessons, the teacher should be actively responding to any emergent needs or situation-specific requirements. These teacher tasks could impose a significant cognitive load on the teacher and would require the teacher to be trained to perform these roles. In addition, a screening process might be needed for teachers to ensure that they have the basic skills and abilities required by the role, including the ability to operate and interact with the swarm, make decisions in cognitively demanding environments and have excellent scanning strategies to maintain situation awareness about swarm members. Inspecting this list may suggest that teachers will need to go through similar screening processes to those air traffic controllers go through.

15.8 Blended Curriculum Design

Learner-centred curriculum design cares mainly about the learner's needs and aims to design a curriculum to accelerate learning or attain high learner performance, or both. This approach to curriculum design is useful when the task can be learnt autonomously or when a teacher's knowledge can be easily accessible in the required form. However, teacher-centred curriculum design aims to improve teachers' ability to deliver their knowledge in a usable form given their skills and abilities. This approach does not explicitly consider the learner's needs or how to ensure the delivered lessons meet these needs. Blended curriculum design seeks to take the characteristics of both the learner and teacher into consideration to inform the design of a curriculum that makes the best use of the teacher and ensures their teaching addresses the learner's needs. Below is a case study for blended curriculum design in learning to shepherd swarms with various behaviours.

15.8.1 Case Study 5: Shepherding Swarms with Various Behavioural Responses

Similar to Sect. 15.7.1, this case study is in the domain of swarm shepherding. However, the objective here was to learn to shepherd simulated sheep swarms that exhibited different behavioural responses to the shepherding agent. The case study was developed in a 2D simulated environment such that in each simulation run, the shepherding agent could have a different type of swarm to guide. Swarm behavioural responses were controlled by changing the sensitivity of a swarm member to the shepherding agent. These behavioural responses mimic different levels of familiarity of sheep flocks to their shepherd. Having a domain expert to supervise the learning of the required shepherding behaviours was important to ensure the learnt behaviours complied with animal ethics and welfare standards [854].

Allowing interactive learning experiences can enable a teacher to monitor the learner's performance and provide lessons to address its learning needs. However, it was necessary to be careful about budgeting the limited human time in this interaction, particularly because past studies have shown that shepherding can take millions of time steps to learn [406]. We designed a system that allowed for interactive teaching-learning experiences while considering the limited teaching budget. The system is shown in Fig. 15.6.

Initially, the teacher provided a few demonstrations to show the learner how to shepherd responsive swarms that had high levels of sensitivity to the shepherding agent. The demonstrations were used by the learning from demonstration component to learn how an action should be selected in the demonstrated swarm states. The performance of the agent was then evaluated using different swarms to identify swarm states the agent could handle successfully and those where the agent struggled in selecting an appropriate action. The agent then proceeded with self-learning to improve its identified weaknesses. The duration of self-learning was controlled by the self-learning budget B_{SL}, which was set according to the available computation resources and capped by the length of the interval before the teacher became available again.

After self-learning was complete, the teaching budget B_T was checked to see whether more demonstrations could be requested from the teacher. If so, a number of swarm states where the shepherding agent struggled were selected and presented to the teacher to give demonstrations for them. The selection of those struggle states has to be done carefully to ensure the teaching budget is best used for learning in sufficiently diverse struggle states to maximise the benefits of the requested demonstrations. The proper distribution of the teaching budget along the learning process is also important to ensure the student can make use of the teacher's knowledge when it is most needed.

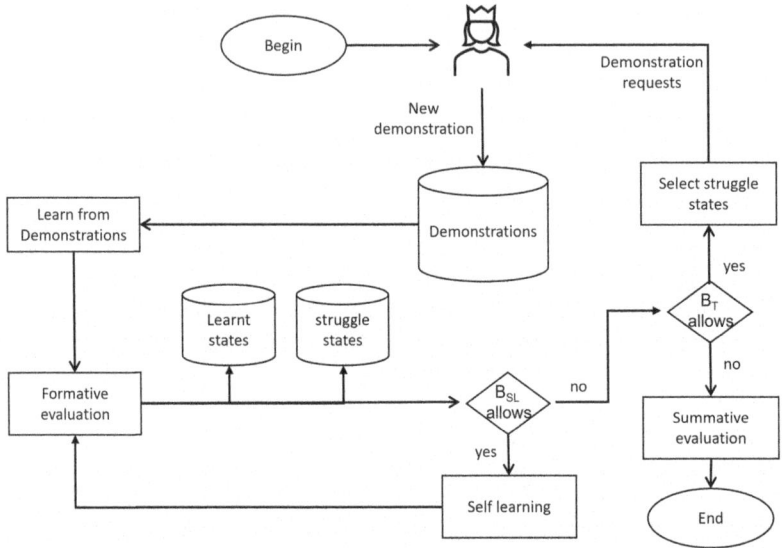

Fig. 15.6 Learning from a teacher with a limited teaching budget. B_{SL} and B_T refer to self-learning budget and teacher budget, respectively

The curriculum designed in this case study not only enabled simple-to-complex lesson sequencing but also incorporated various learning strategies (learning from demonstrations and self-learning), enabled the learner to take an active role in assessing its own performance and requesting lessons to help with its weaknesses and offered a way to maximise the benefit from the teacher within their limited availability.

15.9 Lessons Learnt and Recommendations

ME is a new, but natural, revolution in AI. ME systematises the thought processes that ML experts go through when designing a ML algorithm for a particular task and when faced with challenges in achieving a good learning outcome. Moreover, ME extends this systematisation process all the way to the need phase, where stakeholder analysis occurs to establish the need and requirements that call on ML and AI experts to design and develop autonomous agents for a particular problem. In short, ME transforms the ad hoc design of intelligent machines to a systematic process using theories from human education. One could see ME as the system engineering for cognition.

In this chapter, we presented diverse case studies ranging from a single agent design (Case Study 2) and agent-on-agent interaction (Case Study 3) to agent on swarm interaction (Case Studies 4 and 5) and multi-agent swarm interaction (Case

Study 1). ME was a deliberate choice from the start in Case Studies 1, 4 and 5. In Case Studies 2 and 3, the use of ME was not deliberate; that is, ME came after the fact, when the researchers were faced with low accuracy only to discover that the task was difficult and needed some form of a decomposition. This formed the first learning lesson in this chapter: ME needs to be a deliberate choice from the start.

ML researchers could easily be driven by accuracy alone, whereas they normally innovate multiple approaches and algorithmic choices to engineer success by trying multiple ideas. This lack of a methodological approach has multiple drawbacks: (1) lessons learnt are not recorded, (2) there is a lack of reflective learning by the designer and (3) it could waste significant computational times unnecessarily. Our first recommendation, therefore, is that ME needs to be a deliberate choice from the start.

Case Study 3 demonstrates that autonomous learning can hit a complexity wall. Most successful stories in ML today use supervised learning, in which massive amounts of data are used to train the AI. The alternative is to design clever reward functions to guide the AI using RL. The latter normally relies on a massive computational cost, as the search algorithm attempts to wait for these few tiny tipping points to improve its performance when it reaches a plateau. Case Studies 2, 3 and 4 hit a complexity wall. In one, it was necessary to decompose the problem and fall back on ME to guide the process. In another, it was necessary to call on the expertise of a human. These choices were made when the model failed to perform. ME overcomes this risk by hedging against it from the start. By analysing the complexity of a task, we gain more insight into the design of more efficient ML agents, and we develop understanding to escape, or at least manage, the complexity wall with which a task is faced. Our second recommendation, therefore, is that task complexity analysis is a risk mitigation strategy.

The above two recommendations apply to ME for any problem. For a swarm system, the case studies teach us that ME is necessary because of the overwhelming complexity the swarm problem space imposes on any ML algorithm. Case Study 3 shows that we need to augment the learning process with inputs from humans, and although we may lack human experts to demonstrate particular tasks to the machine, the availability of data from a good-enough demonstrator could help the machine to jump-start and bootstrap to a better performance level. We recommend, therefore, that learning in a swarm context is complex and calls for early decomposition and analysis of the problem space and adoption of an ME methodology.

15.10 Summary and Future Directions

Research on robot swarm systems has recently proliferated because of their ability to perform tasks that exceed the performance limits of single robots. One of the main questions that limit the wider adoption of robot swarms in real-life application is how to design local behaviours that result in the intended global swarm behaviour. AI techniques, specifically, EAs and RL, have demonstrated the significant role they can play in answering this design question. However, the literature has been missing methodological guidance to systematically conduct learning. Moreover, the literature falls short on methodologies capable of incorporating human domain knowledge in a way that enables humans to influence the operations of AI techniques and ensure that they result in behaviours that truly meet humans requirements.

ME offers a methodology for bringing humans and machines together in a holistic education process with the objective of designing curricula for humans to effectively interact with machines and for machines to use these interactions in learning the desirable behaviours. In this chapter, we focused on designing a curriculum for teaching a machine a set of required behaviours. We presented the main factors that need to be considered while designing a curriculum for the machine. We then identified three approaches to curriculum design according to whether the design is driven mainly by the needs and characteristics of the learner, the teacher or both. We presented a few case studies for these curriculum design approaches in single-robot platforms, as well as in the control of robot swarms.

Future researchers can extend on this work in multiple directions. First, we focused mainly on learner-centred, teacher-centred and blended curriculum design approaches. Future work could expand on these approaches by identifying how other aspects, for example, task complexity and educational resources, can have practical implications on the curriculum design process. Second, future work can study the use of ME in designing swarm behaviours with different levels of centralisation, including fully distributed and hierarchical interactions. This will help reveal whether the curriculum needs to consider factors specific to swarm behaviour types. Third, we focused only on the role of the ME agent while assuming the teacher already comes with the relevant skills. Future work could study the roles of other agents in the machine education framework and identify how to qualify a human to perform the teacher role or how to assess the teacher's skill level.

Acknowledgments This research was funded by the Australian Research Council Discovery Grant DP200101211.

Open Access This chapter is licensed under the terms of the Creative Commons Attribution 4.0 International License (http://creativecommons.org/licenses/by/4.0/), which permits use, sharing, adaptation, distribution and reproduction in any medium or format, as long as you give appropriate credit to the original author(s) and the source, provide a link to the Creative Commons license and indicate if changes were made.

The images or other third party material in this chapter are included in the chapter's Creative Commons license, unless indicated otherwise in a credit line to the material. If material is not included in the chapter's Creative Commons license and your intended use is not permitted by statutory regulation or exceeds the permitted use, you will need to obtain permission directly from the copyright holder.

Chapter 16
Hyper-Teaming

Adaptive Teaming and Coordination of Multi-domain Autonomous Robotic Systems

David Johnson and Felix Kong

This chapter describes a multi-agent thinking swarms operating collaboratively with a human team on the ground. It is the result of a multi-year project funded by the Australian Defence Force, led by Mission Systems, to determine the "art of the possible" in regards to what can be achieved with today's technology, as applied to human-machine teaming at the tactical level. We show that robots and humans need to guide one another to achieve compelling results and that although none of this is easy, there are techniques that allow us to build a scalable system of systems to meet real-world goals.

16.1 Introduction

No plan survives first contact with the enemy. When future wars are fought with teams of autonomous systems and humans (potentially on both sides), the ability to communicate command and intent between various elements of the team will be a critical component of a robust and resilient force.

Mission Systems, in conjunction with University of Technology Sydney, AMSL Aero and Defence Science and Technology Group (platform division), is working with the Australian Army's Robotics and Autonomous Systems Implementation and Coordination Office to develop hyper-teaming. This "first contact" capability prototype will demonstrate that mixed robot-human teams can perform sophisti-

D. Johnson (✉)
Mission Systems, Lane Cove West, NSW, Australia
e-mail: david.johnson@missionsystems.com.au

F. Kong
The Robotics Institute, School of Mechanical and Mechatronic Engineering, The University of Technology Sydney, Ultimo, NSW, Australia
e-mail: Felix.Kong@uts.edu.au

© The Author(s) 2025
S. Ng et al. (eds.), *Thinking Swarms*, https://doi.org/10.1007/978-3-031-82790-7_16

cated, multiphase, coupled intelligence, surveillance and reconnaissance (ISR) tasks driven by high-level commands provided by a human on the loop. This project builds on our prior work on decentralised ISR with a heterogeneous multi-robot system [509, 510] and is also applicable to other applications, such as search and rescue and firefighting.

The hyper-teaming system combines decentralised exploratory behaviours and task-driven behaviour trees to accomplish various phases of a mission as defined for the human members of the team. To provide an adaptive and collaborative human-robot teaming framework, the system must be as follows:

- *Explainable*: robot actions should be readily understood.
- *Modular*: the system should be adaptable and upgradable.
- *Generic*: open architecture, to work on a variety of platforms.
- *Scalable*: works on many robots over wide areas.
- *Low cognitive demand*: to support humans with minimum intervention
- *Intuitive to control*: using a combination of natural language, haptic and visual interfaces.

Robots then coordinate among themselves autonomously to assist humans on the ground, providing information through the Civilian Team Awareness Kit and responding to tactical guidance ("avoid this area") in addition to direct ("give me the joystick") or indirect ("go and look at XXX") commands.

In order for the system to operate effectively as a whole, each robot has a number of ancillary requirements and functions:

- Sensing, processing and communications
- Collision avoidance
- A common operating picture
- A mission: **What is it there to do?**

The framework for delivering this system is depicted in Fig. 16.1. Although not all of the elements shown in this diagram are discussed here, given that many remain a work in progress as implied by the technology readiness level, this diagram goes some way to describing the complexity of this system of systems, and the level of effort required to build, deploy and maintain a truly intelligent swarm that is capable of operational utility.

16.2 Related Work

16.2.1 Civilian Context

Although the most obvious example of autonomous "field" robots in the civilian/commercial domain is probably self-driving cars, which need to operate within the human environment, these do not specifically demonstrate teaming behaviour.

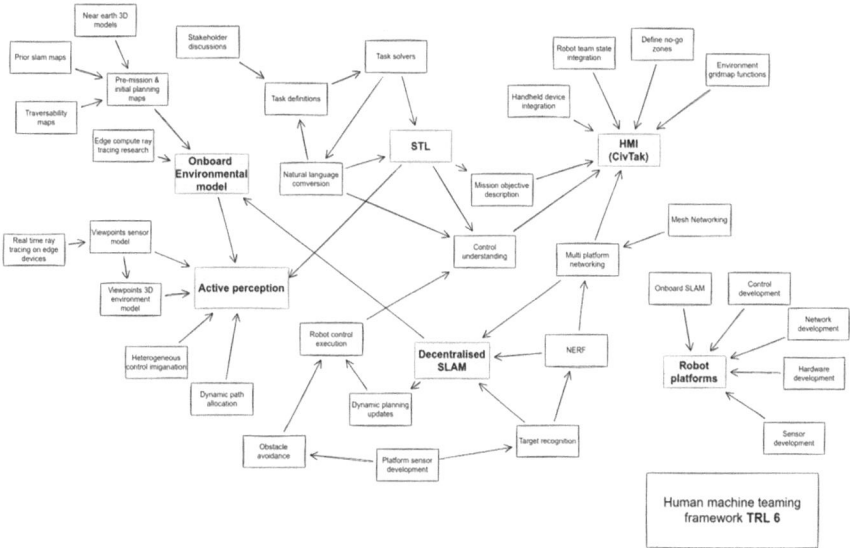

Fig. 16.1 Hyper-teaming functional network diagram

There is, however, significant work underway in vehicle-to-vehicle and vehicle-to-infrastructure communications to ensure cooperative behaviour for improved road safety and efficiency of operation.

Self-driving cars have now reached a point at which they can handle the vast majority of on-road circumstances. It is the remaining "corner cases" that have caused progress to slow in recent years. The limiting factor is safety. These systems are designed to carry civilian passengers, and the criterion for success is that they must be safer than human-driven systems. Unfortunately, it is the very small percentage of challenging conditions that cause most of the accidents, requiring millions of kilometres of training data to provide assurance in system performance. They must also interact with nonautonomous road users that may not strictly follow the road rules, and it is this unpredictable aspect that makes things especially difficult. The Waymo/Google approach to training autonomous systems using pre-acquired and mapped data is currently leading the field but does not easily scale to unknown areas.

In controlled industrial scenarios, such as mining and agriculture, progress has been steadier, and a number of fielded systems are now operating around the world. The defining factor here is the controlled environment. Indeed, the patented Rio Tinto model for an integrated mine automation system employs "islands of automation" to provide both known geometry and well-understood features, along with guarantees relating to communications and isolation from external unexpected influences. One point to note is that unlike the military and self-driving passenger-vehicle environments, it is both possible and necessary to halt mining activities in bad weather, to avoid damage to costly equipment.

16.2.2 The Need for Data Abstraction

The idea of heterogeneous teams of military robots acting in a coordinated manner is not new and was demonstrated to a limited extent in activities such as Unmanned Warrior (2016) and Autonomous Warrior (2018). In Unmanned Warrior (2016), an autonomous underwater vessel equipped with a long-range synthetic aperture sonar was used to detect a mine-like object and directly task a second smaller, more dynamic autonomous underwater vessel with shorter-range sensing capability to perform a reacquisition of the target as a prelude to neutralisation. In Autonomous Warrior (2018), the idea of multi-robot teaming was extended to achieve cooperative behaviour in four domains (underwater, surface-water, air and ground) simultaneously, using a "single" command operator interface and control framework (MAPLE). The two activities required years of planning and rehearsing to achieve their eventual aims, and a plethora of technicians, developers and support personnel was in attendance to ensure the final demonstration was successful. However, much of that effort went into ensuring the hardware (both in terms of robotic platforms and backbone infrastructure) could operate robustly in its own right.

Since 2018, advances in hardware have led to more resilient and operational-ready platforms, while perception and mapping algorithms based on deep-learning and information theoretic methods have continued to improve contextual understanding. In control and planning, Monte Carlo tree search (MCTS) and its decentralised equivalent (Dec-MCTS) have enabled optimal control of small-scale teams for specific paths. Digital twin simulations have also found an important place in the offline training and evaluation of these algorithms.

The limiting factor for both crewed and uncrewed systems is bandwidth, whether for processing or communicating information effectively. For humans processing data this bandwidth limitation is known as operator overload; therefore, presenting only the most relevant and timely information on a need-to-know basis becomes critical across many aspects of modern technology. For robots, the initial bottleneck is the processing capacity of raw sensor data, which must then be fused to extract meaningful information about the environment, for example, what is static versus dynamic and how the world can be segmented into various objects of known class. It is this high-level data that can then be processed, onboard the robot ideally, to determine what is relevant within the context of the mission. Only then should that relevant data be passed through the next bottleneck of the communications channel, which (according to range, environment and other operational parameters) may be extremely limited. Information at this still higher level of abstraction (depending on platform capability) is then typically transmitted back to a centralised processing system to be aggregated from multiple sources to extract additional meaning before being passed down the chain of command as required.

16.3 Justification for Defence Capability

A review of the openly available military doctrine from Australia, the US and the UK showed a steady move towards agile control, multi-domain integration and an emphasis on high-tempo, rapid response to changing circumstances in the battlespace. The US Army's *Operations Field Manual* describes any operational mission as having the following prerequisites:

- Understanding the operating environment
- Understanding the mission objective
- Gaining relative positional advantage

Command is then applied to a team of combat, combat support and combat service support soldiers and their equipment to orchestrate effects. Many of the terms that define this doctrine show the need for adaptation and anticipation in the face of enemy movement to maintain operational and information advantage. Autonomous systems that are proactive and operate collaboratively with synchronised sensors, actuators and networks will be critical to the next generation of human-machine teams, particularly in near-peer adversarial environments.

A recent article from West Point [757] described "The Eight Rules of Urban Warfare and Why We Must Work to Change Them" and concluded that "the combination of humans and robotics (manned/unmanned teaming) arguably holds the greatest potential for enabling swarm tactics, since it could allow rapid massing of a force that has identified any enemy strongpoint during an attack".

The concepts of sustained high-tempo movement, manoeuvring and intelligence sharing, with rapid transitions between tasks and contingent operations according to anticipated risks, allow for momentum, pressure and initiative to be maintained over an enemy force. The use of first contact autonomous systems would alleviate force-protection requirements for frontline personnel, while with greater knowledge of the geometry of the environment, combat power could also be directed with greater precision.

The challenge is then to deliver sufficient mass with the necessary speed and stealth to maintain the element of surprise, particularly as peer-on-peer confrontations would likely involve an opposing force of similarly capable autonomous systems and sensors.

16.4 System Description

The hyper-teaming construct is driven by three core principles that define the interface between command, the human-robot team and the operating environment:

- Mission command: high-level command intent is provided at the mission-task level, in the same form as it is done today. This framework allows integration of autonomous agents with human counterparts.

- Common operating picture: all agents are provided with a common picture of the area of operations before manoeuvring to the objective. In the case of autonomous systems, this common picture is likely to be in the form of a 3-D map, with which the operation can be rehearsed and analysed from all conceivable angles (in an accelerated manner) before execution.
- Accuracy: information shared between agents (and reported up the command hierarchy) during mission execution must be timely and relevant. Information sharing will convey new information about the environment, such as discrepancies in the a priori map, or the location of dynamic threats or obstacles.

16.4.1 Mission Command and Control

Our approach to the human-robot command interface is to inject commands at a higher level of abstraction, transforming it to coupled tasks that may include nonspatio-temporal constraints and interdependent subtasks. The planning and execution of behaviour to achieve tasks must involve intrinsic capabilities to actively search for and identify objects, coordinate team members to achieve this common goal and monitor the teams' progress towards the goal. We focus on the representation, planning and execution elements and address them by representing missions using a form of temporal logic, a model-checking tool from formal methods [202]. These mission task specifications generate a probabilistic representation of intent, which subsequently define an objective function maximised using the Dec-MCTS algorithm [118]. A schematic representation of the overall control framework is shown in Fig. 16.2.

High-level tasks and subtasks are expressed using signal temporal logic (STL) [234, 511], a time-bounded temporal logic defined over continuous space and time capable of specifying complex robotic tasks accurately and explicitly by avoiding the need for discretisation (gridding) of spatio-temporal values. We can represent any task that can be expressed in STL, which includes all mission command examples discussed, and we can formally verify the validity of the STL model. Examples of task categories include sequencing ("go to areas p1, p2 and p3 in order"), coverage ("go to areas p1, p2 and p3 every 10 minutes in any order", i.e. patrol), avoidance ("avoid all hostile agents h until reaching target t") and condition ("whenever hostile agent h is seen, move to safe area and report enemy position within 1 minute"). Missions can be constructed as combinations of tasks that are coupled logically. The power of this representation is apparent when considering traditional representations in robotics, in which all conditional behaviour needs to be manually spelled out as a long list of rules or as an intricate cost function. STL is compact and exact. STL expressions can be combined with typical optimisation objectives, such as finding a time-minimal path satisfying the STL specification, finding the safest path in terms of avoiding hostile agents and so on. Dec-MCTS is used to find solutions to expected categories of STL specifications efficiently and optimisation algorithms to allow the system to communicate necessary information.

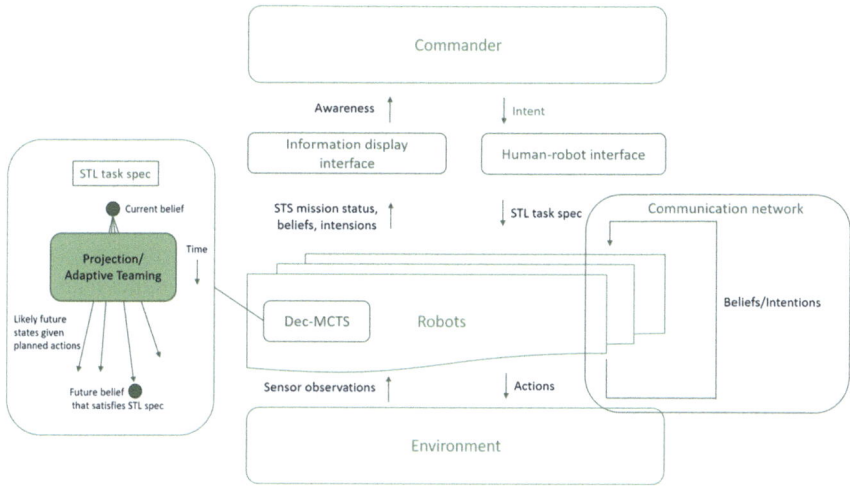

Fig. 16.2 Hyper-teaming framework

16.4.2 Goal-Oriented Cooperative Action in a Common Operating Picture

In this stage of the project, we aimed to develop the capability to:

- Adaptively reassign robotic agents to human units to support high-level goals
- Coordinate actions between human-robot teams though intelligent communication
- Understand the conditions for reorganising the robotic agent teams to reduce risk to the mission

Adaptive teaming refers to the process of forming and reforming subgroups within the overall robotic team dynamically in response to context. Coordination of individual robot actions occurs within each unit towards achieving the unit's objective. This distinction between intra-unit and inter-unit coordination allows for a hierarchical command structure within the overall team that can change over time as tasks are completed and as new information is obtained. It is consistent with the Australian Defence Force's Future C2 Concept vision for artificial intelligence. Our work builds on the advantageous properties of the Dec-MCTS algorithm to define a measure for the level of coupling between plans of individual robots and uses this measure to decide when subgroupings should change. Logical representations of tasks are then required to structure teaming at the functional level.

16.4.3 Accuracy

As part of the Dec-MCTS perception loop, robots must be able to evaluate independently an objective measure of their actions (and others) in relation to its impact on mission success. This requires the ability to project future actions into the common operating picture as it evolves using online simulation, allowing them to act quickly according to the most likely explanation of the uncertain situation observed at that point in time. By communicating only semantic information relating to potential obstacles, threats and other changes in the environment, rather than low-level sensor data, bandwidth requirements can be greatly reduced. Similarly, limited bandwidth is required to distribute robot plans both to other robots and human teammates, allowing the framework to scale effectively. Future work would extend to projecting likely patterns of behaviour by both adversaries and other actors in the scenario and arriving at a jointly optimal set of actions quickly. This approach would allow us to explore and potentially deploy learning strategies to enable systems to evolve their ability to project over time.

16.4.4 Decentralized Monte Carlo Tree Search for Multi-robot Coordination

16.4.4.1 Overview

The *multi-robot coordination module* is responsible for coordinating the behaviour of multiple robots in a way that achieves the objectives of the robot team. In this project, the robot team's objectives were to support the human team by providing ISR. Given estimates from the STL inference module of the human team's likely future movements, the multi-robot coordination module was responsible for tasking each robot in the team in a way that provided useful ISR to the human team, that is, to move the robots in such a way that their sensors covered as much of the area from which the human team would be visible as possible.

Our system makes use of Dec-MCTS [118, 119] to solve the multi-robot task allocation problem. Dec-MCTS is an online, decentralised algorithm that has been demonstrated in a wide variety of multi-robot applications including underwater [567], aerial [748], agricultural [768] and other domains.

Consider a team of R robots $1, 2, \ldots, R$, where each robot r plans its own sequence of actions $x_r = (x_1^r, x_2^r, \ldots)$. Each action x_n^r has an associated cost c_n^r, and each robot has a cost budget B^r such that the sum of the costs is constrained to be less than the budget, that is, $\sum_n c_n^r \leq B^r$. In our scenario, B^r is the distance travelled by performing the action x_n^r, but can be time, energy or some other limited quantity that is depleted by performing x_n^r. Additional constraints on the feasibility of an action x_n^r are also easily incorporated, such as robot-specific traversability constraint and collision avoidance with other robots on the team. Thus, there is a set of feasible action sequences for each robot $\mathbf{x}^r \in \mathscr{X}^r$. We denote $\mathbf{x}$ as the set of action sequences selected by all robots $\mathbf{x} := \{\mathbf{x}^1, \mathbf{x}^2, \ldots, \mathbf{x}^R\}$ and $\mathbf{x}^{(r)}$ as the set of action sequences for all robots except r.

In addition to a robot-specific budget B^r and action feasibility constraints, we defined a reward function $g(\mathbf{x})$ over the sequence of actions of all robots in the team. This reward function served as an objective function that Dec-MCTS attempted to maximise within a (possibly previously unknown) time period. In general, $g(\mathbf{x})$ encoded the "usefulness" of the team performing their actions $\mathbf{x}$ and was chosen to maximise information gain.

In summary, the problem that Dec-MCTS solved is the following: allocate task sequences $\mathbf{x}$ to all robots $r \in \{1, \ldots, R\}$ that satisfy the constraints and maximise the reward $g(\mathbf{x})$.

16.4.4.2 Decentralised Monte Carlo Tree Search for Year 1 of Hyper-Teaming

In our initial demonstration scenario (the first of three demonstrations), several simplifying assumptions were made to keep development reasonable while still showing novel and relevant system capabilities. A key assumption was that objects and threats in the environment are static. Another assumption was that the probability of successfully detecting a threat is 100%. These two assumptions meant that if a location was seen by a robot, it did not need to be revisited. Finally, we represented the planning problem and sensor models in 2-D using grid-based methods. These assumptions describe a highly idealised situation; in future demonstrations, these assumptions are relaxed to provide more realistic scenarios (see Sect. 16.5).

To compute the reward function $g(\mathbf{x})$, we made use of the output of the STL inference module, which was a (relatively large) collection of possible paths P the human team might take. The density of such paths in a particular location (e.g. grid cells on a map) was interpreted to be proportional to the probability that the human team would pass through that location. An example of such paths as the human team move towards the mission goal is shown in Fig. 16.3.

It can be seen at the start of the mission (see Fig. 16.3) that there was a high density of paths leading directly towards the goal and a lower density region along the road. As the mission progressed, if the human team followed the road, as shown in Fig. 16.3, the prediction would still be biased towards thinking the humans would make for the "ford" rather than continuing along the road.

Under the aforementioned assumptions and given a human team intent distribution, we can now express $g(\mathbf{x})$. With a small abuse of notation, the action-wise reward $g(\mathbf{x}_n^r)$ for a single action x_n^r is calculated using the sensor model $S(x)$, which represents the field of view of the vehicle. For each grid cell in $S(x)$ (i.e. "seen" by the vehicle's sensors), the reward per cell seen by $S(x_n^r)$ is:

- Zero if the grid cell has previously been seen by any vehicle
- One if no paths p from the human intent distribution P pass through it
- Proportional to the number of paths p otherwise

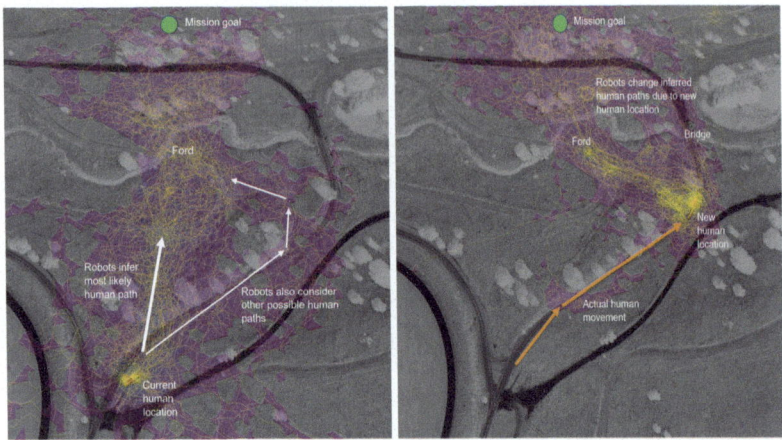

Fig. 16.3 Left: human intent prediction at mission-start. Right: evolution of human intent prediction over time

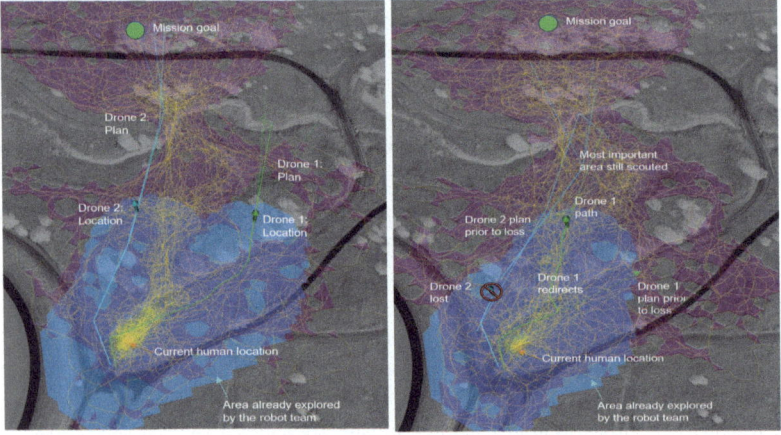

Fig. 16.4 Left: human intent prediction at mission-start. Right: evolution of human intent prediction over time

The overall behaviour encouraged by this formulation of $g(\mathbf{x})$ is the robots using their sensors to cover the area that will be traversed by the human team, prioritising the likely locations first. The decentralised nature of the system allows for graceful degradation, so that if one or more drones are removed from the team, the remainder will automatically plan to cover the terrain according to $g(\mathbf{x})$, as seen in Fig. 16.4.

This is a major simplification of useful ISR, in which it would be much more useful for the robots to search areas from which the human team can be observed rather than observing the path, and consider the regions that the human team can observe. This requires an additional line-of-sight calculation during reward function evaluation. We detail this and other improvements that are being addressed in the next phase of the project in the following subsection.

Despite some simplifications, the Year 1 demonstration is a useful proof of concept. Although the sensor modelling in the reward function could have been (and is currently being) improved, this demonstration clearly shows that Dec-MCTS can be successfully used to coordinate multiple robots to perform an ISR-like task.

16.4.4.3 Decentralised Monte Carlo Tree Search for Year 2 of Hyper-Teaming

In this section, we present the results and key findings from our second-year field trials, which extended on the capabilities of our decentralised motion planning system demonstrated in the first year.

One of the primary objectives of our second-year field trials was to assess the scalability of our decentralised motion planner. In the first year, we successfully demonstrated its effectiveness in managing a relatively small area (200 cm × 200 m). In the second year, we validated its performance when scaled up to a much larger operational area encompassing a full grid square of 1000 × 1000 m. In addition to the larger area, the terrain was also quite different, consisting of vegetation of varying densities (from tall grass to tightly packed copses of trees with occasional tracks and cuttings).

Given that we learnt a lesson from Year 1, the planner for airborne agents was also updated to operate at a height above ground, rather than in a fixed plane, allowing for variations in terrain height to be accounted for. Traversability maps were then created for each class of agent (human, ground-vehicle and drone) to account for the varying obstructions and dynamics of each class.

Given the processing power onboard the robots, the density of the graph nodes for the full grid square could not remain the same as the density for the 200 × 200 m version. Instead of merely rejection sampling uniformly random points against a traversability grid map, which was feasible in the smaller scale first-year trials, a Voronoi graph was constructed. This led to much more useful samples being created, maintaining a good performance while achieving the required decrease in density. Avenues for future work include further increasing the area to multiple grid squares and adding new members to the robot team to allow this to happen effectively.

An essential aspect of modern mission planning and execution is the integration of named areas of interest (NAI), which are designated areas to scout in advance of executing an action. In our second-year field trials, we integrated an NAI interface from Civilian Team Awareness Kit software running on a handheld tablet to our decentralised motion planner.

This integration allowed for dynamic updates and sharing of NAI data across the network, enabling mission planners and operators to mark and prioritise specific areas for observation, exploration or action. Multiple flight tests with AMSL's Vulcan unmanned aircraft system verified that the planner tasked robots to visit NAIs without the operator individually assigning robots to NAIs. This scalability is important to break the single-operator-to-robot constraint that currently limits the use of unmanned aircraft systems in real-world scenarios. Future testing required here would be to scale up the number of robots and NAIs.

To further enhance the decision-making capabilities of our decentralised motion planner, we integrated 3D ray-tracing software into the system according to NVIDIA's GPU Voxel database. This software uses a 3-D common operating picture shared by robots, providing an estimate of what would be seen in any given pose in the environment. The decentralised motion planner was interfaced with viewpoints to calculate rewards and evaluate potential paths for mission objectives. The planner was improved to plan over poses instead of merely positions, given that what a robot sees is pose dependent rather than just position dependent.

An example of such an exploratory mission is shown in Fig. 16.5. This shows the increased area of operations, the Voronoi graph-based probabilistic roadmap (PRM) around known obstructions in the form of tall trees, mapped in a previous flight by onboard LiDAR (purple), NAIs (red circles) and observed viewpoints (green dots).

This integration led to improved performance, because the planner now had an understanding of what the sensors onboard the robot could see; however, the computational cost was significant, especially given the limited computing resources onboard Vulcan. There are efficiency improvements to be made in slimming down the back end calculations of the ray-tracing software, as well as in the interface.

Finally, although this feature was not tested in the field, the planner is capable of incorporating kinematic motion constraints, such as Dubins models, for fixed-wing aircraft and kinematically constrained ground robots.

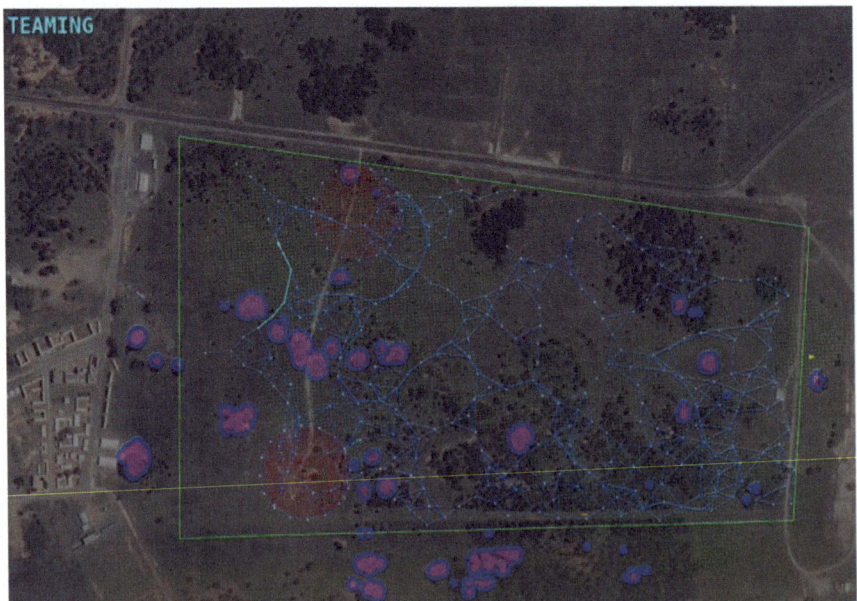

Fig. 16.5 A snapshot of Vulcan flight data from Year 2

16.5 Summary and Future Work

We described the beginnings of a framework designed to allow humans and robots to work collaboratively, with an emphasis on robots coordinating themselves around the human mission. A key learning during this process, in addition to how hard this was to achieve in practice, is that providing timely guidance to robots is just as important as receiving timely and non-disruptive feedback. Timely, further, does not always mean real time because the capabilities, endurance and processing power of these heterogeneous systems (including humans) vary considerably.

Although there has been significant progress in scaling up the area that the robot team covers, the next frontier for scalability is scaling up the number of robotic agents. This will require efficient computation from the planner, especially if the new robotic agents are lightweight quadcopters that have relatively little computing power. The lessons learnt in scaling up are critical for the future as we scale up from 10s to 100s to 1000s of robotic agents.

Finding the appropriate time to stop, think and regroup can also be of benefit, before the next high-tempo coordinated activity is implemented. In addition, computation-heavy processes are better suited to larger or ground-based robots (that have more power), which act as team leaders for smaller systems such as drones. This then allows the hyper-teaming framework to scale more readily to advanced missions through a system of systems or, more broadly, a team of teams command hierarchy.

Acknowledgments This research was funded by the Robotics and Autonomous Systems Implementation and Coordination Office through the Defence Cooperative Research Centre for Trusted Autonomous Systems.

Open Access This chapter is licensed under the terms of the Creative Commons Attribution 4.0 International License (http://creativecommons.org/licenses/by/4.0/), which permits use, sharing, adaptation, distribution and reproduction in any medium or format, as long as you give appropriate credit to the original author(s) and the source, provide a link to the Creative Commons license and indicate if changes were made.

The images or other third party material in this chapter are included in the chapter's Creative Commons license, unless indicated otherwise in a credit line to the material. If material is not included in the chapter's Creative Commons license and your intended use is not permitted by statutory regulation or exceeds the permitted use, you will need to obtain permission directly from the copyright holder.

Part V
Thinking Swarm: Way Forward

Chapter 17
Future Directions

Simon Ng

Thinking Swarms is a multidisciplinary exploration of swarming robotics. The breadth of discussion in the preceding chapters, and in particular the exploratory nature of some, makes writing any concluding chapter a challenge. It would be easy to assemble the proposed future work from each individual chapter and reproduce it here or to forge a completely separate path. This concluding chapter takes a middle ground. We synthesise recommendations and lessons from earlier chapters into a set of opportunities, but through a lens that resonates with this particular author. Without claiming to be definitive, we offer avenues for further exploration: a continual refinement of concepts to enrich the conversation in coherent ways; deeper investigation into social expectations; a greater focus on regulators as partners in the domain; expanding open world applications of our technology so they are resilient to both stochastic and epistemological uncertainty; exploiting large language models as critical "semantic" partners within a broader autonomous system; and strengthening and extending our simulation toolbox to support all of the above.

17.1 Introduction

Swarm robotics [353, 694] promises to open society to a world of possibilities, from environmental monitoring and search and rescue operations through to military capability and space exploration [152, 632]. Development of these complex systems presents unique challenges spanning multiple domains: engineering, regulation,

S. Ng (✉)
Trusted Autonomous Systems (TAS), Toowong, QLD, Australia
e-mail: simon.ng@tasdcrc.com.au

computational science, philosophy and social science. This monograph, "Thinking Swarms", assembles diverse papers that explore this multifaceted nature of swarm robotics, covering topics from conceptual philosophy to practical implementation. "Thinking Swarms" is a call to action for researchers, engineers and policymakers to engage in the interdisciplinary collaboration needed to realise the benefits of swarm robotics while mitigating potential risks and unintended consequences.

The chapters within this volume offer a perspective on the current state of swarm robotics research and insights into potential future directions. Presenting a wide range of perspectives in this monograph will, we hope, stimulate a rich dialogue among engaged communities, including the general public. The provocative title, "Thinking Swarms", reflects our goal of encouraging readers to think critically about the nature of swarm intelligence and its potential impact on our world.

It is crucial that we develop a shared understanding of the key concepts and challenges as the field evolves. We present a framework for categorising swarm constructs according to their physical appearance, behaviour, topology and cognition. These categories are not exhaustive or definitive but do provide a useful launching point for further discussion.

"Thinking Swarms" delves into the social and ethical dimensions of swarm robotics, exploring how public perception and trust in these systems can and must shape their development and deployment. The book highlights the complex and often conflicting ways in which swarms are viewed. We must work towards developing swarm robotic systems that are not only technologically advanced but also socially responsible and ethically sound.

Each chapter offers the thoughts of its authors on "where to next", and there is no value in repeating these verbatim here. We, as editors, respect each author's expertise; they are best placed to direct the next phase of their enquiries. However, particular ideas emerged as we digested, discussed, commented on and assembled this monograph, and we offer some of these here while making no claim that they are the best ideas.

We start with encouraging more work to refine the core concepts offered in the early chapters; we explore social factors governing acceptance, trust and regulation; we raise the challenge of open worlds; and, finally, we consider some emerging practices and technologies that could have profound effects on how swarm robotics will be realised beyond the laboratory, focusing on new neuro-symbolic approaches and simulation.

17.2 Refine the Concepts to Support Shared Understanding

Chapter 1 presents the results of a 1-day workshop held with professionals working in robotics and autonomous systems. Analysis of the collected ideas provides an initial insight into the conceptual landscape that the workshop participants used when thinking about robotic systems generally and swarming systems more specifically.

The concepts naturally collapse into one of four categories (physical appearance, behaviour, topology and cognition) under which personal constructs can be organised. These are not the only or even necessarily the best categories or constructs and they are, by definition, personal. The four categories can be further aggregated into form (physical appearances and behaviour; see Table 1.1) and function (topological relationships between entities and the underlying cognitive processes that form the foundation of "thinking" swarms; see Table 1.2).

The personal nature of these constructs may be construed as limitations, but this in our minds is a naive view: personal interpretation is central to better understanding of the issues related to social licence and interpretability. Considering these divergent personal constructs remains central to the practice of thinking about swarms.

Work remains to be done on these concepts. We need to better understand how they interrelate, to what extent they describe a swarm or team collectively and as individuals and the relationship between these ideas and other functional, conceptual and philosophical frameworks for talking about "thinking swarms". We need to establish how stable they are as concepts and how functional design choices shift the interpretative needle along these dipoles. Further analysis should also be done to refine and test the validity of these constructs using a broader audience. The constructs themselves could be considered for use as a more explicit framework within which to undertake sentiment analysis by allowing researchers to track sentiment along structured and grounded dimensions.

17.3 Continue to Explore Social Expectations

Drone swarms used in entertainment, no matter how orchestrated and centralised control might be, can be decidedly prosaic in comparison with a starling murmuration or a self-organising school of fish, and yet we are engrossed by them. Even so, the idea of swarming robots as things that exist for our entertainment is thoroughly normalised.

We are still learning about how drone and swarm behaviour influences public perception, but we know that inexplicable or unexpected behaviour can affect trust. Trust in human-machine interaction, defined as an expectation that the machine will act as intended [556], is a complex area of investigation that has a long history, and it has an important role in informing regulation [617]. The panic at drone shows in Chongqing, or in crowds at Mecca and Pamplona, is an acute example of what happens when swarms (natural or artificial) start behaving erratically.[1]

The "slaughterbots" campaign tapped into trust issues [826], surfacing awareness of the threat of drones and drone swarms and contributing to the movement against

---

[1] Exploring the effect of heterogenous swarms on panic in crowds would be an interesting extension to the work of [364] and others.

artificial intelligence (AI) in broader society [621] and (in part) driving concerns about how to control AI and autonomous weapons [27]. The social implications of the "slaughterbots" campaign are illustrated in extracts from public media collected by Robertson and Boshuijzen-van Burken (see Chap. 5).

Robertson and Boshuijzen-van Burken (see Chap. 5) provide insights into perception, sentiment and trust in the context of swarming robotics in their study, but the landscape remains partly mapped. If terms such as "artificial intelligence" and "autonomy" are value-laden now when we are only really understanding them as fields of technical and social endeavours, then we are at risk of throwing the baby out with the bath water if that "value-laden" perception is not informed by reality. An alternate mapping by Klarin et al. (see Chap. 2), philosophical discussions by Boshuijzen-van Burken (see Chap. 4) and the opening introduction by the editors make it clear that more work needs to be done to understand trust and its mediating factors in this field.

Longitudinal public sentiment analysis on this issue would be a good first step, followed by a deliberate attempt to engineer the right values into our designs and language. Linking the constructs from Chap. 1 to measures of sentiment in a more formal way would provide structure for evaluating sentiment. [454] used constructs to assess psychological state and psychological triggers, and a similar approach might be used here. We can imagine, for example, evaluations of systems being performed through the lens of these constructs to better understand the relationship between these qualities and positive or negative sentiment.

Mount and Beesley (see Chap. 6) propose a simulation-based scenario-driven approach to better understand how people interpret a "swarming" system and even expand the conceptual boundary of swarming by bringing to the discussion "slow swarms", which interact proximally and distally in space and time. How do we perceive of, interpret and establish trust in "slow swarms"? The work conducted by the authors in this book presents an opportunity to follow through on some of these hermeneutics experiments (see Johnson, Chap. 16; Smith, Chap. 14; and Moses et al., Chap. 12), but simulation tools may prove central to comprehensive studies of this sort, as discussed later in this chapter.

The concepts in Tables 1.1 and 1.2 could aid us in drawing links between what we "see" and how we "feel" about what we "see", lending structure to experiments intended to elucidate how design relates to trust. Similarly, they might help to refine our understanding of what observable properties play into how autonomous systems influence human behaviour. Imagine, if you will, an experiment where subjects exposed to combinations of manifestations of a robotic system consistent with the ontological concepts in Tables 1.1 and 1.2 are asked to verbally assess where the system fits within those constructs while independent measures are made of their physiological markers. Suddenly, we have a link between manifested properties that can be targeted in design and physiological references that tell us how those properties engender positive or negative responses. The concepts could also be used to support value-based design assessments, social licence studies and even our development of interventions for influencing the behaviour of humans when they are interacting with swarms of robotic systems.

17.4 Focusing on Regulatory Partnership

In the world of swarm robotics, it is important to recognise that we are navigating uncharted territory. Horne (see Chap. 9) and Guihen (see Chap. 13) make strong initial contributions to considerations of swarming robotics and regulation, but like all good researchers, they raise as many questions as they answer. The regulatory landscape for autonomous systems continues to evolve, and swarm robotics presents some unique challenges for developing effective governance frameworks and ensuring ethical application [396].

Swarm robotics systems exhibit emergent properties—behaviours that arise from the interactions between individual robots and their environment, rather than being explicitly programmed—and this makes it difficult to predict and control the behaviour of swarms or to satisfactorily claim that a swarm abides by regulations under all conditions. Layered control mechanisms and a degree of self-awareness and self-reflection may provide engineering solutions, but assigning liability and accountability when things go wrong remains critical to proper regulation [708]. Enforcing compliance with regulations and laws through an ethical black box [836] (especially in the military context where swarms may be operating in complex, dynamic and unknown environments) is made especially difficult when swarm robotics systems rely on decision and control processes often distributed and decentralised [702].

Testing and evaluation is part of the solution. But we must design swarm robotics systems from the ground up to be more amenable to regulatory and ethical oversight.

Can we develop swarm robotics systems that can reason about regulations and laws in real time, interpreting their own behaviour in the context of legal and ethical constraints, using approaches such as GenEth [50]?

Goal-driven or intent-driven reasoning is a topic of considerable interest already [33, 151], and it relies on new methods for representing abstract concepts (the intent behind laws and regulations) so that the outcomes are consistent with the underlying motives for the rules that should govern the system. Do we need ethical, legal or regulatory governors that ensure swarms operate within predefined boundaries [59]? How might this sort of reasoning be implemented within a swarm in a way that is computationally tractable and accommodating of the fundamental characteristics of swarms (distributed, amorphous and loosely bounded in time and space)? It surely is not enough to rely on a human to intervene. Minimum AI can enhance legal compliance [703], but this sort of "safety check" relies itself on epistemological assumptions that allow a system to know when something is "wrong".

Regulators and innovators must work closely together to develop new approaches to conceptualising, designing, testing, evaluation and certification of swarm robotics systems—yes, regulators need to be at the start and not simply the end. Including regulators in the world of swarm robotics requires a sustained effort by all stakeholders. We need to work together to develop new technical solutions, testing and evaluation methodologies and governance frameworks that can keep pace with the rapid advancement of AI and swarm robotics technology [616]. Close collaboration between regulators and innovators, and designing swarm robotics systems with regulation and oversight in mind from the start, will improve our chances of realising swarms that are not only technologically impressive but also legally compliant and ethically sound [60]. It is an exciting challenge, and one that will require ongoing dialogue, experimentation and iteration.

17.5 Extend Open World Applications

Complex systems are characterised by the emergence of global order from local interactions and circular causality. Emergent properties can be difficult to predict or even comprehend. According to [675], "the tolerable variation of relationships between the elements can be exceeded and thus lead to massive instabilities". Open, dynamic systems interacting over temporal and spatial scales ("slow swarms") can become something unexpected and unwanted through simple virtue of selective dynamic forces that we do not fully understand. This is a semi-formal definition that might be applied to swarms, but it is not necessarily what people understand when they hear the term.

The Israeli Defence Force's "Lavender" decision-support system combines large time-scale training protocols with medium time-scale policy setting and short time-scale target prioritisation within an open world of machine classifiers interacting with priorities and pressures set by military commanders and shifting social dynamics within the affected population, with profound implications for reliability [416]. The notion of a "slow swarm" is not often explored, but it remains critical because a "slow swarm" is almost invariably functioning within a non-stationary/nonergodic open world. How do we hedge against this, enabling machines to recognise when they are out of bounds or even test bound assumptions deliberately through active hypothesis generation?

Open worlds present a range of fundamental problems for robotics. Approaches for dealing with open world problems include category theory [230] and methods for out-of-distribution detection (see a review by Yang et al. [847]). Work in partially observable Markov decision processes provide another avenue for dealing with epistemological *and* stochastic uncertainty, but a robust general solution remains elusive.

Back's (see Chap. 7) computationally light SLAIT-AI algorithm appears to have the capacity to detect anomalies in sequential data with no prior exposure. It is being investigated for use in isolating abnormal speech in a stream of text, detecting previously unobserved shifts in sentiment in an ongoing dialogue and even unexpected or novel motions among ensembles. Could it offer a real-time sensor for potential "unexpected or previously unobserved" situations? Combined with machine learning, perhaps it could: reveal shifts in the dynamics of large-scale robotic swarms too subtle for humans to detect in near real time and without significant data or computational pipeline overheads; provide a formalism for discovering a behavioural "language" for swarms; and provide a technique for a swarm to reflect on itself, offering it a means to understand whether its interplay with the environment has resulted in unexpected patterns in its own behaviour.

Applying SLAIT-AI to a system such as Lavender could help to determine when the pattern of life is shifting from the a priori statistics in the training set, introducing a safeguard into the system. This metacognitive process is central to allowing machine systems to reason about themselves. Cardier et al. (see Chap. 8) concentrated on the value of a system, such as SLAIT-AI, for building synthetic languages that can describe "semantic" narrative elements by analogy (or likeness) without having a priori knowledge of what these elements might be. This evolving-model approach is a particularly attractive line of enquiry and should be pursued further.

Smith (see Chap. 14) explores this in a complimentary way by extending prior work in the field of lifelong machine learning, capitalising on robotic hierarchical graph neurons (R-HGN) and behaviour creation algorithms to allow swarming systems to learn by combining experiences across the ensemble and then selecting for new behaviours or new combinations of existing behaviours to adapt. The application to network management provides a good controlled test case, and extending this to other salient problems presents a significant opportunity. Coupling this with a perceptual anomaly technique such as SLAIT-AI, we may have a means for not only learning from experience but also seeing and codifying unexpected new experiences.

17.6 Consider Integration of Swarm Robotics with Large Language Models

Large language models (LLMs) built on transformer networks have upended the conventional view of what AI is and what it might be capable of doing [399, 764]. This monograph was started well before the LLM revolution, and consequently has little to say on these types of models, but that opens a new possibility for future work.

LLMs are transforming the way we interact with machines, and a number of commercial LLMs are now available free or via subscription on mobile and desktop devices. These models are being extended to support molecular modelling for drug discovery, chemistry research [25, 133] and a range of other industries. More recently, transformer models that combine language and motion into their architecture and interface with the external world by exploiting visual question answering systems are showing promise in solving heretofore intractable problems in robot real-world interaction [154, 867].

However, LLMs are not without limits [103]. The training data and computational power required to build the largest of these models are substantial, and the most sophisticated models, such as GPT4 and Opus3, use large collections of publicly available text and video. They are increasingly effective at performing well against benchmark tests of intelligence, reasoning and even "theory of mind" [261]. However, they continue to be inscrutable and irrational (although considerable effort is being expended to overcome the explanatory gap, see [780]) and they are prone to hallucination [589] and abuse [777], with significant implications for scientific veracity and commercial and civil applications. Despite these shortcomings, it is not unreasonable to think that they can be exploited to enhance swarming robotics practices as well, either by offering alternative approaches to solving issues related to control, perception and coordination or by complementing existing formal methods, such as those discussed in this book.

One area of potential research might be to borrow from robot-transformer models the idea of encoding within the network not only semantic information (language tokens) but also kinetic information (pose-related tokens) and even to consider the extent to which behaviours (or behavioural elements) might be tokenised so that sequences of discrete behaviours could be interpreted and appropriate responses enacted.

Another avenue is using LLMs to extend and enhance current approaches, including agent-based systems (see, e.g. [397] and [166]) and formal reasoning systems (see, e.g. [792]), the wider semantic Internet, human users, other large language models and even perceptual systems. How this "neuro-symbolic" architecture can be applied in the world of distributed, amorphous, heterogeneous and dynamic systems is an open question, but worth exploring. In fact, recent announcements [583] suggest that the next breakthrough in large-scale cooperative AI is through this means of coupling agency with human-like semantic language engines.

17.7 Developing a Simulation and Analysis Toolbox

Simulation plays a crucial role in the development and testing of swarm robotic systems, allowing researchers and engineers to explore various designs and control strategies and scenarios in a fast, safe and cost-effective manner before deploying physical robots [145]. However, it also provides an important mechanism for sharing knowledge and advances, for testing social factors governing the perception and employment of these systems and even for extending pedagogical practice (see, e.g. Chap. 15). For swarming robotics in particular, it allows us to explore scale in ways that would prove challenging in the real world.

To realise the potential of simulation for swarm robotics, a number of enhancements are needed. Simulation environments must, of course, provide appropriately abstracted models of the physical and social worlds [260]. The tools should also support large-scale swarms that have hundreds or thousands of individual robots [168], requiring efficient algorithms that can simulate distributed and potentially highly asynchronous interactions and processes. Scalability will be essential.

Researchers need to set up experiments, visualise the results and iterate designs, and the environment must support this workflow. Standard approaches to interfacing with tools for specifying robot behaviours, defining experimental metrics and supporting analysis require new pedagogical methods, new forms of data and so on. User-friendly simulation tools will lower the barrier to entry for swarm robotics research and enable more rapid prototyping and experimentation and increase the robustness of developmental outcomes.

Greater standardisation and interoperability across the swarm robotics community and an open approach to the deployment of this infrastructure will be critical. This includes common frameworks for describing robot capabilities, environments and tasks, as well as shared libraries of reusable components and modules. Standardisation efforts should also address the challenge of repeatability in swarm robotics research by providing guidelines for documenting and sharing simulation setups and results and allowing more effective bootstrapping across development groups.

Another area in which simulation can be used is in the testing and validation of swarm robotics systems. By subjecting simulated swarms to a wide range of scenarios and edge cases, researchers can identify potential failure modes and vulnerabilities before deploying physical robots [290]. Simulation-based testing can also be used to evaluate the compliance of swarm algorithms with regulatory requirements and ethical principles, such as safety constraints or fairness criteria.

Finally, simulation tools must be integrated in a scalable way with real-world test beds and software and hardware platforms to enable seamless transfer of swarm algorithms from virtual to physical environments. This will support calibrating simulation models according to real-world data and adapting control strategies to account for discrepancies between simulated and actual robot behaviours [410].

Developing enhanced simulation tools and promoting standardisation and best practices, the swarm robotics community can accelerate progress towards more capable, reliable and beneficial swarm systems. A comprehensive simulation and analysis toolbox will enable researchers to explore new frontiers in swarm intelligence while ensuring the safety and trustworthiness of the technology.

17.8 Conclusion

This monograph explores the diverse landscape of swarm robotics, from theoretical foundations to practical applications and societal implications, and bringing together experts from various fields has helped to highlight the interdisciplinary nature of swarm robotics research and the need for collaboration across domains to address the complex challenges involved. Key themes have emerged.

Refining concepts and terminology should enhance effective communication among researchers, engineers and policymakers. A coherent framework for categorising swarm constructs according to their physical appearance, behaviour, topology and cognition, as well as exploring how these categories relate to other conceptual and philosophical constructs, has been started here but remains preliminary.

The social and ethical dimensions of swarm robotics need ongoing attention, despite the significant work done in this area to date, because swarm robotics potentially add further considerations. Public perceptions of and trust in AI and robotics are not assured, and swarm robotics presents unique properties that can affect acceptance and adoption, making it crucial that we continue to explore how people view swarms, both in nature and in technological contexts, and work towards designing systems that are not only effective but also transparent, explainable and aligned with societal values.

The integration of swarm robotics into real-world applications will require close collaboration between researchers, engineers and policymakers to develop appropriate regulatory frameworks and governance structures. This includes establishing standards for testing and evaluation, defining liability and accountability measures and ensuring that swarm systems operate within legal and ethical boundaries. Simulation tools and test beds will play a critical role in this process, enabling researchers to explore the behaviour of swarms in realistic scenarios and identify potential risks and unintended consequences.

Rapid advancements in AI and machine learning, particularly in the realm of neuro-symbolic methods, present both opportunities and challenges for swarm robotics. Although these technologies have the potential to enhance the capabilities of swarm systems, supporting natural language communication and reasoning about abstract concepts, they also raise concerns about transparency, accountability and the alignment of AI systems with human values. It is essential to explore how these emerging technologies can be responsibly integrated into swarm robotics while addressing the fundamental challenges of swarm coordination, control and resilience in dynamic environments.

We hope that this monograph provides a valuable contribution to the field of swarm robotics by bringing together diverse perspectives and highlighting the key challenges and opportunities that lie ahead. A commitment to interdisciplinary collaboration, public engagement and responsible innovation is central to unlocking the full potential of this technology for society.

Acknowledgments The author acknowledges the use of Claude 3.0 as a preliminary drafting tool in the introductory and concluding portions of this chapter.

Open Access This chapter is licensed under the terms of the Creative Commons Attribution 4.0 International License (http://creativecommons.org/licenses/by/4.0/), which permits use, sharing, adaptation, distribution and reproduction in any medium or format, as long as you give appropriate credit to the original author(s) and the source, provide a link to the Creative Commons license and indicate if changes were made.

The images or other third party material in this chapter are included in the chapter's Creative Commons license, unless indicated otherwise in a credit line to the material. If material is not included in the chapter's Creative Commons license and your intended use is not permitted by statutory regulation or exceeds the permitted use, you will need to obtain permission directly from the copyright holder.

References

1. https://ardupilot.org. Accessed: 28th November 2023
2. https://oceanai.mit.edu/moos-ivp/pmwiki/pmwiki.php?n=Main.HomePage. Accessed: 28th November 2023
3. https://www.ros.org/blog/community/. Accessed: 28th November 2023
4. https://oceaninfinity.com/. Accessed: 28th November 2023
5. https://oceanai.mit.edu/swarmbox/pdfs/memo_swarm_overview.pdf. Accessed: 28th November 2023
6. http://wiki.ros.org/trex_executive. Accessed: 28th November 2023
7. https://kcl-planning.github.io/ROSPlan/. Accessed: 28th November 2023
8. https://www.nasa.gov/directorates/heo/scan/engineering/technology/technology_readiness_level/. Accessed: 28th November 2023
9. http://mavlink.io/en/. Accessed: 28th November 2023
10. https://bluerobotics.com/about/. Accessed: 28th November 2023
11. https://www.kickstarter.com/projects/847478159/913072768. Accessed: 28th November 2023
12. https://xkcd.com/927/. Accessed: 28th November 2023
13. AMC Search Educat. https://www.amcsearch.com.au/amsl. Accessed: 28th November 2023
14. Inspector-general of the Australian defence force. Afghanistan inquiry report 103. https://afghanistaninquiry.defence.gov.au/
15. Protocol Additional to the Geneva Conventions of 12 August 1949, and relating to the Protection of Victims of International Armed Conflicts. Tech. Rep. Article 36 (Protocol I) (1977)
16. OpenAI Five Defeats Dota 2 World Champions. https://openai.com/blog/openai-five-defeats-dota-2-world-champions/ (2019)
17. Australian code of practice for the design, construction, survey and operation of autonomous and remotely operated vessels. Trusted Autonomous Systems. https://www.rasgateway.com.au/resource-hub/australian-code-of-practice-for-the-design-construction-survey-and-operation-of-autonomous-remotely-operated-vessels (2022)
18. Abbass, H., Harvey, J., Yaxley, K.: Lifelong testing of smart autonomous systems by shepherding a swarm of watchdog artificial intelligence agents. ArXiv Preprint p. 1812.08960 (2018). http://arxiv.org/abs/1812.08960
19. Abbass, H., Petraki, E., Hussein, A., McCall, F., Elsawah, S.: A model of symbiomemesis: machine education and communication as pillars for human-autonomy symbiosis. Phil. Trans. R. Soc. A. **379**(2207), 20200,364 (2021). https://doi.org/10.1098/rsta.2020.0364

20. Abbass, H., Petraki, E., Hussein, A., McCall, F., Elsawah, S.: A model of symbiomemesis: machine education and communication as pillars for human-autonomy symbiosis. Philosophical Transactions of the Royal Society A **379**(2207), 20200,364 (2021)
21. Abbass, H.A.: MBO: Marriage in honey bees optimization-a haplometrosis polygynous swarming approach. pp. 207–214 (2001)
22. Abbass, H.A., Elsawah, S., Petraki, E., Hunjet, R.: Machine Education: Designing semantically ordered and ontologically guided modular neural networks, pp. 948–955. IEEE (2019). https://doi.org/10.1109/SSCI44817.2019.9003083
23. Abdelli, A., Amamra, A., Yachir, A.: Swarm robotics: A survey. In: International conference on computing systems and applications, pp. 153–164. Springer (2022)
24. Abe, K., Watanabe, D.: Songbirds possess the spontaneous ability to discriminate syntactic rules. Nature neuroscience **14**(8), 1067–1074 (2011)
25. Abramson, J., Adler, J., Dunger, J., Evans, R., Green, T., Pritzel, A., Ronneberger, O., Willmore, L., Ballard, A.J., Bambrick, J., Bodenstein, S.W., Evans, D.A., Hung, C.C., O'Neill, M., Reiman, D., Tunyasuvunakool, K., Wu, Z., Žemgulytė, A., Arvaniti, E., Beattie, C., Bertolli, O., Bridgland, A., Cherepanov, A., Congreve, M., Cowen-Rivers, A.I., Cowie, A., Figurnov, M., Fuchs, F.B., Gladman, H., Jain, R., Khan, Y.A., Low, C.M.R., Perlin, K., Potapenko, A., Savy, P., Singh, S., Stecula, A., Thillaisundaram, A., Tong, C., Yakneen, S., Zhong, E.D., Zielinski, M., Žídek, A., Bapst, V., Kohli, P., Jaderberg, M., Hassabis, D., Jumper, J.M.: Accurate structure prediction of biomolecular interactions with alphafold 3. Nature **630**(8016), 493–500 (2024). https://doi.org/10.1038/s41586-024-07487-w
26. Acconcjaioco, M., Ntalampiras, S.: One-shot learning for acoustic identification of bird species in non-stationary environments. In: 2020 25th International Conference on Pattern Recognition (ICPR), pp. 755–762. IEEE (2021)
27. Adam, D.: Lethal AI weapons are here: how can we control them?—nature.com. https://www.nature.com/articles/d41586-024-01029-0 (2024). [Accessed 6-06-2024]
28. Adams, T.K.: Future warfare and the decline of human decisionmaking. The US Army War College Quarterly: Parameters **41**(4), 1 (2011)
29. Adelman, R., Kieran, D.: Remote warfare: new cultures of violence. University of Minnesota Press
30. Adhikari, A., Wenink, E., van der Waa, J., Bouter, C., Tolios, I., Raaijmakers, S.: Towards fair explainable AI: a standardized ontology for mapping XAI solutions to use cases, explanations, and AI systems. In: Proceedings of the 15th International Conference on PErvasive Technologies Related to Assistive Environments, pp. 562–568 (2022)
31. Agarwal, S., Tripathy, T., Borkar, A., Sinha, A.: Relative heading based pattern generation. IFAC-PapersOnLine **50**(1), 12,759–12,764 (2017)
32. Agency, E.S.: Navipedia. https://gssc.esa.int/navipedia/index.php/Main_Page. Accessed: 28th November 2023
33. Aha, D.W.: Goal reasoning: Foundations, emerging applications, and prospects. AI Magazine **39**(2), 3–24 (2018). https://doi.org/10.1609/aimag.v39i2.2800. https://onlinelibrary.wiley.com/doi/abs/10.1609/aimag.v39i2.2800
34. Ahmad, M.: The use of drones in Pakistan: An inquiry into the ethical and legal issues. Political Quarterly **85**(1), 65–74 (2014). https://doi.org/10.1111/j.1467-923X.2014.12068.x
35. Ahmi, A., Elbardan, H., Raja Mohd Ali, R.H.: Bibliometric analysis of published literature on industry 4.0. In: ICEIC 2019—International Conference on Electronics, Information, and Communication, pp. 1–6. https://doi.org/10.23919/ELINFOCOM.2019.8706445 (2019)
36. Al-Jumeily, D., Hussain, A., Mallucci, C., Oliver, C.: Assured Decision and Meta-Governance for Mobile Medical Support Systems. In: Applied Computing in Medicine and Health, pp. 166–182. Morgan Kaufman Publishers, Cambridge Massachusetts (2015). P 168
37. Albiero, D., Garcia, A.P., Umezu, C.K., de Paulo, R.L.: Swarm robots in mechanized agricultural operations: a review about challenges for research. Computers and Electronics in Agriculture **193**, 106,608 (2022)

38. Alboaie, L., Alboaie, S., Panu, A.: Swarm communication-a messaging pattern proposal for dynamic scalability in cloud. In: 2013 IEEE 10th International Conference on High Performance Computing and Communications & 2013 IEEE International Conference on Embedded and Ubiquitous Computing, pp. 1930–1937. IEEE (2013)
39. Alhudhaif, A., Saeed, A., Imran, T., Kamran, M., Alghamdi, A.S., Aseeri, A.O., Alsubai, S.: A particle swarm optimization based deep learning model for vehicle classification. Computer Systems Science and Engineering **40**(1), 223–235 (2022). https://doi.org/10.32604/CSSE.2022.018430
40. Alkinani, M.H., Almazroi, A.A., Adhikari, M., Menon, V.G.: Design and analysis of logistic agent-based swarm-neural network for intelligent transportation system. Alexandria Engineering Journal **61**(10), 8325–8334 (2022). https://doi.org/10.1016/j.aej.2022.01.046
41. Allen, C.H.: The seabots are coming here: Should they be treated as 'vessels'? The Journal of Navigation **65**(4), 749–752 (2012)
42. Allen, G.: Understanding AI technology. Tech. rep., Joint Artificial Intelligence Center (JAIC) The Pentagon United States (2020)
43. Altshuler, Y., Pentland, A., Bruckstein, A.: In: Swarms and Network Intelligence in Search. Springer
44. Amari, S.I.: Information Geometry and Its Applications, *Applied Mathematical Sciences*, vol. 194. Springer Japan, New York (2016)
45. Amigó, J., Szczepanski, J., Wajnryb, E., Sanchez-Vives, M.: Estimating the entropy rate of spike trains via Lempel-Ziv complexity. Neural computation **16**, 717–36 (2004)
46. Amjady, N., Keynia, F., Zareipour, H.: Wind power prediction by a new forecast engine composed of modified hybrid neural network and enhanced particle swarm optimization. IEEE transactions on sustainable energy **2**(3), 265–276 (2011)
47. Andersen, J.: Understanding and interpreting algorithms: toward a hermeneutics of algorithms. Media, Culture & Society **42**(7–8), 1479–1494. https://doi.org/10.1177/0163443720919373
48. Anderson, A.J., Lalor, E.C., Lin, F., Binder, J.R., Fernandino, L., Humphries, C.J., Conant, L.L., Raizada, R.D., Grimm, S., Wang, X.: Multiple regions of a cortical network commonly encode the meaning of words in multiple grammatical positions of read sentences. Cerebral cortex **29**(6), 2396–2411 (2019)
49. Anderson, J.R.: How Can the Human Mind Occur in the Physical Universe? Oxford University Press, New York (2007)
50. Anderson, M., Anderson, S.L.: GenEth: a general ethical dilemma analyzer. Paladyn, Journal of Behavioral Robotics **9**(1), 337–357 (2018). https://doi.org/10.1515/pjbr-2018-0024
51. Anderson, P.W.: More is different. Science **177**(4047), 393–396 (1972). https://doi.org/10.1126/science.177.4047.393. https://www.science.org/doi/abs/10.1126/science.177.4047.393
52. Andras, P.: Social learning, environmental adversity and the evolution of cooperation. Artificial Life Conference Proceedings **ALIFE 2016, the Fifteenth International Conference on the Synthesis and Simulation of Living Systems**(28), 290–297 (2016). https://doi.org/10.1162/978-0-262-33936-0-ch051. https://www.mitpressjournals.org/doi/abs/10.1162/978-0-262-33936-0-ch051
53. Ang, J.C., Mirzal, A., Haron, H., Hamed, H.N.A.: Supervised, unsupervised, and semi-supervised feature selection: a review on gene selection. IEEE/ACM transactions on computational biology and bioinformatics **13**(5), 971–989 (2015)
54. Angelis, E.: Images of the invisible war: An interview with Trevor Paglen. The RUSI journal **160**(6), 78–83. https://doi.org/10.1080/03071847.2015.1124627
55. Ansart, A., Juang, J.C.: Generalized cyclic pursuit: A model-reference adaptive control approach. In: 2020 5th International Conference on Control and Robotics Engineering (ICCRE), pp. 89–94. IEEE (2020)
56. Ansart, A., Juang, J.C.: Generalized cyclic pursuit: An estimator-based model-reference adaptive control approach. In: 2020 28th Mediterranean Conference on Control and Automation (MED), pp. 598–604. IEEE (2020)

57. Ansart, A., Juang, J.C.: Integral sliding control approach for generalized cyclic pursuit formation maintenance. Electronics **10**(10), 1217 (2021)
58. Apostolidis, G.K., Hadjileontiadis, L.J.: Swarm decomposition: A novel signal analysis using swarm intelligence. Signal Processing **132**, 40–50 (2017)
59. Arkin, R.: Governing lethal behavior in autonomous robots. Chapman and Hall/CRC (2009)
60. Arkin, R.C.: Ethics and autonomous systems: Perils and promises [point of view]. Proceedings of the IEEE **104**(10), 1779–1781 (2016)
61. Arnold, R., Carey, K., Abruzzo, B., Korpela, C.: What is a robot swarm: a definition for swarming robotics. In: 2019 IEEE 10th annual ubiquitous computing, electronics & mobile communication conference (uemcon), pp. 0074–0081. IEEE (2019)
62. Arnon, I., Snider, N.: More than words: Frequency effects for multi-word phrases. Journal of memory and language **62**(1), 67–82 (2010)
63. Arriazu Muñoz, R.: A research methodology in the service of critical thinking: Hermeneutic approach in the post-truth era. Education policy analysis archives **26**(148)
64. Article36: Structuring Debate on Autonomous Weapons Systems. In: Memorandum for delegates to the Convention on Certain Conventional Weapons (CCW). Article36.org, Geneva (2022). https://article36.org/wp-content/uploads/2020/12/Autonomous-weapons-memo-for-CCW.pdf
65. Arvidsson, M.: The swarm that we already are: artificially intelligent (AI) swarming 'insect drones', targeting and international humanitarian law in a posthuman ecology. Journal of human rights and the environment **11**(1), 114–137. https://doi.org/10.4337/jhre.2020.01.05
66. Asia Pacific Security Magazine: China demonstrates swarm technology capabilities. https://www.asiapacificsecuritymagazine.com/china-demonstrates-swarm-technology-capabilities/ (2021). Accessed: 30 May 2022
67. Associated Press: Buzzing, silence, then an explosion: How killer drones swarm skies over Ukraine. The Sydney Morning Herald, https://www.smh.com.au/world/europe/swarms-of-killer-drones-buzz-skies-over-ukraine-20221018-p5bqk5.html (2022). Accessed: 18 October 2022
68. Atınç, G.M., Stipanović, D.M., Voulgaris, P.G., Karkoub, M.: Collision-free trajectory tracking while preserving connectivity in unicycle multi-agent systems. In: 2013 American control conference, pp. 5392–5397. IEEE (2013)
69. Austen, J.: Pride and Prejudice. T. Egerton, Whitehall (1813)
70. Australia: Australia's system of control and applications for autonomous weapon systems
71. Australian Army Research Centre: Looking to 2040: The swarm advantage. https://researchcentre.army.gov.au/library/land-power-forum/looking-2040-swarm-advantage (2021). Accessed: 26 March 2021
72. Australian Maritime Safety Authority: Domestic commercial vessel requirements. https://www.amsa.gov.au/vessels-operators/domestic-commercial-vessels/domestic-commercial-vessel-requirements (2020). Accessed: 30 May 2022
73. Australian Maritime Safety Authority: Flag state administration. https://www.amsa.gov.au/vessels-operators/flag-state-administration (2020). Accessed: 20 Oct 2022
74. Australian Maritime Safety Authority: What is port state control. https://www.amsa.gov.au/vessels-operators/port-state-control/what-port-state-control (2020). Accessed: 20 Oct 2022
75. Babić, A., Mandić, F., Mišković, N.: Development of visual servoing-based autonomous docking capabilities in a heterogeneous swarm of marine robots. Applied Sciences **10**(20), 7124 (2020)
76. Back, A., Wiles, J.: Determining the number of samples required to estimate entropy in natural sequences. IEEE Trans. on Information Theory **65**(7, July), 4345–4352 (2019)
77. Back, A., Wiles, J.: Transitive entropy—a rank ordered approach for natural sequences. IEEE Journal of Selected Topics in Signal Processing **14**(2), 312–321 (2020)
78. Back, A., Wiles, J.: Entropy estimation using a linguistic Zipf-Mandelbrot-Li model for natural sequences. Entropy **23**(9), 1100 (2021)
79. Back, A., Wiles, J.: An Information Theoretic Approach to Symbolic Learning in Synthetic Languages. Entropy **24**(2), 259 (2022)

80. Back, A.D., Angus, D., Wiles, J.: Determining the number of samples required to estimate entropy in natural sequences. IEEE Trans. on Information Theory **65**(7), 4345–4352 (2019)
81. Back, A.D., Angus, D., Wiles, J.: Transitive entropy—a rank ordered approach for natural sequences. IEEE Journal of Selected Topics in Signal Processing **14**(2), 312–321 (2020)
82. Back, A.D., Cardier, B.: Predicting complex agent outcomes based on narrative swarm models. In: S. Ng, J. Scholz, H.A. Abbass (eds.) Thinking Swarms. Springer (2025)
83. Back, A.D., Trappenberg, T.P.: Selecting inputs for modeling using normalized higher order statistics and independent component analysis. IEEE Transactions on Neural Networks **12**(3), 612–617 (2001)
84. Back, A.D., Weigend, A.S.: A first application of independent component analysis to extracting structure from stock returns. International journal of neural systems **8**(04), 473–484 (1997)
85. Back, A.D., Wiles, J.: Entropy estimation using a linguistic Zipf-Mandelbrot-Li model for natural sequences. Entropy **23**(9) (2021). 1100
86. Back, A.D., Wiles, J.: Estimating sentence-like structure in synthetic languages using information topology. Entropy **24**(7), 859 (2022)
87. Back, A.D., Wiles, J.: An information theoretic approach to symbolic learning in synthetic languages. Entropy **24**(2) (2022). 259
88. Bäck, T., Schwefel, H.P.: An overview of evolutionary algorithms for parameter optimization. Evolutionary computation **1**(1), 1–23 (1993)
89. Backstrom, A., Henderson, I.: New capabilities in warfare: an overview of contemporary technological developments and the associated legal and engineering issues in article 36 weapons reviews. International Review of the Red Cross **94**(886), 483–514 (2012)
90. Bagares, R.R.: Enkapsis and the development of customary international law: An encyclopedic approach to inter-legality (2019)
91. Bagares, R.R.: Enkapsis and the development of customary international law: An encyclopedic approach to inter-legality. The Theory, Practice, and Interpretation of Customary International Law (2022). https://doi.org/10.2139/ssrn.3506461. https://www.ssrn.com/abstract=3506461. [Online; accessed 2023-09-06]
92. Baig, Z.A., Szewczyk, P., Valli, C., Rabadia, P., Hannay, P., Chernyshev, M., Johnstone, M., Kerai, P., Ibrahim, A., Sansurooah, K., Syed, N., Peacock, M.: Future challenges for smart cities: Cyber-security and digital forensics. Digital Investigation **22**, 3–13 (2017). https://doi.org/10.1016/j.diin.2017.06.015
93. Baldazo, D., Parras, J., Zazo, S.: Decentralized multi-agent deep reinforcement learning in swarms of drones for flood monitoring. In: 2019 27th European Signal Processing Conference (EUSIPCO), pp. 1–5 (2019). https://doi.org/10.23919/EUSIPCO.2019.8903067
94. Baldwin, R., Cave, M., Lodge, M.: Understanding regulation: theory, strategy, and practice. Oxford university press (2011)
95. Ballerini, M., Cabibbo, N., Candelier, R., Cavagna, A., Cisbani, E., Giardina, I., Lecomte, V., Orlandi, A., Parisi, G., Procaccini, A., Viale, M., Zdravkovic, V.: Interaction ruling animal collective behavior depends on topological rather than metric distance: Evidence from a field study. Proceedings of the National Academy of Sciences **105**(4), 1232–1237 (2008). https://doi.org/10.1073/pnas.0711437105. https://www.pnas.org/doi/abs/10.1073/pnas.0711437105
96. Bansal, J.C., Sharma, H., Jadon, S.S., Clerc, M.: Spider monkey optimization algorithm for numerical optimization. Memetic computing **6**, 31–47 (2014). Publisher: Springer
97. Bari, A., Zhao, R., Swetha Pothineni, J., Saravanan, D.: Swarm Intelligence Algorithms and Applications: An Experimental Survey. In: Y. Tan, Y. Shi, W. Luo (eds.) Advances in Swarm Intelligence, vol. 1, pp. 3–17. Springer, Shenzhen, China (2023)
98. Barthes, R.: The Pleasure of the Text. Hill and Wang, New York (1975)
99. Basden, A.: Enkapsis (2003). https://www.dooy.info/enkapsis.html
100. Batycka, D.: Bot battle in the sky? a saboteur unleashed enemy drones to attack studio drift's latest aerial drone ballet in Hamburg. https://news.artnet.com/art-world/studio-drift-drone-attack-2108561. Retrieved from

101. BBC: Mh370: Four-year hunt ends after private search is completed. https://www.bbc.com/news/world-asia-44285241. Accessed: 28th November 2023
102. Beesley, D., Mount, G.: Digital ethnography redux: Interpreting drone cultures and microtargeting in an era of digital transformation. In: 4th International Conference on Advanced Research Methods and Analytics (CARMA 2022. https://doi.org/10.4995/CARMA2022.2022.15083
103. Bender, E.M., Gebru, T., McMillan-Major, A., Shmitchell, S.: On the dangers of stochastic parrots: Can language models be too big? In: Proceedings of the 2021 ACM Conference on Fairness, Accountability, and Transparency, FAccT '21, p. 610–623. Association for Computing Machinery, New York, NY, USA (2021). https://doi.org/10.1145/3442188.3445922
104. Bengio, Y., Ducharme, R., Vincent, P.: A neural probabilistic language model. Advances in neural information processing systems **13** (2000)
105. Beni, G.: From swarm intelligence to swarm robotics. Proceedings of the 2004 international conference on Swarm Robotics (2020)
106. Beni, G., Meyers, R.A.: Swarm intelligence. In: R.A. Meyers (ed.) Encyclopedia of Complexity and Systems Science, pp. 8869–8888. New York, NY (2009). https://doi.org/10.1007/978-0-387-30440-3_530
107. Beni, G., Wang, J.: Swarm intelligence. In: Proc. 7th Ann. Meeting of the Robotics Society of Japan, pp. 425–428. RSJ Press, Japan (1989)
108. Beni, G., Wang, J.: Swarm intelligence in cellular robotic systems. In: Robots and biological systems: towards a new bionics?, pp. 703–712 (1993)
109. Beni, G., Şahin, E., Spears, W.M.: From swarm intelligence to swarm robotics. Lecture Notes in Computer Science, pp. 1–9. Berlin, Heidelberg (2005). https://doi.org/10.1007/978-3-540-30552-1_1
110. Benjamin, M.: Drone warfare: killing by remote control. OR Books, New York
111. Benjamin M. R., S.H.N.P.M..L.J.J.: Nested autonomy for unmanned marine vehicles with moos-ivp. Journal of Field Robotics **27**(6), 834–875 (2010)
112. Berlinger, F., Wulkop, P., Nagpal, R.: Self-organized evasive fountain maneuvers with a bioinspired underwater robot collective. In: IEEE International Conference on Robotics and Automation (ICRA), pp. 9204–9211. IEEE (2021)
113. Bernus, P., Martin, R., Noran, O., Molina, A.: IFIP WG5. 12 architectures for enterprise integration: Twenty-five years of the GERAM framework. In: M. Goedicke (ed.) Advancing Research in Information and Communication Technology AICT 600, pp. 245–268. Springer, Cham (2021)
114. Bernus, P., Noran, O.: Static vs dynamic architecture of aware cyber physical systems of systems. In: L. Pufahl, D. Karastoyanova, A. Gill (eds.) Proc. IEEE 25th International Enterprise Distributed Object Computing Workshop (EDOCW), pp. 186–193. IEEE (2021)
115. von Bertalanffy, L.: General System Theory. George Braziller (1968)
116. Bertalanffy, L.v.: General system theory: Foundations, development, applications. G. Braziller, New York (1968). https://systemika.g-i.cz/record/1570/files/Bertalanffy,%20Ludwig%20von%20-%20The%20Meaning%20of%20General%20System%20Theory.pdf
117. Besnier, V., Bursuc, A., Picard, D., Briot, A.: Triggering failures: Out-of-distribution detection by learning from local adversarial attacks in semantic segmentation. In: Proceedings of the IEEE/CVF International Conference on Computer Vision, pp. 15,701–15,710 (2021)
118. Best, G., Cliff, O.M., Patten, T., Mettu, R.R., Fitch, R.: Dec-MCTS: Decentralized planning for multi-robot active perception. The International Journal of Robotics Research **38**(2–3), 316–37 (2019)
119. Best, G., Cliff, O.M., Patten, T., Mettu, R.R., Fitch, R.: Decentralised monte carlo tree search for active perception. In: Algorithmic Foundations of Robotics XII: Proceedings of the Twelfth Workshop on the Algorithmic Foundations of Robotics, pp. 864–879. Springer (2020)

120. Bhalerao, S.V., Pachori, R.B.: Sparse spectrum based swarm decomposition for robust nonstationary signal analysis with application to sleep apnea detection from EEG. Biomedical Signal Processing and Control **77**, 103,792 (2022)
121. Bharti, V., Biswas, B., Shukla, K.K.: Swarm intelligence for deep learning: concepts, challenges and recent trends. In: B. Biswas, C. Kalayci, S. Mirjalili (eds.) Advances in Swarm Intelligence: Variations and Adaptations for Optimization Problems, pp. 37–58. Springer (2023)
122. Bhatia, S., Richie, R., Zou, W.: Distributed semantic representations for modeling human judgment. Current Opinion in Behavioral Sciences **29**, 31–36 (2019)
123. Bigo, D., Isin, E., Ruppert, E. (eds.): Data politics: worlds, subjects. Routledge, rights
124. Bimber, B., Zúñiga, H.: The unedited public sphere. New media & society **22**(4), 700–715. https://doi.org/10.1177/1461444819893980
125. Binder, J.R., Desai, R.H.: The neurobiology of semantic memory. Trends in cognitive sciences **15**(11), 527–536 (2011)
126. Binder, J.R., Desai, R.H., Graves, W.W., Conant, L.L.: Where is the semantic system? a critical review and meta-analysis of 120 functional neuroimaging studies. Cerebral cortex **19**(12), 2767–2796 (2009)
127. Binney, R.J., Ramsey, R.: Social semantics: The role of conceptual knowledge and cognitive control in a neurobiological model of the social brain. Neuroscience & Biobehavioral Reviews **112**, 28–38 (2020)
128. Bleakney, W., Hipple Jr, J.A.: A new mass spectrometer with improved focusing properties. Physical Review **53**(7), 521 (1938)
129. Blum, C., Li, X.: Swarm intelligence in optimization. In: Swarm intelligence, pp. 43–85. Springer (2008)
130. Bodinger-DeUriarte, C. (ed.): Interfacing ourselves: living in the digital age. Routledge, New York
131. Boden, M.: The creative mind: Myths and mechanisms. New York: HarperCollins Publishers (1990)
132. Bogost, I.: Persuasive games: the expressive power of videogames. MIT Press, Cambridge, MA
133. Boiko, D.A., MacKnight, R., Kline, B., Gomes, G.: Autonomous chemical research with large language models. Nature **624**(7992), 570–578 (2023). https://doi.org/10.1038/s41586-023-06792-0
134. Bonabeau, E., Dorigo, M., Theraulaz, G.: Swarm intelligence: from natural to artificial systems (No. 1). Oxford University Press (1999)
135. Bonabeau, E., Theraulaz, G., Dorigo, M., Theraulaz, G., Marco, D.d.R.D.F., et al.: Swarm intelligence: from natural to artificial systems. 1. Oxford university press (1999)
136. Boner, J.: Philosophy and Political Economy, 0 edn. Routledge (2018). https://www.taylorfrancis.com/books/9781351316439. https://doi.org/10.4324/9781351316446
137. Boothby, W.H.: Weapons and the law of armed conflict. Oxford University Press (2016)
138. Bothell, D.: ACT-R 7.21 reference manual. Tech. rep., Department of Psychology, Carnegie Mellon University (2021). http://act-r.psy.cmu.edu/actr7.x/reference-manual.pdf
139. Bouktif, S., Fiaz, A., Ouni, A., Serhani, M.A.: Multi-sequence LSTM-RNN deep learning and metaheuristics for electric load forecasting. Energies **3**, 1–21 (2020)
140. Boulanin, V.: Implementing Article 36 weapon reviews in the light of increasing autonomy in weapon systems. Stockholm International Peace Research Institute [Stockholms internationella ... (2015)
141. Bousquet, A.: The eye of war: military perception from the telescope to the drone. University of Minnesota Press, Minneapolis
142. Bowden, R., Windridge, D., Kadir, T., Zisserman, A., Brady, M.: A linguistic feature vector for the visual interpretation of sign language. In: European Conference on Computer Vision, pp. 390–401. Springer (2004)
143. Boyack, K.W.: Mapping knowledge domains: Characterizing PNAS. Proceedings of the National Academy of Sciences of the United States of America **101**, 5192–5199 (2004). https://doi.org/10.1073/pnas.0307509100

144. Brambilla, M., Brutschy, A., Dorigo, M., Birattari, M.: Property-driven design for robot swarms: A design method based on prescriptive modeling and model checking. ACM Transactions on Autonomous and Adaptive Systems (TAAS) **9**(4), 1–28 (2014)
145. Brambilla, M., Ferrante, E., Birattari, M. and Dorigo, M., 2013. Swarm robotics: a review from the swarm engineering perspective. Swarm Intelligence, 7, pp.1–41.
146. Brambilla, M., Ferrante, E., Birattari, M., Dorigo, M.: Swarm robotics: A review from the swarm engineering perspective. Swarm Intelligence **7**(1), 1–41 (2013). https://doi.org/10.1007/s11721-012-0075-2
147. Brambilla, M., Ferrante, E., Birattari, M., Dorigo, M.: Swarm robotics: a review from the swarm engineering perspective. Swarm Intelligence **7**(1), 1–41 (2013)
148. Branch, J.: What's in a name? metaphors and cybersecurity. International Organization **75**(1), 39–70 (2021). https://doi.org/10.1017/S002081832000051X
149. Bremner, N.: From learner-centred to learning-centred: Becoming a 'hybrid' practitioner. International Journal of Educational Research **97**, 53–64 (2019). https://doi.org/10.1016/j.ijer.2019.06.012. https://www.sciencedirect.com/science/article/pii/S0883035519306871
150. Bresnan, J., Asudeh, A., Toivonen, I., Wechsler, S.: Lexical-functional syntax. John Wiley & Sons (2015)
151. Bride, H., Dong, J.S., Green, R., Hóu, Z., Mahony, B., Oxenham, M.: Gravitas: A model checking based planning and goal reasoning framework for autonomous systems. Engineering Applications of Artificial Intelligence **97**, 104,091 (2021). https://doi.org/10.1016/j.engappai.2020.104091. https://www.sciencedirect.com/science/article/pii/S0952197620303432
152. Briscoe, S.: Drone swarms: The good, the bad, and the terrifying future. Briscoe, S. (2023, September 19). Drone swarms: The good, the bad, and the terrifying future. ASIS Homepage. https://www.asisonline.org/security-management-magazine/latest-news/today-in-security/2023/september/drone-swarms-good-bad-and-terrifying/. [Accessed 20-06-2024]
153. Brislin, R.W.: Cross-cultural research methods: Strategies, problems, applications. Springer (1980)
154. Brohan, A., Brown, N., Carbajal, J., Chebotar, Y., Chen, X., Choromanski, K., Ding, T., Driess, D., Dubey, A., Finn, C., Florence, P., Fu, C., Arenas, M.G., Gopalakrishnan, K., Han, K., Hausman, K., Herzog, A., Hsu, J., Ichter, B., Irpan, A., Joshi, N., Julian, R., Kalashnikov, D., Kuang, Y., Leal, I., Lee, L., Lee, T.W.E., Levine, S., Lu, Y., Michalewski, H., Mordatch, I., Pertsch, K., Rao, K., Reymann, K., Ryoo, M., Salazar, G., Sanketi, P., Sermanet, P., Singh, J., Singh, A., Soricut, R., Tran, H., Vanhoucke, V., Vuong, Q., Wahid, A., Welker, P., Wohlhart, P., Wu, J., Xia, F., Xiao, T., Xu, P., Xu, S., Yu, T., Zitkovich, B.: Rt-2: Vision-language-action models transfer web knowledge to robotic control (2023)
155. Bruner, J.: The Narrative Construction of Reality. Critical Inquiry **18**, 1–21 (1991)
156. Bruvik, E.M., Fer, I., Våge, K., Haugan, P.M.: A revised ocean glider concept to realize stommel's vision and supplement argo floats. Ocean Science Discussions pp. 1–27 (2019)
157. Bruvik, E.M., Fer, I., Våge, K., Haugan, P.M.: A revised ocean glider concept to realize stommel's vision and supplement argo floats. Ocean Science Discussions pp. 1–27 (2019)
158. Buckle, J.R., Knox, A., Siviter, J., Montecucco, A.: Autonomous underwater vehicle thermoelectric power generation. Journal of Electronic Materials **42**(7) (2013)
159. Bui, L., Barlow, M., Abbass, H.: A multi-objective risk-based framework for mission capability planning. New Mathematics and Natural Computation **5**, 459–485
160. Burke, E.K., Petrovic, S., Qu, R.: Case-based heuristic selection for timetabling problems. Journal of Scheduling **9**(2), 115–132 (2006)
161. Boshuijzen-van Burken, C.: Value sensitive design for autonomous weapon systems—a primer. Ethics and Information Technology **25**(1), 1–14 (2023). https://doi.org/10.1007/s10676-023-09687-w
162. Boshuijzen-van Burken, C., Gore, R., Dignum, F., Royakkers, L., Wozny, P., Shults, F.L.: Agent-based modelling of values: The case of value sensitive design for refugee logistics. Journal of Artificial Societies and Social Simulation **23**(4), 6 (2020). https://doi.org/10.18564/jasss.4411. http://jasss.soc.surrey.ac.uk/23/4/6.html

163. Boshuijzen van Burken, C., Haftor, D.M.: Complexities and dilemmas in the sharing economy : The uber case. Llinnaeus University, Vaxjo, Sweden (2017). https://doi.org/10.15626/dirc.2015.04. https://open.lnu.se/index.php/dilemmas/article/view/542. [Online; accessed 2017-06-19]
164. Boshuijzen-van Burken, C.G.: Definitions, sources and categorisations for thinking swarms. In: S. Ng, J. Scholz, H.A. Abbass (eds.) Thinking Swarms. Springer (2025)
165. Boshuijzen-van Burken, C.G.: A philosophical analysis of "thinking swarms". In: S. Ng, J. Scholz, H.A. Abbass (eds.) Thinking Swarms. Springer (2025)
166. Buşoniu, L., Babuška, R., De Schutter, B.: Multi-agent Reinforcement Learning: An Overview, pp. 183–221. Springer Berlin Heidelberg, Berlin, Heidelberg (2010). https://doi.org/10.1007/978-3-642-14435-6_7
167. B.W., H., B., K., B., R., Y., Z., J., B., J.P., R., F.P., C.: An autonomous vehicle based open ocean Lagrangian observatory. In: Proceedings of IEEE AUV'2018, p. 1–5. IEEE (2018)
168. Bădică, A., Bădică, C., Ivanović, M., Dănciulescu, D.: Multi-agent modelling and simulation of graph-based predator-prey dynamic systems: A BDI approach. Expert Systems **35**(5), e12,263 (2018). https://doi.org/10.1111/exsy.12263. https://onlinelibrary.wiley.com/doi/abs/10.1111/exsy.12263. E12263 10.1111/exsy.12263
169. Cardier, B.: Unputdownable: How the agencies of compelling story assembly can be modelled using formalisable methods from knowledge representation and in a fictional tale about seduction. Ph.D. thesis, University of Melbourne (2013)
170. Cardier, B.: Unputdownable: How the Agencies of Compelling Story Assembly Can Be Modelled Using Formalisable Methods From Knowledge Representation, and in a Fictional Tale About Seduction. Ph.D. thesis, University of Melbourne, Melbourne, Australia (2013)
171. Cardier, B.: Narrative causal impetus: Governance through situational shift in game of thrones. In: Seventh Intelligent Narrative Technologies Workshop (2014)
172. Cardier, B.: The Evolution of Interpretive Contexts in Stories. In: M. Finlayson, B. Miller, A. Lieto, R. Ronfard (eds.) Sixth International Workshop on Computational Models of Narrative, vol. 45. Dagstuhl Publishing, Saarbrücken/Wadern (2015)
173. Cardier, B.: The shared story-narrative principles for innovative collaboration: Narrative principles for innovative collaboration beth cardier. In: Computational Context, pp. 209–234. CRC Press (2018)
174. Cardier, B., Back, A.D., Korte, J., Pounds, P.: Understanding swarms of sensory drone information as synthetic narratives. In: S. Ng, J. Scholz, H.A. Abbass (eds.) Thinking Swarms. Springer (2025)
175. Cardier, B., Nieslen, A.C., Shull, J., Sanford, L.D.: Outside the lines: Visualizing influence across heterogeneous contexts in PTSD. In: Systems Engineering and Artificial Intelligence, pp. 535–569. Springer (2021)
176. Cardier, B., Sanford, L.D., Goranson, H.T., Lundberg, P.S., Ciavarra, R.P., Devlin, K., Cassas, N., Erioli, A.: Modeling the resituation of memory in neurobiology and narrative. In: 2017 AAAI Spring Symposium Series (2017)
177. Cardona, G.A., Calderon, J.M.: Robot swarm navigation and victim detection using rendezvous consensus in search and rescue operations. Applied Sciences **9**(8), 1702 (2019). https://doi.org/10.3390/app9081702
178. Carroll, M., Shah, R., Ho, M.K., Griffiths, T., Seshia, S., Abbeel, P., Dragan, A.: On the utility of learning about humans for human-AI coordination. Advances in Neural Information Processing Systems **32** (2019)
179. Cashmore, M., Fox, M., Long, D., Magazzeni, D., Ridder, B., Carrera, A., Palomeras, N., Hurtos, N., Carreras, M.: ROSPlan: Planning in the robot operating system. In: Proceedings of the International Conference on Automated Planning and Scheduling, vol. 25, pp. 333–341 (2015). https://doi.org/10.1609/icaps.v25i1.13699
180. Chadwick, K.: Unmanned maritime systems will shape the future of naval operations: is international law ready? In: Maritime Security and the Law of the Sea, pp. 132–156. Edward Elgar Publishing (2020)

181. Chakraborty, A., Kar, A.K.: Swarm intelligence: A review of algorithms. Nature-inspired computing and optimization pp. 475–494 (2017)
182. Chamayou, G.: Drone Theory. Penguin Books Limited, United Kingdom
183. Chambers, N., Jurafsky, D.: Unsupervised learning of narrative event chains. In: Proceedings of ACL-08: HLT, pp. 789–797 (2008)
184. Chandler, D., Fuchs, C. (eds.): Digital objects, digital subjects: interdisciplinary perspectives on capitalism, labour and politics in the age of big data. Penguin, London
185. Chang, C.M., Toda, K., Gui, X., Seo, S.H., Igarashi, T.: Can Eyes on a Car Reduce Traffic Accidents? In: Proceedings of the 14th International Conference on Automotive User Interfaces and Interactive Vehicular Applications, pp. 349–359. Association for Computing Machinery, New York (2022). https://dl.acm.org/doi/10.1145/3543174.3546841
186. Chaplin, J.: Dooyeweerd's notion of societal structural principles. Philosophia reformata **60**(1), 16–36 (1995). Publisher: Brill
187. Chaplin, J.: Herman Dooyeweerd, Christian Philosopher of State and Civil Society. University of Notre Dame Press, Notre Dame (2011)
188. Charniak, E.: Toward a model of children's story comprehension. Ph.D. thesis, Massachusetts Institute of Technology (1972)
189. Chaturvedi, S., Peng, H., Roth, D.: Story comprehension for predicting what happens next. In: Proceedings of the 2017 Conference on Empirical Methods in Natural Language Processing, pp. 1603–1614 (2017)
190. Checkland, P.: Systems thinking. In: W.L. Currie, B. Galliers (eds.) Rethinking Management Information Systems, pp. 45–56. Oxford University Press (1999)
191. Chen, S.F., Goodman, J.: An empirical study of smoothing techniques for language modeling. Computer Speech & Language **13**(4), 359–394 (1999)
192. Chen, X., Xu, L., Liu, Z., Sun, M., Luan, H.: Joint learning of character and word embeddings. In: Twenty-fourth international joint conference on artificial intelligence (2015)
193. Chen, Z., Liu, B., Brachman, R., Stone, P., Rossi, F.: Lifelong Machine Learning: Second Edition. Synthesis Lectures on Artificial Intelligence and Machine Learning. Morgan & Claypool Publishers (2018). https://books.google.com.au/books?id=JQ5pDwAAQBAJ
194. Chen, F., & Ren, W. (2019). On the control of multi-agent systems: A survey. Foundations and Trends® in Systems and Control, 6(4), 339–499.
195. Cheraghi, A.R., Shahzad, S., Graffi, K.: Past, present, and future of swarm robotics. In: Proceedings of SAI Intelligent Systems Conference, pp. 190–233 (2021). https://doi.org/10.1007/978-3-030-82199-9_13
196. Christian, B.: The alignment problem: How can artificial intelligence learn human values?
197. Christian, K., Allen, C.: 'territorial' ravens disrupt surge in wing drone deliveries under Canberra's lockdown (2021). https://www.abc.net.au/news/2021-09-22/territorial-ravens-disrupt-canberra-drone-deliveries/100480470
198. Christiano, P.F., Leike, J., Brown, T., Martic, M., Legg, S., Amodei, D.: Deep reinforcement learning from human preferences. Advances in neural information processing systems **30** (2017)
199. Chu, S.C., Tsai, P.w., Pan, J.S., Yang, Q., Webb, G.: Cat swarm optimization. In: Q. Yang, G. Webb (eds.) PRICAI 2006: Trends in Artificial Intelligence, vol. 4099, pp. 854–858. Berlin, Heidelberg (2006). http://link.springer.com/10.1007/978-3-540-36668-3_94
200. Cichocki, A., Amari, S.I.: Families of alpha- beta- and gamma- divergences: Flexible and robust measures of similarities. Entropy **12**(6), 1532–1568 (2010)
201. Clark, B., Patt, D., Schramm, H.: Mosaic warfare: Exploiting artificial intelligence and autonomous systems to implement decision-centric operations
202. Clarke, E.M., Henzinger, T.A., Veith, H., Bloem, R.: Handbook of model checking. Springer (2018)
203. Clothier, R., Palmer, J., Walker, R., Fulton, N.: Definition of airworthiness categories for civil unmanned aircraft systems (UAS). In: ICAS 2010: 27th International Congress of the Aeronautical Sciences. Nice

204. Clouser, R.: A Brief Sketch of the Philosophy of Herman Dooyeweerd. Axiomathes (Journal Article) (2010). http://dx.doi.org/10.1007/s10516-009-9075-2
205. Coeckelbergh, M.: From killer machines to doctrines and swarms, or why ethics of military robotics is not (necessarily) about robots. Philosophy & Technology **24**, 269–278 (2011). https://doi.org/10.1007/s13347-011-0019-6
206. Colbourne, J.: The geometry of trochoid envelopes and their application in rotary pumps. Mechanism and Machine Theory **9**(3–4), 421–435 (1974)
207. Collar, P.G., McPhail, S.D.: Autosub—an autonomous unmanned submersible for ocean data-collection. Electronics & Communication Engineering Journal **7**(3), 105–114 (1995)
208. Collins, M.: Head-driven statistical models for natural language parsing. Computational linguistics **29**(4), 589–637 (2003)
209. Coombes, M., Chen, W.H., Liu, C.: Boustrophedon coverage path planning for UAV aerial surveys in wind. In: 2017 International Conference on Unmanned Aircraft Systems (ICUAS), pp. 1563–1571. IEEE (2017)
210. Copeland, D., Liivoja, R., Sanders, L.: The utility of weapons reviews in addressing concerns raised by autonomous weapon systems. Journal of Conflict and Security Law **28**(2), 285–316 (2023)
211. Corporation, I.: Drone light shows powered by intel. https://www.intel.com/content/www/us/en/technology-innovation/aerial-technology-light-show.html. Retrieved from
212. Cox, M.T.: Metacognition in computation: a selected research review. Artificial Intelligence **169**(2), 104–141 (2005)
213. Cox, M.T.: Perpetual self-aware cognitive agents. AI Mag **28**(1), 32 (2007)
214. Crandall, J.W., Anderson, N., Ashcraft, C., Grosh, J., Henderson, J., McClellan, J., Neupane, A., Goodrich, M.A.: Human-swarm interaction as shared control: Achieving flexible fault-tolerant systems. In: International Conference on Engineering Psychology and Cognitive Ergonomics, pp. 266–284. Springer (2017). https://doi.org/10.1007/978-3-319-58472-0_21
215. Croft, W.: Radical construction grammar: Syntactic theory in typological perspective. Oxford University Press on Demand (2001)
216. Crogan, P.: Visions of Swarming Robots: Artificial Intelligence and Stupidity in the Military-Industrial Projection of the Future of Warfare. Palgrave Macmillan, Cham (2019). https://doi.org/10.1007/978-3-030-21836-2_5
217. Cui, C., Lin, P., Nie, X., Yin, Y., Zhu, Q.: Hybrid textual-visual relevance learning for content-based image retrieval. Journal of Visual Communication and Image Representation **48**, 367–374 (2017)
218. Culler, J.: Structuralist Poetics: Structuralism, Linguistics and the Study of Literature. Routedge, London, New York (1975)
219. Cunningham, D., Noakes, J.: "what if she's from the FBI?" the effects of covert forms of social control on social movements. In: Surveillance and governance: Crime control and beyond. Emerald Group Publishing Limited (2008)
220. Curtin, T.B., Bellingham, J.G., Catipovic, J., Webb, D.: Autonomous oceanographic sampling networks. Oceanography **6**(3), 86–94 (1993)
221. Daixin, T., Hongwei, X., Huijuan, Y., Hao, Y., Wen, H.: Optimization of group control strategy and analysis of energy saving in refrigeration plant. Energy and Built Environment **3**(4), 525–535 (2022). https://doi.org/10.1016/j.enbenv.2021.05.006
222. Damasio, H., Tranel, D., Grabowski, T., Adolphs, R., Damasio, A.: Neural systems behind word and concept retrieval. Cognition **92**(1–2), 179–229 (2004)
223. Dautriche, I., Mahowald, K., Gibson, E., Piantadosi, S.T.: Wordform similarity increases with semantic similarity: An analysis of 100 languages. Cognitive science **41**(8), 2149–2169 (2017)
224. Davis, E., Pounds, P.: Direct Sensing of Thrust and Velocity for a Quadrotor Rotor Array. IEEE Robotics and Automation Letters **2**(3, July) (2017)
225. Davis, M.: The role of autonomous systems in Australia's defence. After Covid-19: Australia and the world rebuild (Volume 1) pp. 106–109 (2020)

226. Deer, W., Pounds, P.: Light-Weight Whiskers for Contact. IEEE Robotics and Automation Letters **4**(2, April) (2019)
227. Delageniere, B.: Intelligence analysis as cryptic hermeneutics. Intelligence and national security **36**(4), 541–554. https://doi.org/10.1080/02684527.2021.1893075
228. Deleuze, G.: Difference and repetition. Columbia University Press, New York
229. DeLillo, D.: White Noise. New York: Viking Press (1985)
230. Delvenne, J.C.: Category theory for autonomous and networked dynamical systems. Entropy **21**(3) (2019). https://doi.org/10.3390/e21030302. https://www.mdpi.com/1099-4300/21/3/302
231. Demazeau, Y., Holvoet, T., Corchado, J., Costantini, S.: Advances in practical applications of agents, multi-agent systems, and trustworthiness. In: The PAAMS Collection: 18th International Conference (October 7–9, 2020. L'Aquila, Italy
232. Department of Prime Minister and Cabinet, Regulation: Note the office of best practice regulation (OBPR) was set up in Australia to assist in implementing best practice regulation. https://www.pmc.gov.au/domestic-policy/regulation (2022). Accessed: 19 October 2022
233. DeRoo, N.: Meaning, Being, and Time: The Phenomenological Significance of Dooyeweerd's Thought. In: J.A. Simmons, J.E. Hackett (eds.) Phenomenology for the Twenty-First Century, pp. 77–96. Palgrave Macmillan UK, London (2016). https://doi.org/10.1057/978-1-137-55039-2_5. http://link.springer.com/10.1057/978-1-137-55039-2_5
234. Deshmukh, J.V., Donze, A., Ghosh, S., Jin, X., Juniwal, G., Seshia, S.A.: Robust online monitoring of signal temporal logic. Formal Methods in System Design **51**(1), 5–30 (2017)
235. Devitt, K., Horne, R., Assaad, Z., Broad, E., Kurniawati, H., Cardier, B., Scott, A., Lazar, S., Gould, M., Adamson, C., Karl, C., Schrever, F., Keay, S., Tranter, K., Shellshear, E., Hunter, D., Brady, M., Putland, T.: A Robotics Roadmap for Australia, chap. Trust and Safety. Robotics Australia Group (2022). https://roboausnet.com.au/wp-content/uploads/2021/11/Robotics-Roadmap-for-Australia-2022_compressed-1.pdf
236. Devitt, S.: Trustworthiness of autonomous systems. In: H. Abbass, J. Scholz, D. Reid (eds.) Foundations of Trusted Autonomous Systems, p. 161–184 (2018)
237. Devitt, S.: Trustworthiness of autonomous systems. Foundations of trusted autonomy (Studies in Systems, Decision and Control, Volume 117) pp. 161–184 (2018)
238. Devlin, K.J.: A Uniform Framework for Describing and Analyzing the Modern Battlefield; Feasibility Study p. 19 (2011)
239. DeVries, L., Paley, D.A.: Multivehicle control in a strong flowfield with application to hurricane sampling. Journal of Guidance, Control, and Dynamics **35**(3), 794–806 (2012)
240. Dick, W., Carey, L., Carey, J.O.: The systematic design of instruction. Pearson (2009)
241. Ding, W., Yan, G., Lin, Z.: Formations on two-layer pursuit systems. In: 2009 IEEE International Conference on Robotics and Automation, pp. 3496–3501. IEEE (2009)
242. Ding, W., Yan, G., Lin, Z.: Collective motions and formations under pursuit strategies on directed acyclic graphs. Automatica **46**(1), 174–181 (2010)
243. Diplomat, T.: Ready for service by 2027. https://thediplomat.com/2019/03/us-intelligence-russias-nuclear-capable-poseidon-underwater-drone-ready-for-service-by-2027/ (2022). Accessed: 28th November 2023
244. Dong, Y., Su, H., Zhu, J., Zhang, B.: Improving interpretability of deep neural networks with semantic information. In: Proceedings of the IEEE conference on computer vision and pattern recognition, pp. 4306–4314 (2017)
245. Donthu, N., Kumar, S., Mukherjee, D., Pandey, N., Lim, W.M.: How to conduct a bibliometric analysis: An overview and guidelines. Journal of Business Research **133**, 285–296 (2021). https://doi.org/10.1016/j.jbusres.2021.04.070
246. Dooyeweerd, H.: A new critique of theoretical thought, vol. I-V. The Presbyterian and Reformed Publishing Company, Amsterdam/Philadelphia (1953)
247. Dooyeweerd, H.: A new critique of theoretical thought, vol. I-V. The Presbyterian and Reformed Publishing Company, Amsterdam/Philadelphia (1953)
248. Dooyeweerd, H.: A New Critique of Theoretical Thought,The General Theory of the Modal Spheres, vol. 2, 2 edn. The Presbyterian and Reformed Publishing Company (1969)

249. Dorigo, M.: Optimization, learning and natural algorithms. Ph. D. Thesis, Politecnico di Milano (1992)
250. Dorigo, M., Birattari, M., Stutzle, T.: Ant colony optimization. IEEE Computational Intelligence Magazine **1**(4), 28–39 (2006). https://doi.org/10.4249/scholarpedia.1461
251. Dorigo, M., Theraulaz, G., & Trianni, V. (2020). Reflections on the future of swarm robotics. Science robotics, 5(49), eabe4385.
252. Dorigo, M., Theraulaz, G., & Trianni, V. (2021). Swarm robotics: Past, present, and future [point of view]. Proceedings of the IEEE, 109(7), 1152–1165.
253. Douglas, D.M., Howard, D., Lacey, J.: Moral responsibility for computationally designed products. AI and Ethics **1**(3), 273–281 (2021)
254. Dressler, A.: Poetics of conspiracy and hermeneutics of suspicion in tacitus's dialogus de oratoribus. Classical antiquity **32**(1), 1–34. https://doi.org/10.1525/ca.2013.32.1.1
255. DronesAreSuperb: Drone show failed. https://www.youtube.com/watch?app=desktop&v=btD0GKFXtbkt%3D160. Retrieved from
256. DroneSec: Drone hacking analysis: Drones drop from sky during light show. https://dronesec.com/blog/drone-hacking-analysis-drones-drop-from-sky-during-light-show. Retrieved from
257. Du Gay, P., Hall, S., Janes, L., Madsen, A.K., Mackay, H., Negus, K.: Doing cultural studies: The story of the Sony Walkman. In: P. Vannini (ed.) Handbook of Ethnographic Film and Video, first edn., p. 247–255. Sage Publications.Fish, London
258. Duan, H., Qiao, P.: Pigeon-inspired optimization: a new swarm intelligence optimizer for air robot path planning. International Journal of Intelligent Computing and Cybernetics **7**(1), 24–37 (2014). https://doi.org/10.1108/IJICC-02-2014-0005
259. Duarte, M., Gomes, J., Oliveira, S.M., Christensen, A.L.: Evolution of repertoire-based control for robots with complex locomotor systems. IEEE Transactions on Evolutionary Computation **22**(2), 314–328 (2018). https://doi.org/10.1109/TEVC.2017.2722101
260. Duarte, M., Gomes, J., Oliveira, S.M., Christensen, A.L.: Evolution of repertoire-based control for robots with complex locomotor systems. IEEE Transactions on Evolutionary Computation **22**(2), 314–328 (2018)
261. van Duijn, M.J., van Dijk, B.M.A., Kouwenhoven, T., de Valk, W., Spruit, M.R., van der Putten, P.: Theory of mind in large language models: Examining performance of 11 state-of-the-art models vs. children aged 7–10 on advanced tests (2023)
262. van Eck, N.J., Waltman, L.: Software survey: Vosviewer, a computer program for bibliometric mapping. Scientometrics **84**(2), 523–538 (2010). https://doi.org/10.1007/s11192-009-0146-3
263. Edwards, S.: Swarming and the future of warfare (2005)
264. Edwin Hutchins: How a Cockpit Remembers Its Speeds. Cognitive Science **19**, 265–288 (1995)
265. Einhorn, H., Hogarth, R.: Judging Probable Cause. Psychological Bulletin **99**, 3–19 (1986)
266. El-Fiqi, H., Campbell, B., Elsayed, S., Perry, A., Singh, H.K., Hunjet, R., Abbass, H.A.: The limits of reactive shepherding approaches for swarm guidance. IEEE Access **8**, 214,658–214,671 (2020)
267. Electronica, A.: Spaxels / Klangwolke—Quadcopter. Ars Electronica Blog
268. Electronica, A.: Throwback: Spaxels. Ars Electronica Blog
269. Endsley, M.R.: Situation awareness: operationally necessary and scientifically grounded. Cogn. Tech Work **17**, 163–167 (2015)
270. Engebraaten, S.A., Moen, J., Yakimenko, O.A., Glette, K.: A framework for automatic behavior generation in multi-function swarms. Frontiers in Robotics and AI **7** (2020). https://doi.org/10.3389/frobt.2020.579403
271. Eusuff, M.M., Lansey, K.E.: Optimization of water distribution network design using the shuffled frog leaping algorithm. J. Water Resour. Plann. Manage. **129**(3), 210–225 (2003). https://doi.org/10.1061/(ASCE)0733-9496(2003)129:3(210)
272. Evans, N., Levinson, S.C.: The myth of language universals: Language diversity and its importance for cognitive science. Behavioral and brain sciences **32**(5), 429–448 (2009)
273. Evans, V.: Cognitive linguistics. Edinburgh University Press (2006)

274. Evered, M., Burling, P., Trotter, M., et al.: An investigation of predator response in robotic herding of sheep. International Proceedings of Chemical, Biological and Environmental Engineering **63**, 49–54 (2014)
275. Faint, C., Harris, M.: F3ead: Ops/intel fusion 'feeds' the sof targeting process. Small Wars Journal **31**(7) (2012)
276. Faraji, M., Preuschoff, K., Gerstner, W.: Balancing new against old information: the role of puzzlement surprise in learning. Neural computation **30**(1), 34–83 (2018)
277. Farrington, E.: Parametric equations at the circus: Trochoids and poi flowers. The College Mathematics Journal **46**(3), 173–177 (2015)
278. Fauconnier, G., Turner, M.: The Way We Think: Conceptual Blending and the Mind's Hidden Complexities. Basic Books, New York (2002)
279. Fei, G., Wang, S., Liu, B.: Learning cumulatively to become more knowledgeable. In: Proceedings of the 22nd ACM SIGKDD International Conference on Knowledge Discovery and Data Mining, pp. 1565–1574. ACM (2016)
280. Fernández, F., Veloso, M.: Probabilistic policy reuse in a reinforcement learning agent. In: Proceedings of the fifth international joint conference on Autonomous agents and multiagent systems, pp. 720–727. ACM (2006)
281. Ferrante, E., Turgut, A.E., Huepe, C., Stranieri, A., Pinciroli, C., Dorigo, M.: Self-organized flocking with a mobile robot swarm: a novel motion control method. Adaptive Behavior **20**(6), 460–477 (2012). https://doi.org/10.1177/1059712312462248
282. Ferreira, F., Henderson, J.M.: Recovery from misanalyses of garden-path sentences. Journal of Memory and Language **30**(6), 725–745 (1991)
283. Feynman, R.: Report of the presidential commission on the space shuttle challenger accident,. Volume 2: Appendix F
284. Filminquiry: 10 films that utilize a frame narrative. Filminquiry Magazine (2020). https://www.filminquiry.com/10-films-frame-narrative/
285. Fish, A.R.: Drones: visual anthropology from the air (2020)
286. Forbes: U.S. navy destroys target with drone swarm—and sends a message to china (2021). Accessed: 30 May 2022
287. Forster, E.: Aspects of the Novel. Mariner Books, Boston (1956)
288. Foundation, R.P.: Raspberry Pi 4 Product Brief (2019). https://static.raspberrypi.org/files/product-briefs/Raspberry-Pi-4-Product-Brief.pdf
289. Francesca, G., Birattari, M.: Automatic design of robot swarms: Achievements and challenges. Frontiers in Robotics and AI **3** (2016). https://doi.org/10.3389/frobt.2016.00029
290. Francesca, G., Birattari, M.: Automatic design of robot swarms: Achievements and challenges. Frontiers in Robotics and AI **3** (2016). https://doi.org/10.3389/frobt.2016.00029
291. Francesca, G., Brambilla, M., Brutschy, A., Trianni, V., Birattari, M.: Automode: A novel approach to the automatic design of control software for robot swarms. Swarm Intelligence **8**(2), 89–112 (2014)
292. Frantzman, S.: Drone wars: pioneers, killing machines, artificial intelligence, and the battle for the future. Bombardier Books, New York
293. Freitas, D., Lopes L.G., M.D., F.: Particle swarm optimisation: A historical review up to the current developments. Entropy **22**, 262 (2020)
294. Friedman, B., Hendry, D.G.: Value Sensitive Design: Shaping Technology with Moral Imagination. MIT Press, MIT (2019)
295. Friedman, B., Kahn Jr., P.H., Borning, A.: Value Sensitive Design and Information Systems, pp. 69–101 (2008). https://doi.org/10.1002/9780470281819.ch4. https://onlinelibrary.wiley.com/doi/abs/10.1002/9780470281819.ch4
296. Frigg, R.: Models and fiction. Synthese **172**(2), 251–268 (2010)
297. Froede, W.G.: The NSU-Wankel rotating combustion engine. SAE Transactions **69**, 179–193 (1961)
298. Frow, J.: Genre: The New Critical Idiom. Routledge, London & New York (2014)

299. Fu, Y., Xiang, T., Jiang, Y.G., Xue, X., Sigal, L., Gong, S.: Recent advances in zero-shot recognition: Toward data-efficient understanding of visual content. IEEE Signal Processing Magazine **35**(1), 112–125 (2018)
300. Fuhrman, S., Cunningham, M.J., Wen, X., Zweiger, G., Seilhamer, J.J., Somogyi, R.: The application of Shannon entropy in the identification of putative drug targets. Biosystems (A6E) **55**(1–3), 5–14 (2000)
301. Fuller, M., O'Kane, M.: NSW independent flood inquiry (volume 2: Full report). Tech. rep., New South Wales Government, Sydney, Australia (2022). https://www.nsw.gov.au/nsw-government/projects-and-initiatives/floodinquiry
302. Furstenberg, H.: Stationary processes and prediction theory. 44. Princeton University Press (1960)
303. Gadamer, H.G. (ed.): Truth and method, 2nd edn. Seabury Press, New York
304. Galindo, J.: Authentic learning (simulation, lab, field. In: ABLConnect.harvard.edu.
305. Gallie, W.: Essentially contested concepts. Proceedings of the Aristotelian Society **56**, 167–198
306. Galloway, K.S., Justh, E.W., Krishnaprasad, P.: Portraits of cyclic pursuit. In: 2011 50th IEEE Conference on Decision and Control and European Control Conference, pp. 2724–2731. IEEE (2011)
307. Galloway, K.S., Justh, E.W., Krishnaprasad, P.S.: Symmetry and reduction in collectives: cyclic pursuit strategies. Proceedings of the Royal Society of London A: Mathematical, Physical and Engineering Sciences **469**(2158) (2013)
308. Galvin, P., Klarin, A., Nyuur, R., Burton, N.: A bibliometric content analysis of do-it-yourself (DIY) science: where to from here for management research? Technology Analysis & Strategic Management **33**(10), 1255–1266 (2021). https://doi.org/10.1080/09537325.2021.1959031
309. García, J., Fernández, F.: Probabilistic policy reuse for safe reinforcement learning. ACM Transactions on Autonomous and Adaptive Systems (TAAS) **13**(3), 14 (2019)
310. Garrett-Glaser, B.: Securing airports from drone threats is a 'wicked problem. In: FAA and Industry Agree—Avionics International. Aviationtoday.com
311. Gawron, J.M., Stephens, K.: Sparsity and normalization in word similarity systems. Natural Language Engineering **22**(3), 351–395 (2016)
312. Gennari, S.P., MacDonald, M.C., Postle, B.R., Seidenberg, M.S.: Context-dependent interpretation of words: Evidence for interactive neural processes. Neuroimage **35**(3), 1278–1286 (2007)
313. Gentner, D.: Structure-mapping: A theoretical framework for analogy. Cognitive Science **7**, 55–170 (1951)
314. Gentner, D.: Structure-Mapping: A Theoretical Framework for Analogy. Cognitive Science **7**, 155–170 (1983)
315. Genz, A., Bretz, F.: Comparison of methods for the computation of multivariate t probabilities. Journal of Computational and Graphical Statistics **11**(4), 950–971 (2002)
316. George, T.: Hermeneutics. Stanford Encyclopedia of Philosophy
317. Gersho, A., Cuperman, V.: Vector quantization: A pattern-matching technique for speech coding. IEEE Communications Magazine **21**(9), 15–21 (1983)
318. Gerz, M., Schwarze, A., Stuch, H.: Connecting the dots—enhancing the information processing chain for the detection of hybrid threats for host nation support and territorial operations. https://www.sto.nato.int/publications/STO%20Meeting%20Proceedings/STO-MP-IST-190/MP-IST-190-02.pdf. Retrieved from
319. Gianvecchio, S., Wang, H.: An entropy-based approach to detecting covert timing channels. IEEE Trans. Dependable Secur. Comput. **8**(6), 785–797 (2011)
320. Gibbs, N.: Aust warned of complacency as AI infiltrates workplaces. https://www.aap.com.au/news/aust-warned-of-complacency-as-ai-infiltrates-workplaces/. Retrieved from
321. Gibbs, R.W.: The Cambridge Handbook of Metaphor and Thought. Cambridge University Press, Cambridge (2008)

322. Gibson, E., Futrell, R., Piantadosi, S.P., Dautriche, I., Mahowald, K., Bergen, L., Levy, R.: How efficiency shapes human language. Trends in cognitive sciences **23**(5), 389–407 (2019)
323. Glas, G.: Churchland, kandel and dooyeweerd on the reducibility of mind states. Philosophia Reformata **67**(2), 148–172 (2002). Publisher: Brill
324. Glas, G.: Christian philosophical anthropology. a reformation perspective. Philosophia Reformata **75** (2010). https://doi.org/10.1163/22116117-90000493
325. Glomsrud, J.A., Ødegårdstuen, A., Clair, A.L.S., Smogeli, Ø.: Trustworthy versus explainable AI in autonomous vessels. In: Proceedings of the International Seminar on Safety and Security of Autonomous Vessels (ISSAV) and European STAMP Workshop and Conference (ESWC), vol. 37 (2019)
326. Goldberg, Y., Levy, O.: word2vec explained: deriving mikolov et al.'s negative-sampling word-embedding method. arXiv preprint arXiv:1402.3722 (2014)
327. Goldfarb, A., Lindsay, J.R.: Prediction and judgment: Why artificial intelligence increases the importance of humans in war. International Security **46**(3), 7–50. https://doi.org/10.1162/isec_a_00425
328. Gomes, J., Christensen, A.L.: Task-agnostic evolution of diverse repertoires of swarm behaviours. In: M. Dorigo, M. Birattari, C. Blum, A.L. Christensen, A. Reina, V. Trianni (eds.) Swarm Intelligence, pp. 225–238. Springer International Publishing, Cham (2018)
329. Gomes, J., Mariano, P., Christensen, A.L.: Cooperative coevolution of partially heterogeneous multiagent systems. In: Proceedings of the 2015 International Conference on Autonomous Agents and Multiagent Systems, pp. 297–305. International Foundation for Autonomous Agents and Multiagent Systems (2015)
330. Goranson, H.T., Cardier, B.: A two-sorted logic for structurally modeling systems. Progress in biophysics and molecular biology **113**, 141–178 (2013)
331. Goussac, N., Jevglevskaja, N., Liivoja, R., Sanders, L.: Enhancing the legal review of autonomous weapon systems: Report of an expert meeting (2023)
332. Graesser, A.C., Kreuz, R.J.: A Theory of Inference Generation During Text Comprehension. Discourse Processes **16**, 145–160 (1993). https://doi.org/10.1080/01638539309544833
333. Granroth-Wilding, M., Clark, S.: What happens next? event prediction using a compositional neural network model. In: Proceedings of the AAAI Conference on Artificial Intelligence, vol. 30 (2016)
334. Graphika, Observatory, S.I.: Unheard voice: Evaluating five years of pro-western covert influence operations
335. Gray, R., Gray, A., Rebolledo, G., Shore, J.: Rate-distortion speech coding with a minimum discrimination information distortion measure. IEEE Transactions on Information Theory **27**(6), 708–721 (1981)
336. Greenwald, G.: How covert agents infiltrate the internet to manipulate, deceive, and destroy reputations. The Intercept **24** (2014)
337. Gregory, C., Vardy, A.: microusv: A low-cost platform for indoor marine swarm robotics research. HardwareX **7**, e00,105 (2020)
338. Grimal, F., Sundaram, J.: Combat drones: Hives, swarms, and autonomous action? Journal of Conflict and Security Law **23**(1), 105–135 (2018)
339. Grois, E., Wilkins, D.C.: Learning strategies for story comprehension: a reinforcement learning approach. In: Proceedings of the 22nd international conference on Machine learning, pp. 257–264 (2005)
340. Gronauer, S., Diepold, K.: Multi-agent deep reinforcement learning: a survey. Artificial Intelligence Review **55**(2), 895–943 (2022)
341. Guan, J., Wang, Y., Huang, M.: Story ending generation with incremental encoding and commonsense knowledge. In: Proceedings of the AAAI Conference on Artificial Intelligence, vol. 33, pp. 6473–6480 (2019)
342. Guihen, D.: The barriers and opportunities of effective underwater autonomous swarms. In: S. Ng, J. Scholz, H.A. Abbass (eds.) Thinking Swarms. Springer (2025)
343. Gunningham, N., Grabosky, P., Sinclair, D.: Smart regulation: Designing environmental policy. Oxford University Press (1998)

344. Günther, F., Rinaldi, L., Marelli, M.: Vector-space models of semantic representation from a cognitive perspective: A discussion of common misconceptions. Perspectives on Psychological Science **14**(6), 1006–1033 (2019)
345. Günther, F., Rinaldi, L., Marelli, M.: Vector-space models of semantic representation from a cognitive perspective: A discussion of common misconceptions. Perspectives on Psychological Science **14**(6), 1006–1033 (2019)
346. Guo, S., Guan, Y., Tan, H., Li, R., Li, X.: Frame-based neural network for machine reading comprehension. Knowledge-Based Systems **219**, 106,889 (2021)
347. Guthrie, D., Allison, B., Liu, W., Guthrie, L., Wilks, Y.: A closer look at skip-gram modelling. In: Proceedings of the Fifth International Conference on Language Resources and Evaluation (LREC'06). European Language Resources Association (ELRA), Genoa, Italy (2006)
348. Gutzwiller, R.S., Reeder, J.: Dancing with algorithms: Interaction creates greater preference and trust in machine-learned behavior. Human Factors **63**(5), 854–867 (2021)
349. Ha, D., Tang, Y.: Collective intelligence for deep learning: a survey of recent developments. Collective Intelligence **1**(1), 1–16 (2022). https://doi.org/10.1177/26339137221114874
350. Hadfield-Menell, D., Milli, S., Abbeel, P., Russell, S.J., Dragan, A.: Inverse reward design. Advances in Neural Information Processing Systems **30** (2017)
351. Hagenauer, J., Helbich, M.: A comparative study of machine learning classifiers for modeling travel mode choice. Expert Systems with Applications **78**, 273–282 (2017)
352. Hall, L.M.: Trochoids, roses, and thorns-beyond the spirograph. The College Mathematics Journal **23**(1), 20–35 (1992)
353. Hamann, H.: Swarm Robotics: A Formal Approach (2018). https://doi.org/10.1007/978-3-319-74528-2
354. Hambling, D: What are drone swarms and why does every military suddenly want one? Forbes (Website) https://www.forbes.com/sites/davidhambling/2021/03/01/what-are-drone-swarms-and-why-does-everyone-suddenly-want-one/?sh=5978fea2f5c6 (2021). Accessed: 1 Mar 21
355. Hammes, T.X.: The future of warfare: Small, many, smart vs. few & ecquisite? War on the Rocks Commentary, https://warontherocks.com/2014/07/the-future-of-warfare-small-many-smart-vs-few-exquisite/ (2014)
356. Harzing, A.W., Alakangas, S.: Google scholar, scopus and the web of science: A longitudinal and cross-disciplinary comparison. Scientometrics **106**(2), 787–804 (2016). https://doi.org/10.1007/s11192-015-1798-9
357. Hauert, S., Zufferey, J.C., Floreano, D.: Evolved swarming without positioning information: an application in aerial communication relay. Autonomous Robots **26**(1), 21–32 (2009)
358. Hausser, J., Strimmer, K.: Entropy inference and the James-Stein estimator, with application to nonlinear gene association networks. Journal of Machine Learning Research **10**(Jul), 1469–1484 (2009)
359. Héder, M.: From NASA to EU: the evolution of the TRL scale in public sector innovation. The Innovation Journal **22**(2), 1–23 (2017)
360. Héder, M.: From NASA to EU: the evolution of the TRL scale in public sector innovation. The Innovation Journal **22**(2), 1–23 (2017)
361. Heidegger, M.: The age of the world picture. In: Y. J., K. Haynes (eds.) Off the Beaten Track. Cambridge University Press, Cambridge
362. Heinerman, J., Drupsteen, D., Eiben, A.E.: Three-fold adaptivity in groups of robots: The effect of social learning. In: Proceedings of the 2015 Annual Conference on Genetic and Evolutionary Computation, pp. 177–183. ACM (2015)
363. Heinerman, J., Zonta, A., Haasdijk, E., Eiben, A.E.: On-line evolution of foraging behaviour in a population of real robots. In: European Conference on the Applications of Evolutionary Computation, pp. 198–212. Springer (2016)
364. Helbing, D., Farkas, I., Vicsek, T.: Simulating dynamical features of escape panic. Nature **407**(6803), 487–490 (2000). https://doi.org/10.1038/35035023
365. Hellsten, I., Dawson, J., Leydesdorff, L.: Implicit media frames: Automated analysis of public debate on artificial sweeteners (2012)

366. Hellsten, I., Porter, A.J., Nerlich, B.: Imaging the future at the global and national scale: A comparative study of British and Dutch press coverage of Rio 1992 and Rio 2012. Environmental Communication **8**(4), 468–488 (2014). https://doi.org/10.1080/17524032.2014.911197
367. Henkel, D., Brown, T.X.: Delay-tolerant communication using mobile robotic helper nodes. In: Modeling and Optimization in Mobile, Ad Hoc, and Wireless Networks and Workshops, 2008. WiOPT 2008. 6th International Symposium on, pp. 657–666. IEEE (2008)
368. Hepworth, A., Yaxley, K., Baxter, D., Keene, J.: Swarming and counterswarming: Report on applied research directions and future opportunities for swarm systems in defence. Tech. Rep. Australian Army Occasional Paper No. 11, Australian Army Research Centre (2020)
369. Herman, D.: Genette Meets Vygotsky: Narrative Embedding and Distributed Intelligence. Language and Literature **15**, 357–380 (2006)
370. Herman, D.: Genette meets Vygotsky: Narrative embedding and distributed intelligence. Language and Literature **15**, 357–380 (2006)
371. Herrera, J., Pury, P.: Statistical keyword detection in literary corpora. Eur. Phys. J. B **63**, 135–146 (2008)
372. Hettiarachchige, Y., Khan, A., Barca, J.C.: Multi-object tracking of swarms with active target avoidance. In: 2018 15th International Conference on Control, Automation, Robotics and Vision (ICARCV), pp. 1204–1209. IEEE (2018)
373. Hildebrand, J.: Situating hobby drone practices. Digital Culture & Society **3**(2), 207–218
374. Hildenbrandt, H., Carere, C., Hemelrijk, C.: Self-organized aerial displays of thousands of starlings: a model. Behavioral Ecology **21**(6), 1349–1359 (2010). https://doi.org/10.1093/beheco/arq149
375. Hinchey, M.G., Sterritt, R., Rouff, C.: Swarms and swarm intelligence. Computer **40**(4), 111–113 (2007). https://doi.org/10.1109/MC.2007.144
376. Hirschman, L., Light, M., Breck, E., Burger, J.D.: Deep read: A reading comprehension system. In: Proceedings of the 37th annual meeting of the Association for Computational Linguistics, pp. 325–332 (1999)
377. Hoffman, P.: An individual differences approach to semantic cognition: Divergent effects of age on representation, retrieval and selection. Scientific reports **8**(1), 1–13 (2018)
378. Hoffman, P., McClelland, J.L., Lambon Ralph, M.A.: Concepts, control, and context: A connectionist account of normal and disordered semantic cognition. Psychological review **125**(3), 293 (2018)
379. Homayounnejad, M., Overill, R.: Preventing autonomous weapon systems from being used to perpetrate intentional violations of the laws of war. TLI Think (2018)
380. Hong, J.S., Chang, P.L.: The trochoid-like track in Typhoon Dujuan (2003). Geophysical research letters **32**(16) (2005)
381. Hong, M.T., Benjamin, J.J., Muller-Birn, C.: Coordinating agents: Promoting shared situational awareness in collaborative sensemaking. In: CSCW '18 ACM Conference on Computer Supported Cooperative Work and Social Computing, pp. 217–220. ACM (2018)
382. Honkela, T., Hyvarinen, A.: Linguistic feature extraction using independent component analysis. In: 2004 IEEE International Joint Conference on Neural Networks (IEEE Cat. No. 04CH37541), vol. 1, pp. 279–284. IEEE (2004)
383. Hornborg, A.: The ontology of technology. In: A. Hornborg (ed.) Nature, Society, and Justice in the Anthropocene: Unravelling the Money-Energy-Technology Complex, p. 93–113. Cambridge University Press, Cambridge
384. Horne, R.: Maritime autonomous swarms: legal issues and opportunities. In: S. Ng, J. Scholz, H.A. Abbass (eds.) Thinking Swarms. Springer (2025)
385. Horne, R., Deane, F., Joiner, K., Tranter, K.: Navigating to smoother regulatory waters for Australian commercial vessels capable of remote or autonomous operation: a systematic quantitative literature review. Australian Journal of Maritime & Ocean Affairs **15**(4), 496–517 (2023)
386. Horne, R., Putland, T., Brady, M.: Regulating trusted autonomous systems in Australia. arXiv preprint arXiv:2302.03778 (2023)
387. Horne, R., Vanderkooi, M., Guihen, D.: Autonomous vessel regulation in Australia: Why an Australian code of practice is required

388. Horne, R., Vanderkooi, M., Guihen, D.: Introduction to PhD thesis on Australian regulation of autonomous vessels: Navigating to smoother regulatory waters for Australian commercial vessels capable of remote or autonomous operation. In: IndoPacific International Maritime Conference (2023)
389. Horák, V.: Public sociology and hermeneutics. Critical sociology **43**(2), 309–325. https://doi.org/10.1177/0896920515569083
390. Hosang, J.B.: Rules of Engagement and the International Law of Military Operations. Oxford University Press (2020)
391. Hou, K., Yang, Y., Yang, X., Lai, J.: Cooperative control and communication of intelligent swarms: a survey. Control Theory Technol. **18**(2), 114–134 (2020). https://doi.org/10.1007/s11768-020-9195-1
392. Hovy, D., Yang, D.: The importance of modeling social factors of language: Theory and practice. In: The 2021 Conference of the North American Chapter of the Association for Computational Linguistics: Human Language Technologies. Association for Computational Linguistics (2021)
393. Hsu, J.: U.s. navy's drone boat swarm practices harbor defense: A swarm of roboboats shows it can cooperate as a team on a harbor defense mission. IEEE Spectrum (2016). https://spectrum.ieee.org/navy-drone-boat-swarm-practices-harbor-defense
394. Hu, J., Wellman, M.P., et al.: Multiagent reinforcement learning: theoretical framework and an algorithm. In: ICML, vol. 98, pp. 242–250 (1998)
395. Hua, J., Li, Y., Liu, C., Wang, L.: A zero-shot prediction method based on causal inference under non-stationary manufacturing environments for complex manufacturing systems. Robotics and Computer-Integrated Manufacturing **77**, 102,356 (2022)
396. Huang, C., Zhang, Z., Mao, B., Yao, X.: An overview of artificial intelligence ethics. IEEE Transactions on Artificial Intelligence **4**(4), 799–819 (2023). https://doi.org/10.1109/TAI.2022.3194503
397. Huang, X., Liu, W., Chen, X., Wang, X., Wang, H., Lian, D., Wang, Y., Tang, R., Chen, E.: Understanding the planning of LLM agents: A survey (2024)
398. Huang, X., Tian, Y., He, Y., Tong, E., Niu, W., Li, C., Liu, J., Chang, L.: Exposing spoofing attack on flocking-based unmanned aerial vehicle cluster: A threat to swarm intelligence. Security and Communication Networks **2020** (2020). https://doi.org/10.1155/2020/8889122
399. Hubert, K.F., Awa, K.N., Zabelina, D.L.: The current state of artificial intelligence generative language models is more creative than humans on divergent thinking tasks. Scientific Reports **14**(1), 3440 (2024). https://doi.org/10.1038/s41598-024-53303-w
400. Hui, K.P., Phillips, D., Kekirigoda, A.: Beyond line-of-sight range extension with opal using autonomous unmanned aerial vehicles. In: MILCOM 2017, IEEE Military Communications Conference, pp. 279–284. IEEE (2017)
401. Hui, K.P., Pourbeik, P., George, P., Phillips, D., Magrath, S., Kwiatkowski, M.: Opal-a survivability-oriented approach to management of tactical military networks. In: MILCOM 2011, IEEE Military Communications Conference, pp. 1127–1132. IEEE (2011)
402. Hunt, E., Hauert, S.: A checklist for safe robot swarms. Nature Machine Intelligence **2**, 420–422 (2020). https://doi.org/10.1038/s42256-020-0213-2
403. Hunt, E.R., Hauert, S.: A checklist for safe robot swarms. Nature Machine Intelligence **2**, 420–423 (2020). https://doi.org/10.1038/s42256-020-0213-2
404. Hussein, A., Elsawah, S., Abbass, H.A.: Trust mediating reliability–reliance relationship in supervisory control of human–swarm interactions. Human Factors **62**(8), 1237–1248 (2020). https://doi.org/10.1177/0018720819879273
405. Hussein, A., Elsawah, S., Petraki, E., Abbass, H.A.: A machine education approach to swarm decision-making in best-of-n problems. Swarm Intelligence **16**(1), 59–90 (2022). https://doi.org/10.1007/s11721-021-00206-5
406. Hussein, A., Petraki, E., Elsawah, S., Abbass, H.: Autonomous swarm shepherding using curriculum-based reinforcement learning. In: P. Faliszewski, V. Mascardi, C. Pelachaud, M. Taylor (eds.) Proc. of the 21st International Conference on Autonomous Agents and Multiagent Systems (AAMAS 2022). ACM, Auckland, New Zealand (2022)

407. Hüttenrauch, M., Adrian, S., Neumann, G., et al.: Deep reinforcement learning for swarm systems. Journal of Machine Learning Research **20**(54), 1–31 (2019)
408. Hyvärinen, A.: New approximations of differential entropy for independent component analysis and projection pursuit. In: Advances in Neural Information Processing Systems, pp. 273–279. MIT Press, Cambridge, MA, USA (1998)
409. Hyvärinen, A., Oja, E.: Independent component analysis: algorithms and applications. Neural networks **13**(4–5), 411–430 (2000)
410. Hüttenrauch, M., Šošić, A., Neumann, G.: Guided deep reinforcement learning for swarm systems (2017)
411. IBM: Explainable AI. https://www.ibm.com/watson/explainable-ai (2022). Accessed: 15 Nov 2022
412. Ihde, D.: Philosophy of Technology: an Introduction. Paragon House, New York
413. Ikonen, I., Biles, W.E., Kumar, A., Wissel, J.C., Ragade, R.K.: A genetic algorithm for packing three-dimensional non-convex objects having cavities and holes. In: ICGA, pp. 591–598 (1997)
414. International Committee of the Red Cross: A guide to the legal review of new weapons, means or methods of warfare. Geneva, 4 (2006)
415. iPaulCanada: ipaulcaua on twitter. Twitter.com
416. Iraqi, A.: 'Lavender': The AI machine directing Israel's bombing spree in Gaza—972mag.com. https://www.972mag.com/lavender-ai-israeli-army-gaza/ (2024). [Accessed 20-04-2024]
417. Irmer, M.: Bridging inferences: constraining and resolving underspecification in discourse interpretation. Walter de Gruyter, Berlin; Boston (2011)
418. Ismail, Z. H., Sariff, N., & Hurtado, E. G. (2018). A survey and analysis of cooperative multi-agent robot systems: challenges and directions. Applications of Mobile Robots, 5, 8–14.
419. ISO: Systems and software engineering—system life cycle processes. Standard, International Organization for Standardization, Geneva, CH (2018)
420. ISO/IEC: GERAM. in ISO15704-2019, enterprise modelling and architecture—requirements for enterprise-referencing architectures and methodologies. Standard, International Organization for Standardization, Geneva, CH (2019)
421. Ivanov, D., Dolgui, A., Das, A., Sokolov, B.V.: Digital supply chain twins: Managing the ripple effect, resilience, and disruption risks by data-driven optimization, simulation, and visibility. In: Handbook on Ripple Effects in the Supply Chain, pp. 309–332 (2019)
422. Ivanova, K., Gallasch, G.E., Jordans, J.: Automated and autonomous systems for combat service support: scoping study and technology prioritisation. Defence Science and Technology Group Edinburgh SA Australia (2016)
423. J. Badior: Striking in a new way. Defence News, https://www.defence.gov.au/news-events/news/2021-07-22/striking-new-way (2021). Accessed: 22 July 2021
424. Jackson, D.E., Ratnieks, F.L.: Communication in ants. Current biology **16**(15), R570–R574 (2006)
425. James Albus, .: System description and design architecture for multiple autonomous undersea vehicle. NIST Technical Note 1251 (1988)
426. Jamil, T.: Steganography: the art of hiding information in plain sight. IEEE potentials **18**(1), 10–12 (1999)
427. Jans, B., Bethard, S., Vulic, I., Moens, M.F.: Skip n-grams and ranking functions for predicting script events. In: Proceedings of the 13th Conference of the European Chapter of the Association for Computational Linguistics (EACL 2012), pp. 336–344. ACL; East Stroudsburg, PA (2012)
428. Jawhar, I., Mohamed, N., Al-Jaroodi, J.: UAV-based data communication in wireless sensor networks: Models and strategies. In: Unmanned Aircraft Systems (ICUAS), 2015 International Conference on, pp. 687–694. IEEE (2015)
429. Jefferies, E.: The neural basis of semantic cognition: converging evidence from neuropsychology, neuroimaging and TMS. Cortex **49**(3), 611–625 (2013)

430. Jerome, M.M., Sinha, A., Chung, H., et al.: Geometric pattern formation with coordinated double-integrator agents. In: IFAC Symposium on Robot Control 2018, pp. 172–177. Elsevier (2018)
431. Jiang, H., Gai, J., Zhao, S., Chaudhry, P.E., Chaudhry, S.S.: Applications and development of artificial intelligence system from the perspective of system science: A bibliometric review. Systems Research and Behavioral Science **39**(3), 361–378 (2022). https://doi.org/10.1002/sres.2865
432. Johns, B.T.: Disentangling contextual diversity: Communicative need as a lexical organizer. Psychological Review **128**(3), 525 (2021)
433. Johns, B.T.: Distributional social semantics: Inferring word meanings from communication patterns. Cognitive Psychology **131**, 101,441 (2021)
434. Johnson, D., Eyers, A., Lee, K.M.B., Fitch, R.: Hyper-teaming: Adaptive teaming and coordination of multi-domain autonomous robotic systems. In: S. Ng, J. Scholz, H.A. Abbass (eds.) Thinking Swarms. Springer (2025)
435. Johnson, J.: Automating the OODA loop in an age of intelligent machines: reaffirming the role of humans in command-and control decision-making in the digital age. Defence Studies p. 1–25
436. Johnston, T.: Auslan: The Sign Language of the Australian Deaf Community. Ph.D. thesis, University of Sydney, Sydney (1989)
437. Joneidy, Sina, and Andrew Basden. "Exploring Dooyeweerd's Aspects For Understanding Perceived Usefulness Of Information Systems." Re-Integrating Technology And Economy In Human Life And Societys & Preface, vol. 2, Rozenberg Publishers, 2011, https://rozenbergquarterly.com/iide-proceedings-2011-exploring-dooyeweerds-aspects-for-understanding-perceived-usefulness-of-information-systems/
438. Jones, A., Lin, A., Zhang, P., Tall, M.: Development of a steering behavior for decentralized unmanned underwater vehicle swarm shape formation. In: OCEANS 2021: San Diego–Porto, pp. 1–4. IEEE (2021)
439. Jones, D., Snider, C., Aydin Nassehi, A., Yon, J., Hicks, B.: Non-cooperative games. CIRP JMST **29**, 36–52 (2020)
440. Jordahn, S.: Drift's eibphilharmonie installation disrupted by drones. Dezeen.com https://www.dezeen.com/2022/05/03/studio-drift-elbphilharmonie-drone-installation-cancelled-breaking-waves/. Retrieved from
441. Juang, J.C.: On the formation patterns under generalized cyclic pursuit. IEEE Transactions on Automatic Control **58**(9), 2401–2405 (2013)
442. Juang, J.C.: On the formation patterns in cyclic pursuit of double-integrator agents. In: Proc. IEEE Conference on Decision and Control, pp. 5924–5929. IEEE (Dec. 2012)
443. Juang, J.C.: Cyclic pursuit control for dynamic coverage. In: Proc. IEEE Conference on Control Applications, pp. 2147–2152. IEEE (Oct. 2014)
444. Judson, G., Horne, R.: The regulatory approach for vessels capable of autonomous and remote-controlled operation. In: International Maritime Conference, Pacific (2019)
445. Kallenborn, Z.: Swords and shields: Autonomy, AI, and the offense-defense balance. Georgetown Journal of International Affairs https://gjia.georgetown.edu/2021/11/22/swords-and-shields-autonomy-ai-and-the-offense-defense-balance/. Retrieved from
446. Kallenborn Z.: Future warfare series no. 60: Are drone swarms weapons of mass destruction? The Counterproliferation Papers, United States Air Force Center for Strategic Deterrence Studies (2020). Accessed: 6 May 2020
447. Kalman, R.: A new approach to linear filtering and prediction problems. Journal of Basic Engineering **82**(1), 35–45 (1960)
448. Kalpic, B., Bernus, P.: Business process modelling through the knowledge management perspective. Journal of Knowledge Management **10**(3), 40–56 (2006)
449. Kang, T., Kim, K., Kim, S., Lee, D.: Underwater laser communication with sloped pulse modulation in turbid water. International Journal of Distributed Sensor Networks **15**(3), 155014771983,787 (2019)

450. Kar, A.K.: Bio inspired computing–a review of algorithms and scope of applications. Expert Systems with Applications **59**, 20–32 (2016)
451. Karaboga, D.: An idea based on honey bee swarm for numerical optimization. Tech. Rep. Technical report-tr06, (2005)
452. Karlsrud, J., Rósen, F.: In the eye of the beholder? un and the use of drones to protect civilians. Stability: International Journal of Security and Development **2**
453. Keane, J., Joiner, K.F., Arulampalam, S., Webber, R.: Expediting recovery of autonomous underwater vehicles in dynamic mission environments: A system-of-systems challenge for underwater warfare. Journal of Field Robotics **39**(8), 1323–1340 (2022)
454. Kelly, G.A.: The psychology of personal constructs. Vol. 1. A theory of personality. Vol. 2. Clinical diagnosis and psychotherapy. The psychology of personal constructs. Vol. 1. A theory of personality. Vol. 2. Clinical diagnosis and psychotherapy. W. W. Norton, Oxford, England (1955)
455. Kennedy, J.: Swarm intelligence. In: Handbook of nature-inspired and innovative computing, pp. 187–219. Springer (2006)
456. Kennedy, J., Eberhart, R.: Particle swarm optimization. In: Proceedings of ICNN'95-International Conference on Neural Networks, pp. 1942–1948. 4 (1995)
457. Kennedy, J., Eberhart, R.: Particle swarm optimization. pp. 1942–1948 (1995)
458. Kerr, C., Jaradat, R., Ibne Hossain, N.U.: Battlefield mapping by an unmanned aerial vehicle swarm: Applied systems engineering processes and architectural considerations from system of systems. IEEE Access **8**, 20,892–20,903 (2020). https://doi.org/10.1109/ACCESS.2020.2968348
459. Kershenbaum, A.: Entropy rate as a measure of animal vocal complexity. Bioacoustics **23**(3), 195–208 (2014)
460. Khan, A., Ramachandran, V.: A peer-to-peer associative memory network for intelligent information systems. In: ACIS 2002 Proceedings, pp. 6–17 (2002)
461. Khan, R.A., Meyer, A., Konik, H., Bouakaz, S.: Facial Expression Recognition using Entropy and Brightness Features. In: IEEE (ed.) 11th Int. Conf. on Intell. Sys. Design and Applic.(ISDA) (2011)
462. Kholidy, H.A.: An intelligent swarm based prediction approach for predicting cloud computing user resource needs. Computer Communications **151**, 133–144 (2020)
463. Khoo, S., Xie, L., Man, Z.: Robust finite-time consensus tracking algorithm for multirobot systems. IEEE/ASME transactions on mechatronics **14**(2), 219–228 (2009)
464. Kim, J.: Cooperative localization and unknown currents estimation using multiple autonomous underwater vehicles. IEEE Robotics and Automation Letters **5**(2), 2365–2371 (2020)
465. Kim, J., André, E.: Emotion recognition based on physiological changes in music listening. IEEE Trans. Pattern Anal. Mach. Intell. **30**(12), 2067–2083 (2008)
466. Kim, Y., Jernite, Y., Sontag, D., Rush, A.M.: Character-aware neural language models. In: Thirtieth AAAI conference on artificial intelligence (2016)
467. Kimura, M., Moehlis, J.: Novel vehicular trajectories for collective motion from coupled oscillator steering control. Journal on Applied Dynamical Systems **7**(4), 1191–1212 (2008)
468. Klarin, A.: Mapping product and service innovation: A bibliometric analysis and a typology. Technological Forecasting and Social Change **149**, 119,776 (2019). https://doi.org/10.1016/j.techfore.2019.119776
469. Klarin, A.H., A.A., Sharmelly, R.: Professionalism in artificial intelligence: The link between technology and ethics. Systems Research and Behavioral Science pp. 1–24 (2024). https://doi.org/10.1002/sres.2994
470. Klarin, A., Inkizhinov, B., Nazarov, D., Gorenskaia, E.: International business education: What we know and what we have yet to develop. International Business Review **30**(5), 101,833 (2021). https://doi.org/10.1016/j.ibusrev.2021.101833
471. Klarin, A., Seet, P., Jones, J., Dowse, A., Suter, D., Johnstone, M., Cripps, H., Sharafizad, J., Marceddo, T.: Understanding the roots of 'thinking swarms' in defence to find the path forward: A scientometric study of autonomous systems. In: S. Ng, J. Scholz, H.A. Abbass (eds.) Thinking Swarms. Springer (2025)

472. Klarin, A., Suseno, Y.: An integrative literature review of social entrepreneurship research: mapping the literature and future research directions. Business and Society **62**(3), 565–611 (2023). https://doi.org/10.1177/00076503221101611
473. Klarin, A., Suseno, Y., Lajom, J.A.L.: Systematic literature review of convergence: A systems perspective and re-evaluation of the convergence process. IEEE Transactions on Engineering Management **70**(4), 1531–1543 (2023). https://doi.org/10.1109/TEM.2021.3126055
474. Klein, N.: Maritime autonomous vehicles within the international law framework to enhance maritime security. International Law Studies **95**(1), 8 (2019)
475. Klein, N., Guilfoyle, D., Karim, M.S., McLaughlin, R.: Maritime autonomous vehicles: New frontiers in the law of the sea. International & Comparative Law Quarterly **69**(3), 719–734 (2020)
476. Kleinschmidt, D.F., Jaeger, T.F.: Robust speech perception: recognize the familiar, generalize to the similar, and adapt to the novel. Psychological review **122**(2), 148 (2015)
477. for the diffusion of useful knowledge, S., Long, G., cyclopaedia, P.: The penny cyclopædia [ed. by G. Long]. v. 25 (1843)
478. Kolling, A., Walker, P., Chakraborty, N., Sycara, K., Lewis, M.: Human interaction with robot swarms: A survey. IEEE Transactions on Human-Machine Systems **46**(1), 9–26 (2016). https://doi.org/10.1109/THMS.2015.2480801
479. Koop, C., Lodge, M.: What is regulation? an interdisciplinary concept analysis. Regulation & Governance **11**(1), 95–108 (2017)
480. Koos, S., Mouret, J.B., Doncieux, S.: Crossing the reality gap in evolutionary robotics by promoting transferable controllers. In: Proceedings of the 12th annual conference on Genetic and evolutionary computation, pp. 119–126. ACM (2010)
481. Koplenig, A.: Using the parameters of the zipf–mandelbrot law to measure diachronic lexical, syntactical and stylistic changes–a large-scale corpus analysis. Corpus Linguistics and Linguistic Theory **14**(1), 1–34 (2018)
482. Kotseruba, I., Tsotsos, J.K.: 40 years of cognitive architectures: core cognitive abilities and practical applications. Artificial Itelligence Review **53**, 17–94 (2020)
483. Koubâa, A., Allouch, A., Alajlan, M., Javed, Y., Belghith, A., Khalgui, M.: Micro air vehicle link (mavlink) in a nutshell: A survey. IEEE Access **7**, 87,658–87,680 (2019)
484. Krishnanand, K.N., Ghose, D., Prasad, B.: Glowworm swarm based optimization algorithm for multimodal functions with collective robotics applications. MGS **2**(3), 209–222 (2006). https://doi.org/10.3233/MGS-2006-2301
485. Kuhn, T.: The Structure of Scientific Revolutions, [1962, 3rd edn. University of Chicago Press, Chicago
486. Kuhn, T.: The Structure of Scientific Revolutions. University of Chicago Press, Chicago (1973)
487. Kullback, S., Leibler, R.A.: On information and sufficiency. Annals of Mathematical Statistics **22**(1), 79–86 (1951)
488. Kumar, A., Singh, D.: Butterfly optimizer. pp. 1–6 (2015). https://doi.org/10.1109/WCI.2015.7495523
489. Kumar, A., Zhou, A., Tucker, G., Levine, S.: Conservative q-learning for offline reinforcement learning. In: H. Larochelle, M. Ranzato, R. Hadsell, M. Balcan, H. Lin (eds.) Advances in Neural Information Processing Systems, vol. 33, pp. 1179–1191. Curran Associates, Inc. (2020)
490. Kumar, A.A.: Semantic memory: A review of methods, models, and current challenges. Psychonomic Bulletin & Review **28**(1), 40–80 (2021)
491. Kumar, A.A., Steyvers, M., Balota, D.A.: A critical review of network-based and distributional approaches to semantic memory structure and processes. Topics in Cognitive Science (2022)
492. Kuperberg, G.R., Jaeger, T.F.: What do we mean by prediction in language comprehension? Language, cognition and neuroscience **31**(1), 32–59 (2016)
493. Kuzmenko, E., Herbelot, A.: Distributional semantics in the real world: building word vector representations from a truth-theoretic model. In: Proceedings of the 13th International Conference on Computational Semantics-Short Papers, pp. 16–23 (2019)

494. Laird, J.E., Congdon, C.B., Assanie, M., Derbinsky, N., Xu, J.: The SOAR user's manual, 9.6.0. Tech. rep., EECS Univ. of Michigan (2017). https://soar.eecs.umich.edu/downloads/SoarManual.pdf
495. Laird, J.E., Lebiere, C., Rosenbloom, P.S.: A standard model for the mind: Toward a common computational framework across artificial intelligence, cognitive science, neuroscience, and robotics. AI Magazine **38**(4), 13–26 (2017)
496. Lakoff, G.: Metaphors and war: The metaphor system used to justify war in the Gulf. John Benjamins, Amsterdam (1992)
497. Lakoff, G., Johnson, M.: Conceptual metaphor in everyday language. Journal of Philosophy **77**(8), 453–486 (1980)
498. Lanctot, M., Zambaldi, V., Gruslys, A., Lazaridou, A., Tuyls, K., Perolat, J., Silver, D., Graepel, T.: A unified game-theoretic approach to multiagent reinforcement learning. In: I. Guyon, U.V. Luxburg, S. Bengio, H. Wallach, R. Fergus, S. Vishwanathan, R. Garnett (eds.) Advances in Neural Information Processing Systems, vol. 30. Curran Associates, Inc. (2017)
499. Lanzisera, S., Lin, D.T., Pister, K.S.: Rf time of flight ranging for wireless sensor network localization. In: Intelligent Solutions in Embedded Systems, 2006 International Workshop on, pp. 1–12. IEEE (2006)
500. Lappas, V., Shin, H.S., Tsourdos, A., Lindgren, D., Bertrand, S., Marzat, J., Piet-Lahanier, H., Daramouskas, Y., Kostopoulos, V.: Autonomous unmanned heterogeneous vehicles for persistent monitoring. Drones **6**(4), 94 (2022). https://doi.org/10.3390/drones6040094
501. Larochelle, H., Erhan, D., Bengio, Y.: Zero-data learning of new tasks. In: AAAI, vol. 1, p. 3 (2008)
502. Latour, B., Woolgar, S.: Laboratory Life: The Construction of Scientific Facts. Princeton University Press, Princeton
503. Laub, A.: Matrix Analysis for Scientists and Engineers. Society for Industrial and Applied Mathematics (2005)
504. Lawrence, J.: A catalog of special plane curves. Dover Publications (1972)
505. Lawson, C.: An ontology of technology: Artefacts, relations and functions. Techne **12**(1)
506. Lee, J., Back, A.: Subword Entropy Processing for Spoken Language Analysis. Tech. rep., School of Information Technology and Electrical Engineering, The University of Queensland, Brisbane (2019)
507. Lee, J., Bagheri, B., Kao, H.A.: A cyber-physical systems architecture for industry 4.0-based manufacturing systems. Manufacturing Letters **3**, 18–23 (2015)
508. Lee, J.H., Ahn, C.W., An, J.: A honey bee swarm-inspired cooperation algorithm for foraging swarm robots: An empirical analysis. In: 2013 IEEE/ASME International Conference on Advanced Intelligent Mechatronics, pp. 489–493 (2013). https://doi.org/10.1109/AIM.2013.6584139
509. Lee, K.M.B., Kong, F.H., Cannizzaro, R., Palmer, J.L., Johnson, D., Yoo, C., Fitch, R.: Decentralised intelligence, surveillance, and reconnaissance in unknown environments with heterogeneous multi-robot systems. In: Proc. of ICRA2021 Workshop on Robot Swarms in the Real World: From Design to Deployment, p. **Best Poster Award** (2021)
510. Lee, K.M.B., Kong, F.H., Cannizzaro, R., Palmer, J.L., Johnson, D., Yoo, C., Fitch, R.: An upper confidence bound for simultaneous exploration and exploitation in heterogeneous multi-robot systems. In: Proc. of IEEE ICRA, pp. 8685–8691 (2021)
511. Lee, K.M.B., Yoo, C., Fitch, R.: Signal temporal logic synthesis as probabilistic inference. In: Proceedings of IEEE ICRA, pp. 5483–5489 (2021)
512. Lee, S.K., Fekete, S.P., McLurkin, J.: Structured triangulation in multi-robot systems: Coverage, patrolling, voronoi partitions, and geodesic centers. The International Journal of Robotics Research **35**(10), 1234–1260 (2016). https://doi.org/10.1177/0278364915624974
513. Lee, S.S.: Dynamics and control of satellite relative motion: designs and applications. Ph.D. thesis, Virginia Tech (2009)
514. Leguizamon, S.C., Scott, T.F.: Mimicking DNA functions with abiotic, sequence-defined polymers. Polymer Reviews **62**(3), 626–651 (2022)

515. Lehoux, D.: Why does Aristotle think bees are divine? proportion, triplicity and order in the natural world. The British Journal for the History of Science **52**(3), 383–403 (2019). https://doi.org/10.1017/S0007087419000165. Edition: 2019/04/30 publisher: Cambridge University Press
516. Lesne, A., Blanc, J.L., Pezard, L.: Entropy estimation of very short symbolic sequences. Phys. Rev. E **79**, 046,208 (2009)
517. Leu, G., Lakshika, E., Tang, J., Merrick, K., Barlow, M.: Machine education-the way forward for achieving trust-enabled machine agents (2017)
518. Levy, O., Goldberg, Y.: Neural word embedding as implicit matrix factorization. Advances in neural information processing systems **27** (2014)
519. Levy, R.: Expectation-based syntactic comprehension. Cognition **106**(3), 1126–1177 (2008)
520. Levy, R.: Expectation-based syntactic comprehension. Cognition **106**(3), 1126–1177 (2008)
521. Leydesdorff, L., Hellsten, I.: Metaphors and diaphors in science communication: Mapping the case of stem cell research. Science Communication **27**(1), 64–99 (2005)
522. Li, M., Lu, Y.: Angle-of-arrival estimation for localization and communication in wireless networks. In: Signal Processing Conference, 2008 16th European, pp. 1–5. IEEE (2008)
523. Li, Y., & Tan, C. (2019). A survey of the consensus for multi-agent systems. Systems Science & Control Engineering, 7(1), 468-482.
524. Li, M., Xie, Y., Gao, Y., Zhao, Y.: Organization virtualization driven by artificial intelligence. Systems Research and Behavioral Science **39**(3), 633–640 (2022). https://doi.org/10.1002/sres.2863
525. Li, W.: Random texts exhibit Zipf's-law-like word frequency distribution. IEEE Transactions on Information Theory **38**(6), 1842–1845 (1992)
526. Li, X., Jiang, L., Liu, X., Dang, R., Liu, F., Wei, W., Zhang, T., Wang, G.: Modeling and implementation of a novel amphibious robot with multimode motion. Industrial Robot: the international journal of robotics research and application **49**(5), 947–961 (2022)
527. Li, X.l.: An optimizing method based on autonomous animats: fish-swarm algorithm. Systems engineering-theory & practice **22**(11), 32–38 (2002)
528. Li, Y., Xiang, Y., Deng, H., Sun, Z.: An entropy-based index for fine-scale mapping of disease genes. Journal of Genetics and Genomics **34**(7), 661–668 (2007)
529. Li, Z., Ding, X., Liu, T.: Constructing narrative event evolutionary graph for script event prediction. arXiv preprint arXiv:1805.05081 (2018)
530. Licklider, J.C.R.: Man-computer symbiosis. IRE Trans. Hum. Factors Electron. **HFE-1**(1), 4–11 (1960). https://doi.org/10.1109/THFE2.1960.4503259
531. Lien, J.M., Bayazit, O.B., Sowell, R.T., Rodriguez, S., Amato, N.M.: Shepherding behaviors. In: IEEE International Conference on Robotics and Automation, 2004. Proceedings. ICRA'04. 2004, vol. 4, pp. 4159–4164. IEEE (2004)
532. Liivoja, R., Massingham, E., McKenzie, S.: The legal requirement for command and the future of autonomous military platforms. International Law Studies **99**(1), 27 (2022)
533. Lim, K.Y.H., Zheng, P., Chen, C.H.: A state-of-the-art survey of digital twin: techniques, engineering product lifecycle management and business innovation perspectives. J. Intelligent Manufacturing **31**(6), 1313–1337 (2020)
534. Lin, Z., Broucke, M., Francis, B.: Local control strategies for groups of mobile autonomous agents. IEEE Transactions on Automatic Control **49**(4), 622–629 (2004)
535. Ling, H., Luo, H., Chen, H., Bai, L., Zhu, T., Wang, Y.: Modelling and simulation of distributed UAV swarm cooperative planning and perception. International Journal of Aerospace Engineering **2021**, 1–11 (2021)
536. Ling, J.: Is your new car a threat to national security? wired
537. Liu, J., Anavatti, S., Garratt, M., Tan, K.C., Abbass, H.A.: A survey, taxonomy and progress evaluation of three decades of swarm optimisation. Artif Intell Rev **55**(5), 3607–3725 (2022). https://doi.org/10.1007/s10462-021-10095-z
538. Liu, Z., Meyendorf, N., Mrad, N.: The role of data fusion in predictive maintenance using digital twin. AIP Conf. Proc. 020023 **1949**(1) (2018)

539. Llorca, J., Milner, S.D., Davis, C.C.: Mobility control for joint coverage-connectivity optimization in directional wireless backbone networks. In: Military Communications Conference, 2007. MILCOM 2007. IEEE, pp. 1–7. IEEE (2007)
540. Lockwood, E.: Book of Curves. Cambridge University Press (2007)
541. Lodge, R., Zamani, M., Marsh, L., Sims, B., Hunjet, R.: A hybrid multi-modal approach for flocking. In: 2019 12th Asian Control Conference (ASCC), pp. 126–131 (2019)
542. Lu, Y., Morris, K.C., Frechette, S.P.: Current standards landscape for smart manufacturing systems. Report, National Institute of Standards and Technology, Gaithersburg, MD (2016)
543. Luchins, A.S.: Mechanization in problem solving: The effect of einstellung. Psychological Monographs **54**, 1–95 (1942)
544. Luckcuck, M., Farrell, M., Dennis, L.A., Dixon, C., Fisher, M.: Formal specification and verification of autonomous robotic systems: A survey. ACM Computing Surveys (CSUR) **52**(5), 1–41 (2019)
545. Luhmann, N., Clarke, B., Hansen, M.B.N.: Self-organization and autopoiesis. In: B. Clarke, M.B.N. Hansen (eds.) Emergence and Embodiment, pp. 143–156 (2009). http://read.dukeupress.edu/books/book/1324/chapter/164857/SelfOrganization-and-Autopoiesis
546. Luke, S., Cioffi-Revilla, C., Panait, L., Sullivan, K., Balan, G.: Mason: A multiagent simulation environment. Simulation **81**(7), 517–527 (2005)
547. Lundin, D., Gudmundsson, J.T., Minea, T.: High Power Impulse Magnetron Sputtering: Fundamentals, Technologies, Challenges and Applications. Elsevier (2019)
548. Lundquist, E.: Nato's autonomous UUVS are working together to find mines (2021). The Maritime Executive, 29 January 2021
549. Luo, S., Zhang, Z., Wang, S., Zhang, S., Dai, J., Bu, X., An, J.: Network for hypersonic UCAV swarms. Science China Information Sciences **63**(4), 1–28 (2020). https://doi.org/10.1007/s11432-019-2765-7
550. Lupton, D.: The quantified self. Online resource (154 pages)).
551. Lyu, C., Lu, D., Xiong, C., Hu, R., Jin, Y., Wang, J., Zeng, Z., Lian, L.: Toward a gliding hybrid aerial underwater vehicle: Design, fabrication, and experiments. Journal of Field Robotics **39**(5), 543–556 (2022)
552. Ma, S.: Calculation of entropy from data of motion. Journal of Statistical Physics **26**(2), 221–240 (1981)
553. Ma, Y., Zhao, Y., Incecik, A., Yan, X., Wang, Y., Li, Z.: A collision avoidance approach via negotiation protocol for a swarm of USVS. Ocean Engineering **224**, 108,713 (2021)
554. Macknight, V., Medvecky, F.: (google-)knowing economics. Social Epistemology **34**(3), 213–226 (2020). https://doi.org/10.1080/02691728.2019.1702735
555. Mahadevan, K., Somanath, S., Sharlin, E.: Communicating Awareness and Intent in Autonomous Vehicle-Pedestrian Interaction. In: Proceedings of the 2018 CHI Conference on Human Factors in Computing Systems, pp. 1–12. Association for Computing Machinery, New York (2018)
556. Malle, B.F., Ullman, D.: Chapter 1—a multidimensional conception and measure of human-robot trust. In: C.S. Nam, J.B. Lyons (eds.) Trust in Human-Robot Interaction, pp. 3–25. Academic Press (2021). https://doi.org/10.1016/B978-0-12-819472-0.00001-0. https://www.sciencedirect.com/science/article/pii/B9780128194720000010
557. Mancao, C., Hammerschmidt, W.: Epstein-Barr virus latent membrane protein 2A is a B-cell receptor mimic and essential for B-cell survival. Blood, The Journal of the American Society of Hematology **110**(10), 3715–3721 (2007)
558. Mandelbrot, B.: An informational theory of the statistical structure of language. Communication theory **84**, 486–502 (1953)
559. Marslen-Wilson, W., Brown, C.M., Tyler, L.K.: Lexical representations in spoken language comprehension. Language and Cognitive Processes **3**(1), 1–16 (1988)
560. Martin, M.: Local and global processing: The role of sparsity. Memory & Cognition **7**(6), 476–484 (1979)
561. Martínez-Plumed, F., Ferri, C., Hernández-Orallo, J., Ramírez-Quintana, M.J.: Knowledge acquisition with forgetting: an incremental and developmental setting. Adaptive Behavior **23**(5), 283–299 (2015)

562. Maslow, A.H.: Toward a psychology of being. Simon and Schuster (1968)
563. Mathis, C.: Data lakes. Datenbank-Spektrum **17**(3), 289–293 (2017)
564. Maturana, H.R.: The Biological Foundations of Self-Consciousness and the Physical Domain of Existence. In: E. Caianiello (ed.) Physics of Cognitive Processes, pp. 324–379. World Scientific, Singapore (1987)
565. Max, J.: Quantizing for minimum distortion. IRE Transactions on Information Theory **6**(1), 7–12 (1960)
566. McCall, F., Hussein, A., Petraki, E., Elsawah, S., Abbass, H.: Towards a systematic educational framework for human-machine teaming. In: 2021 IEEE International Conference on Engineering, Technology & Education (TALE), pp. 375–382 (2021). https://doi.org/10.1109/TALE52509.2021.9678853
567. McCammon, S., Marcon dos Santos, G., Frantz, M., Welch, T.P., Best, G., Shearman, R.K., Nash, J.D., Barth, J.A., Adams, J.A., Hollinger, G.A.: Ocean front detection and tracking using a team of heterogeneous marine vehicles. Journal of Field Robotics **38**(6), 854–881 (2021)
568. McClelland, J.: The review of weapons in accordance with article 36 of additional protocol i. International Review of the Red Cross **85**(850), 397–420 (2003)
569. McGann, C., Py, F., Rajan, K., Thomas, H., Henthorn, R., McEwen, R.: T-rex: A model-based architecture for AUV control. In: 3rd Workshop on Planning and Plan Execution for Real-World Systems, vol. 2007 (2007)
570. McGreal, S.: The War on Hospital Ships, 1914–1918. Casemate Publishers, Philadelphia, PA. (2009)
571. McKenzie, S.: When is a ship a ship? use by state armed forces of uncrewed maritime vehicles and the united nations convention on the law of the sea. Melbourne Journal of International Law **21**(2), 373–402 (2020)
572. McKenzie, S.: Autonomous technology and dynamic obligations: Uncrewed maritime vehicles and the regulation of maritime military surveillance in the exclusive economic zone. Asian Journal of International Law **11**(1), 146–175 (2021)
573. McKenzie, S.: Law and the future of war (2021)
574. McNabb, M.: Us department of interior grounds DJI drones
575. McNeill, H.: Perth city of light: More than 50 drones plunge into swan river during light show. WAToday.com https://www.watoday.com.au/national/western-australia/an-expensive-event-50-drones-plunge-into-swan-river-during-sky-show-fail-20221121-p5c037.html. Retrieved from
576. Mehta, A.M., Lanzisera, S., Pister, K.S.: Steganography in 802.15. 4 wireless communication. In: 2008 2nd International Symposium on Advanced Networks and Telecommunication Systems, pp. 1–3. IEEE (2008)
577. Meng, X., Liu, Y., Gao, X., Zhang, H., Tan, Y., Shi, Y., Coello, C.A.C.: A new bio-inspired algorithm: Chicken swarm optimization. In: Y. Tan, Y. Shi, C.A.C. Coello (eds.) Advances in Swarm Intelligence, vol. 8794, pp. 86–94. Cham (2014). http://link.springer.com/10.1007/978-3-319-11857-4_10
578. Mero, T.: The martens clause, principles of humanity, and dictates of public conscience. American Journal of International Law **94**(1), 78–89 (2000)
579. Merriam-Webster: "noise" (2022). https://www.merriam-webster.com/dictionary/noise
580. Mesbahi, M., Egerstedt, M.: Graph theoretic methods in multiagent networks. Princeton University Press (2010)
581. Meyerson, E., Miikkulainen, R.: Discovering evolutionary stepping stones through behavior domination. In: Proceedings of the Genetic and Evolutionary Computation Conference, pp. 139–146. ACM (2017)
582. Mikolov, T., Joulin, A., Baroni, M.: A roadmap towards machine intelligence. In: A. Gelbukh (ed.) Proc. 17th Int Conf CICLing (Revised Selected Papers, Part I) LNCS 9623, pp. 29–61. Springer (2018)
583. Miller, R.: Sam Altman: Size of LLMs won't matter as much moving forward | TechCrunch—techcrunch.com. https://techcrunch.com/2023/04/14/sam-altman-size-of-llms-wont-matter-as-much-moving-forward/ (2024). [Accessed 17-04-2024]

584. Mingers, J.: Can social systems be autopoietic? Assessing Luhmann's social theory. The Sociological Review **50**(2), 278–299 (2002). Publisher: Wiley Online Library
585. Mirjalili, S., Mirjalili, S.M., Lewis, A.: Grey wolf optimizer. Advances in engineering software **69**, 46–61 (2014)
586. Mischiati, M., Krishnaprasad, P.: Motion camouflage for coverage. In: Proceedings of the 2010 American Control Conference, pp. 6429–6435. IEEE (2010)
587. Mitcham, C.: Thinking Through Technology: The Path Between Engineering and Philosophy. University of Chicago Press, Chicago
588. Mitchell, W.: Agile sense-making in the battlespace. The Int. C2 Journal **4**(1), 1–33 (2010)
589. Mittelstadt, B., Wachter, S., Russell, C.: To protect science, we must use LLMs as zero-shot translators. Nature Human Behaviour **7**(11), 1830–1832 (2023). https://doi.org/10.1038/s41562-023-01744-0
590. Mizokami, K.: For First Time, Drones Autonomously Attacked Humans: UN Report. Popular Mechanics
591. Mizokami, K.: In a historic first, Ukraine attacked a Russian fleet with autonomous robo-ships: This is the beginning of a new era in naval warfare. https://www.popularmechanics.com/military/navy-ships/a41835387/ukraine-attacks-russian-fleet-with-robo-ships/ (2022). Accessed: 5 Nov 2022
592. Mohamed, R.E., Elsayed, S., Hunjet, R., Abbass, H.: A graph-based approach for shepherding swarms with limited sensing range. In: 2021 IEEE Congress on Evolutionary Computation (CEC), pp. 2315–2322 (2021). https://doi.org/10.1109/CEC45853.2021.9504706
593. Mollica, F., Bacon, G., Zaslavsky, N., Xu, Y., Regier, T., Kemp, C.: The forms and meanings of grammatical markers support efficient communication. Proceedings of the National Academy of Sciences **118**(49), e2025993,118 (2021)
594. Monsingh, J.M., Sinha, A., Chung, H.: Orbital pattern generation for double-integrator agents using consensus strategy. In: Proceedings of ACRA, vol. 2019, pp. 1–8 (2019)
595. Monsingh, J.M., Sinha, A., Chung, H.: Trochoidal patterns generation using generalized consensus strategy for single-integrator kinematic agents. European Journal of Control **Under Revision** (2022)
596. Monsingh, J.M., Sinha, A., Chung, H.: Trochoidal patterns generation using generalized consensus strategy for double-integrator dynamic agents. European Journal of Control **76**, 100,928 (2024)
597. Montalvão, J., Silva, D., Attux, R.: Simple entropy estimator for small datasets. Electronics Letters **48**, 1059–1061 (2012)
598. Moon, T.K.: The expectation-maximization algorithm. IEEE Signal processing magazine **13**(6), 47–60 (1996)
599. Moshfeghi, Y., Piwowarski, B., Jose, J.M.: Handling data sparsity in collaborative filtering using emotion and semantic based features. In: Proceedings of the 34th international ACM SIGIR conference on Research and development in Information Retrieval, pp. 625–634 (2011)
600. Mou, Z., Zhang, Y., Gao, F., Wang, H., Zhang, T., Han, Z.: Deep reinforcement learning based three-dimensional area coverage with UAV swarm. IEEE Journal on Selected Areas in Communications **39**(10), 3160–3176 (2021). https://doi.org/10.1109/JSAC.2021.3088718
601. Moy, G., Shekh, S., Oxenham, M., Ellis-Steinborner, S.: Recent advances in artificial intelligence and their impact on defence. Defence Science and Technology Group, Canberra, 25, (2020)
602. Moyses, H., Palacci, J., Sacanna, S., Grier, D.G.: Trochoidal trajectories of self-propelled Janus particles in a diverging laser beam. Soft Matter **12**(30), 6357–6364 (2016)
603. Mozur, P.: Drone maker D.J.I. may be sending data to china, U.S
604. Mueller, A.: Modern robotics: Mechanics, planning, and control [bookshelf]. IEEE Control Systems Magazine **39**(6), 100–102 (2019)
605. Muhuri, P.K., Ashraf, Z., Goel, S.: A novel image steganographic method based on integer wavelet transformation and particle swarm optimization. Applied Soft Computing **92**, 106,257 (2020)

606. Murayama, I.: Role of agency in causal understanding of natural phenomena. Human Development **37**(4), 198–206 (1994)
607. Murray, R. M. (May 3, 2007). "Recent Research in Cooperative Control of Multivehicle Systems." ASME. J. Dyn. Sys., Meas., Control. September 2007; 129(5): 571–583
608. Muscettola, N., Nayak, P.P., Pell, B., Williams, B.C.: Remote agent: To boldly go where no AI system has gone before. Artificial intelligence **103**(1–2), 5–47 (1998)
609. Nagavalli, S., Chakraborty, N., Sycara, K.: Automated sequencing of swarm behaviors for supervisory control of robotic swarms. In: 2017 IEEE International Conference on Robotics and Automation (ICRA), pp. 2674–2681. IEEE (2017)
610. Narvekar, S., Peng, B., Leonetti, M., Sinapov, J., Taylor, M.E., Stone, P.: Curriculum learning for reinforcement learning domains: A framework and survey. Journal of Machine Learning Research **21**, 1–50 (2020)
611. Nasution, B.B., Khan, A.I.: A hierarchical graph neuron scheme for real-time pattern recognition. IEEE Transactions on Neural Networks **19**(2), 212–229 (2008)
612. NATO: Deep dive recap: Emerging and disruptive technologies, and the gender perspective. https://www.nato.int/cps/en/natohq/news_196562.htm?selectedLocale=en#:~:text=In%20the%20NATO%20AI%20Strategy,%2C%20governability%2C%20and%20bias%20mitigation (2022). Accessed: 7 June 2022
613. Navarro, I., & Matía, F. (2013). An introduction to swarm robotics. International Scholarly Research Notices, 2013(1), 608164
614. Nazarov, D., Klarin, A.: Taxonomy of industry 4.0 research: Mapping scholarship and industry insights. Systems Research and Behavioral Science **37**(4), 535–556 (2020). https://doi.org/10.1002/sres.2700
615. Negri, E., Fumagalli, L., Macchi, M.: "a review of the roles of digital twin in CPS-based production systems". Procedia Manufacturing **11**, 939–948 (2017)
616. Nelson, A.: The Right Way to Regulate AI—foreignaffairs.com. https://www.foreignaffairs.com/united-states/right-way-regulate-artificial-intelligence-alondra-nelson (2024). [Accessed 05-04-2024]
617. Nelson, J., Gorichanaz, T.: Trust as an ethical value in emerging technology governance: The case of drone regulation. Technology in Society **59**, 101,131 (2019). https://doi.org/10.1016/j.techsoc.2019.04.007. https://www.sciencedirect.com/science/article/pii/S0160791X18301854
618. Nemenman, I., Shafee, F., Bialek, W.: Entropy and inference, revisited. In: T.G. Dietterich, S. Becker, Z. Ghahramani (eds.) Advances in Neural Information Processing Systems 14, pp. 471–478. MIT Press, Cambridge, MA (2002)
619. Nguyen, H., Hussein, A., Garratt, M.A., Abbass, H.A.: Swarm metaverse for multi-level autonomy using digital twins. Sensors **23**(10) (2023). https://doi.org/10.3390/s23104892
620. Ni, C.C., Lin, Y.Y., Gao, J., Gu, D., Saucan, E.: Ricci curvature of the internet topology. In: Proceedings of the IEEE Conference on Computer Communications INFOCOM 2015. IEEE Computer Society (2015)
621. Nield, D.: This Horrifying 'Slaughterbot' Video Is The Best Warning Against Autonomous Weapons—sciencealert.com. https://www.sciencealert.com/chilling-drone-video-shows-a-disturbing-vision-of-an-ai-controlled-future (2024). [Accessed 19-06-2024]
622. Nietzsche, F.: The Collected Works of Friedrich Nietzsche. DigiCat (2022)
623. Niklasson, L., Riveiro, M., Johansson, F., Dahlbom, A., Falkman, G., Ziemke, T., Brax, C., Kronhamn, T., Smedberg, M., Warston, H., Gustavsson, P.M.: Extending the scope of situation analysis. In: Proc. 11th Int. Conf. on Information Fusion, pp. 1–8. IEEE (2008)
624. Nikooienejad, N., Maroufi, M., Moheimani, S.R.: Rosette-scan video-rate atomic force microscopy: Trajectory patterning and control design. Review of Scientific Instruments **90**(7), 073,702 (2019)
625. Nimmo, R.: From monarchists to communists: Bees in the socio-political imagination. (2013). https://sociologicalinsect.com/2013/10/10/from-monarchists-to-communists-bees-in-the-socio-political-imagination/. [Online; accessed 2023-06-02]

626. NJ, N.: Teleo-reactive programs for agent control. Journal of Artificial intelligence Research **1**, 139–158 (1994)
627. Nolan, C.: Memento. Summit Entertainment (2001)
628. Noran, O., Romero, D., Zdravkovic, M.: The sensing enterprise: Towards the next generation dynamic virtual organisations. In: IFIP Advances in Information and Communication Technology, 434, pp. 209–216. Springer (2014)
629. Nowak, M., Plotkin, J., Jansen, V.: The evolution of syntactic communication. Nature **404**, 495–498 (2000)
630. Noys, B.: Drone metaphysics. Culture Machine **16**, 1–22
631. Ochs, E., Capps, L.: Living Narrative: Creating Lives in Everyday Storytelling. Harvard University Press, Cambridge Massachusetts (2002)
632. Office, U.G.A.: Science & Tech Spotlight: Drone Swarm Technologies—gao.gov. https://www.gao.gov/products/gao-23-106930. [Accessed 19-06-2024]
633. O'Kane, S.: Parrot exits the toy drone market. Retrieved from.https://www.theverge.com/2019/7/19/20699905/parrot-exit-toy-drone-market-dji-consumers
634. Oksa, S.J., Jalowskia, M., Fritzschea, A., Moslein, K.M.: Cyber-physical modeling and simulation: a reference architecture for designing demonstrators for industrial cyber-physical systems. Procedia CIRP **84**, 257–264 (2019)
635. Olaronke, I., Rhoda, I., Gambo, I., Ojerinde, O.A., Janet, O.: A systematic review of swarm robots. Current Journal of Applied Science and Technology **39**(15), 79–97 (2020)
636. Ollivier, Y.: A visual introduction to Riemannian curvatures and some discrete generalizations. In: G. Dafni, R. McCann, A. Stancu (eds.) Analysis and Geometry of Metric Measure Spaces: Lecture Notes of the 50th Séminaire de Mathématiques Supérieures (SMS), Montréal, 2011, pp. 197–219. AMS (2013)
637. O' shaughnessy, D.: Automatic speech recognition: History, methods and challenges. Pattern Recognition **41**(10), 2965–2979 (2008)
638. Pal, A.: Localization algorithms in wireless sensor networks: Current approaches and future challenges. Network protocols and algorithms **2**(1), 45–73 (2010)
639. Pan, W.T.: A new evolutionary computation approach: fruit fly optimization algorithm. pp. 382–391 (2011)
640. Pan, W.T.: A new fruit fly optimization algorithm: Taking the financial distress model as an example. Knowledge-Based Systems **26**, 69–74 (2012). https://doi.org/10.1016/j.knosys.2011.07.001
641. Pantic, N., Husain, M.I.: Covert botnet command and control using twitter. In: Proceedings of the 31st annual computer security applications conference, pp. 171–180 (2015)
642. Parayil, A., Ratnoo, A.: Bifurcation-based control law for pattern generation. IEEE Control Systems Letters **3**(2), 374–379 (2018)
643. Parisi, G.I., Kemker, R., Part, J.L., Kanan, C., Wermter, S.: Continual lifelong learning with neural networks: A review. Neural Networks (2019)
644. Park, K.T., Son, Y.H., Noh, S.D.: The architectural framework of a cyber physical logistics system for digital-twin-based supply chain control. International Journal of Production Research **59**(19), 5721–5742 (2021). https://doi.org/10.1080/00207543.2020.1788738
645. Pask, G.: Conversation, Cognition and Learning. Elsevier, Amsterdam (1975)
646. Pask, G.: Styles and strategies of learning. British Journal of Educational Psychology **46**, 128–148 (1976)
647. Passino, K.M.: Biomimicry of bacterial foraging for distributed optimization and control. IEEE control systems magazine **22**(3), 52–67 (2002)
648. Patterson, K., Nestor, P.J., Rogers, T.T.: Where do you know what you know? the representation of semantic knowledge in the human brain. Nature reviews neuroscience **8**(12), 976–987 (2007)
649. Pavone, M., Frazzoli, E.: Decentralized policies for geometric pattern formation and path coverage. Journal of Dynamic Systems, Measurement, and Control **129**(5), 633–643 (2007)
650. Pepper, S.: World Hypotheses: A Study in Evidence. California University Press, Berkeley

651. Pérez, I.F., Boumaza, A., Charpillet, F.: Learning collaborative foraging in a swarm of robots using embodied evolution. In: ECAL 2017–14th European Conference on Artificial Life (2017)
652. Petticrew, M., Roberts, H.: Systematic Reviews in the Social Sciences: A Practical Guide. Blackwell Publishing (2006)
653. Piantadosi, S.T., Fedorenko, E.: Infinitely productive language can arise from chance under communicative pressure. Journal of Language Evolution **2**(2), 141–147 (2017)
654. Pinker, S.: The language instinct: How the mind creates language. Penguin UK (2003)
655. Plato: The Timaeus. Macmillan, New York (1888)
656. Porta, A., Guzzetti, S., Montano, N., Furlan, R., Pagani, M., Malliani, A., Cerutti, S.: Entropy, entropy rate, and pattern classification as tools to typify complexity in short heart period variability series. IEEE Transactions on Biomedical Engineering **48**(11), 1282–1291 (2001)
657. Pourpanah, F., Abdar, M., Luo, Y., Zhou, X., Wang, R., Lim, C.P., Wang, X.Z., Wu, Q.J.: A review of generalized zero-shot learning methods. IEEE transactions on pattern analysis and machine intelligence (2022)
658. Prado, C.: Gadamer and Rorty: From interpretation to conversation. In: S. Zabala, J. Malpas (eds.) Consequences of hermeneutics: Fifty years after Gadamer's Truth and method. Northwestern University Press, Evanston
659. Prestes E, e.a.: The first global ontological standard for ethically driven robotics and automation systems [standards]. IEEE Robotics & Automation Magazine **28**(4), 120–124 (2021). https://doi.org/10.1109/MRA.2021.3117414
660. Pulvermüller, F.: The case of cause: neurobiological mechanisms for grounding an abstract concept. Philosophical Transactions of the Royal Society B: Biological Sciences **373**(1752), 20170,129 (2018)
661. Quigley, M., Conley, K., Gerkey, B., Faust, J., Foote, T., Leibs, J., Wheeler, R., Ng., A.Y.: Ros: an open-source robot operating system. In: ICRA workshop on open source software, vol. 3, p. 5 (2009)
662. Rabiner, L., Juang, B.: An introduction to hidden Markov models. IEEE ASSP magazine **3**(1), 4–16 (1986)
663. Rabiner, L.R.: A tutorial on hidden Markov models and selected applications in speech recognition. Proceedings of the IEEE **77**(2), 257–286 (1989)
664. Radford, A., Narasimhan, K., Salimans, T., Sutskever, I., et al.: Improving language understanding by generative pre-training (2018)
665. Rafols, I., Leydesdorff, L., O'Hare, A., Nightingale, P., Stirling, A.: How journal rankings can suppress interdisciplinary research: A comparison between innovation studies and business & management. Research Policy **41**(7), 1262–1282 (2012). https://doi.org/10.1016/j.respol.2012.03.015
666. Ralph, M.A.L., Jefferies, E., Patterson, K., Rogers, T.T.: The neural and computational bases of semantic cognition. Nature Reviews Neuroscience **18**(1), 42–55 (2017)
667. Rappaport, T.S., et al.: Wireless communications: principles and practice, vol. 2. prentice hall PTR New Jersey (1996)
668. Rath, M., Darwish, A., Pati, B., Pattanayak, B.K., Panigrahi, C.R.: Swarm intelligence as a solution for technological problems associated with internet of things. In: Swarm Intelligence for Resource Management in Internet of Things, pp. 21–45 (2020)
669. Raymond, E.S.: How to ask questions the smart way. http://www.catb.org/~esr/faqs/smart-questions.html. Accessed: 28th November 2023
670. Reina, A., Valentini, G., Fernández-Oto, C., Dorigo, M., Trianni, V.: A design pattern for decentralised decision making. PloS one **10**(10), e0140,950 (2015)
671. Ren and Beard, R.W.: Consensus seeking in multiagent systems under dynamically changing interaction topologies. IEEE Transactions on automatic control **50**(5), 655–661 (2005)
672. Ren, W.: Consensus tracking under directed interaction topologies: Algorithms and experiments. In: 2008 American Control Conference, pp. 742–747. IEEE (2008)
673. Ren, W.: Collective motion from consensus with cartesian coordinate coupling. IEEE Transactions on Automatic Control **54**(6), 1330–1335 (2009)

674. Ren, W.: Collective motion from consensus with cartesian coordinate coupling-part I: Single-integrator kinematics. In: Proc. IEEE Conference on Decision and Control, pp. 1006–1011. IEEE (Dec. 2008)
675. Renn, O., Lucas, K.: Systemic risk: The threat to societal diversity and coherence. Risk Analysis **42**(9), 1921–1934 (2022). https://doi.org/10.1111/risa.13654. https://onlinelibrary.wiley.com/doi/abs/10.1111/risa.13654
676. Reynolds, C.W.: Flocks, herds and schools: A distributed behavioral model. In: Proceedings of the 14th Annual Conference on Computer Graphics and Interactive Techniques, pp. 25–34 (1987). https://doi.org/10.1145/37401.37406
677. Reynolds, C.W.: Flocks, herds and schools: A distributed behavioral model. ACM SIGGRAPH computer graphics **21**(4), 25–34 (1987)
678. Riaz, R., Hussein, A., El-Fiqi, H., Abbass, H.: Deep reinforcement learning for autonomous alphadogfight air-to-air combat. under review **3** (2016)
679. Richards, J.C.: Curriculum approaches in language teaching: Forward, central, and backward design. *RELC Journal* **44**(1), 5–33 (2013). https://doi.org/10.1177/0033688212473293
680. Richardson, M., Burges, C.J., Renshaw, E.: Mctest: A challenge dataset for the open-domain machine comprehension of text. In: Proceedings of the 2013 conference on empirical methods in natural language processing, pp. 193–203 (2013)
681. Richardson, V.: Constructivist pedagogy. Teachers College Record **105**(9), 1623–1640
682. Ricœur, P.: Interpretation Theory: Discourse and the Surplus of Meaning. Texas Christian University Press, Fort Worth
683. Roark, B.: Probabilistic top-down parsing and language modeling. Computational linguistics **27**(2), 249–276 (2001)
684. Roberson, T.: On the social shaping of quantum technologies: An analysis of emerging expectations through grant proposals from 2002–2020. Minerva **59**(3), 379–397 (2021). https://doi.org/10.1007/s11024-021-09438-5
685. Roberson, T., Boshuijzen van Burken, C.: Unpacking the metaphors underlying discussions on 'thinking swarms'. In: S. Ng, J. Scholz, H.A. Abbass (eds.) Thinking Swarms. Springer (2025)
686. Rorty, R.: Philosophy and the Mirror of Nature. Princeton University Press, Princeton
687. Rossetto, D.E., Bernardes, R.C., Borini, F.M., Gattaz, C.C.: Structure and evolution of innovation research in the last 60 years: Review and future trends in the field of business through the citations and co-citations analysis. Scientometrics **115**(3), 1329–1363 (2018). https://doi.org/10.1007/s11192-018-2709-7
688. Roy, A.: God of Small Things. Flamingo, London (1997)
689. Royal Australian Navy: RAS-AI strategy 2040 (2020)
690. Rudnick, D.L., Davis, R.E., Eriksen, C.C., Fratantoni, D.M., Perry, M.J.: Underwater gliders for ocean research. Marine Technology Society Journal **38**(2), 73–84 (2004)
691. Sabir, Z., Raja, M.A.Z., Umar, M., Shoaib, M.: Neuro-swarm intelligent computing to solve the second-order singular functional differential model. The European Physical Journal Plus **135**(6), 1–19 (2020)
692. Saha, B., Sharma, S.: Steganographic techniques of data hiding using digital images. Defence Science Journal **62**(1), 11–18 (2012)
693. Şahin, E. (2004, July). Swarm robotics: From sources of inspiration to domains of application. In International workshop on swarm robotics (pp. 10–20). Berlin, Heidelberg: Springer Berlin Heidelberg.
694. Sahin, E., Winfield, A.F.: Special issue on swarm robotics. Swarm Intell. **2**(2–4), 69–72 (2008)
695. Sammons, P.J., Page, J.: Experimentation and validation of vehicle cluster simulator using netlogo. Adelaide, SA, Australia (2009)
696. Sanders, L., Copeland, D.: Developing an approach to the legal review of autonomous weapon systems (2020)
697. Sandhu, R., Georgiou, T., Reznik, E., Zhu, L., Kolesov, I., Senbabaoglu, Y., Tannenbaum, A.: Graph curvature for differentiating cancer networks. Scientific reports **5**(12323) (2015)

698. Sandler, W., Meir, I., Padden, C., Aronoff, M.: The emergence of grammar: Systematic structure in a new language. Proceedings of the National Academy of Sciences **102**(7), 2661–2665 (2005)
699. Sankey, D.W., Storms, R.F., Musters, R.J., Russell, T.W., Hemelrijk, C.K., Portugal, S.J.: Absence of "selfish herd" dynamics in bird flocks under threat. Current Biology **31**(14), 3192–3198 (2021)
700. Saska, M., Vonásek, V., Chudoba, J., Thomas, J., Loianno, G., Kumar, V.: Swarm distribution and deployment for cooperative surveillance by micro-aerial vehicles. Journal of Intelligent & Robotic Systems **84**(1), 469–492 (2016)
701. Schaal, S.: Learning from demonstration. Advances in neural information processing systems **9** (1996)
702. Scharre, P.: Army of none: Autonomous weapons and the future of war. WW Norton & Company (2018)
703. Scholz, J., Galliott, J.: 57The Humanitarian Imperative for Minimally-Just AI in Weapons: Jason Scholz and Jai Galliott, The Humanitarian Imperative for Minimally-Just AI in Weapons In: Lethal Autonomous Weapons. Edited by: Jai Galliott, Duncan MacIntosh and Jens David Ohlin, ©Oxford University Press (2021). https://doi.org/10.1093/oso/9780197546048.003.0005. In: Lethal Autonomous Weapons: Re-Examining the Law and Ethics of Robotic Warfare. Oxford University Press (2021). https://doi.org/10.1093/oso/9780197546048.003.0005
704. Scholz, J., Lambert, D., Gossink, D., Smith, G.: A blueprint for command and control: Automation and interface. In: Proceedings of the 15th International Conference on Information Fusion, pp. 211–217. IEEE (2012)
705. Scholz, J., Ng, S., , Abbass, H.A.: Thinking swarms. In: S. Ng, J. Scholz, H.A. Abbass (eds.) Thinking Swarms. Springer (2025)
706. Schranz, M., Umlauft, M., Sende, M., Elmenreich, W.: Swarm robotic behaviors and current applications. Frontiers in Robotics and AI **7**, 36 (2020)
707. Schranz, M., Umlauft, M., Sende, M., Elmenreich, W.: Swarm robotic behaviors and current applications. Frontiers in Robotics and AI **7** (2020). https://doi.org/10.3389/frobt.2020.00036
708. Schranz, M., Umlauft, M., Sende, M., Elmenreich, W.: Swarm robotic behaviors and current applications. Frontiers in Robotics and AI **7**, 36 (2020)
709. Schürmann, T., Grassberger, P.: Entropy estimation of symbol sequences. Chaos **6**(3), 414–427 (1996)
710. Schwandt, T.: The hermeneutics of suspicion. In: The SAGE dictionary of qualitative inquiry, vol. Vols. 1–0. SAGE Publications, Thousand Oaks, CA
711. Schwartz, M., Adams, D.: Shortest path solution by epitrochoid machine. The College Mathematics Journal **30**(3), 221–225 (1999)
712. of Science, T.W.H.O., Policy, T.: Opportunities and actions for ocean science and technology (2022–2028) (2022)
713. Sechrest, L.: The psychology of personal constructs: George kelly. (1963)
714. Seeja, G., et al.: A survey on swarm robotic modeling, analysis and hardware architecture. Procedia Computer Science **133**, 478–485 (2018)
715. Seet, P.S., Jones, J., Klarin, A., Suseno, Y.: IoT and the future workforce: potential implications. In: Input paper for the Horizon Scanning Project 'The Internet of Things'. Australian Council of Learned Academies (ACOLA), https://acola.org/wp-content/uploads/2021/02/acola-iot-input-paper_iot-and-the-future-workforce-potential-implications_seet-jones-klarin-suseno.pdf (2019)
716. Selkowitz, A., Lakhmani, S., Chen, J.Y., Boyce, M.: The effects of agent transparency on human interaction with an autonomous robotic agent. Proceedings of the Human Factors and Ergonomics Society Annual Meeting **59**(1), 806–810 (2015). https://doi.org/10.1177/1541931215591246
717. Sellars, W.: Philosophy and the scientific image of man. Frontiers of science and philosophy **1**, 35–78 (1962)
718. Semino, E.: Metaphor in Discourse. Cambridge University Press, Cambridge (2008)

719. Sennrich, R., Haddow, B., Birch, A.: Neural machine translation of rare words with subword units. arXiv preprint arXiv:1508.07909 (2015)
720. Serres, M.: Conversations on Science, Culture, and Time. University of Michigan Press, Ann Arbor
721. Shackle, S.: The mystery of the Gatwick drone: Gatwick airport. The Guardian
722. Shami, T.M., El-Saleh, A.A., Alswaitti, M., Al-Tashi, Q., Summakieh, M.A., Mirjalili, S.: Particle swarm optimization: a comprehensive survey. IEEE Access **10**, 10,031–10,061 (2022). https://doi.org/10.1109/ACCESS.2022.3142859
723. Shan, S.: Drones crash during light display at lantern festival. Taipei Times
724. Shannon, C.E.: A mathematical theory of communication. The Bell system technical journal **27**(3), 379–423 (1948)
725. Shannon, C.E.: A mathematical theory of communication (part III). Bell System Technical Journal **XXVII**, 623–656 (1948)
726. Shannon, C.E.: A mathematical theory of communication (parts I and II). Bell System Technical Journal **XXVII**, 379–423 (1948)
727. Sharkey, A.J.C.: Robots, insects and swarm intelligence. Artif Intell Rev **26**(4), 255–268 (2006). https://doi.org/10.1007/s10462-007-9057-y
728. Sharkey, A.J.C.: Robots, insects and swarm intelligence. Artificial Intelligence Review **26**(4), 255–268 (2006). https://doi.org/10.1007/s10462-007-9057-y. http://link.springer.com/10.1007/s10462-007-9057-y
729. Sharma, A., Shoval, S., Sharma, A., Pandey, J.K.: Path planning for multiple targets interception by the swarm of UAVS based on swarm intelligence algorithms: a review. IETE Technical Review (Institution of Electronics and Telecommunication Engineers, India) **39**(3), 675–697 (2022). https://doi.org/10.1080/02564602.2021.1894250
730. Sharma, V., Tripathi, A.K.: A systematic review of meta-heuristic algorithms in IoT based application. Array **14**, 100,164 (2022). https://doi.org/10.1016/j.array.2022.100164
731. Shekar, B.H., Kumari, M.S., Mestetskiy, L., Dyshkant, N.: Face recognition using kernel entropy component analysis. Neurocomputing **74**(6), 1053–1057 (2011)
732. Shelly, M.W.: Frankenstein, or the modern Prometheus. George Routledge and Sons Ltd, London (1891)
733. Sherratt, S.: Multi-level discourse analysis: A feasible approach. Aphasiology **21**(3–4), 375–393 (2007)
734. Shi, L., Zhang, Z., Li, Z., Guo, S., Pan, S., Bao, P., Duan, L.: Design, implementation and control of an amphibious spherical robot. Journal of Bionic Engineering **19**(6), 1736–1757 (2022)
735. Shi, Y.: An optimization algorithm based on brainstorming process:. International Journal of Swarm Intelligence Research **2**(4), 35–62 (2011). https://doi.org/10.4018/ijsir.2011100103
736. Shi, Y., Tan, Y., Shi, Y., Chai, Y., Wang, G.: Brain storm optimization algorithm. In: Y. Tan, Y. Shi, Y. Chai, G. Wang (eds.) Advances in Swarm Intelligence, vol. 6728, pp. 303–309. Berlin, Heidelberg (2011). http://link.springer.com/10.1007/978-3-642-21515-5_36
737. Shore, T., Skantze, G.: Using lexical alignment and referring ability to address data sparsity in situated dialog reference resolution. In: EMNLP, pp. 2288–2297 (2018)
738. Siddaway, A.P., Wood, A.M., Hedges, L.V.: How to do a systematic review: A best practice guide for conducting and reporting narrative reviews, meta-analyses, and meta-syntheses. Annual Review of Psychology **70**, 747–770 (2019). https://doi.org/10.1146/annurev-psych-010418-102803
739. Silva, F., Duarte, M., Correia, L., Oliveira, S.M., Christensen, A.L.: Open issues in evolutionary robotics. Evolutionary computation **24**(2), 205–236 (2016)
740. Simmons, J.: Paintings (1999). http://jeffreysimmonsstudio.com/project/trochoids-paintings-1999/
741. Simonetti, P.: Slocum glider: Design and 1991 field trials. Tech. rep., Report to the Office of Naval Technology (1992)
742. Simoson, A.: An envelope for a spirograph. The College Mathematics Journal **28**(2), 134–139 (1997)

743. Simoson, A.J.: The trochoid as a tack in a bungee cord. Mathematics Magazine **73**(3), 171–184 (2000)
744. Sinha, A., Ghose, D.: Generalization of linear cyclic pursuit with application to rendezvous of multiple autonomous agents. IEEE Transactions on Automatic Control **51**(11), 1819–1824 (2006)
745. Siu, H.C., Peña, J., Chen, E., Zhou, Y., Lopez, V., Palko, K., Chang, K., Allen, R.: Evaluation of human-ai teams for learned and rule-based agents in hanabi. Advances in Neural Information Processing Systems **34** (2021)
746. Slapakova, L., Fusaro, P., Black, J., Dortmans, P.: Supporting the royal Australian navy's campaign plan for robotics and autonomous systems. https://www.rand.org/content/dam/rand/pubs/research_reports/RRA1300/RRA1377-1/RAND_RRA1377-1.pdf (2020). Accessed: 20 Oct 2022
747. Slupska, J.: War, health and ecosystem: Generative metaphors in cybersecurity governance. Philosophy & Technology **34**, 463–482 (2021). https://doi.org/10.1007/s13347-020-00397-5
748. Smith, A.J., Best, G., Yu, J., Hollinger, G.A.: Real-time distributed non-myopic task selection for heterogeneous robotic teams. Autonomous Robots **43**, 789–811 (2019)
749. Smith, P., Aleti, A., Lee, V.C., Hunjet, R., Khan, A.: Robotic hierarchical graph neurons. a novel implementation of HGN for swarm robotic behaviour control. Expert Systems with Applications **186**, 115,675 (2021). https://doi.org/10.1016/j.eswa.2021.115675. https://www.sciencedirect.com/science/article/pii/S0957417421010629
750. Smith, P., Hunjet, R., Aleti, A., Barca, J.C., et al.: Data transfer via UAV swarm behaviours: Rule generation, evolution and learning. Australian Journal of Telecommunications and the Digital Economy **6**(2), 35–58 (2018)
751. Smith, P., Hunjet, R., Khan, A.: Swarm learning in restricted environments: an examination of semi-stochastic action selection. In: 2018 15th International Conference on Control, Automation, Robotics and Vision (ICARCV), pp. 848–855. IEEE (2018)
752. Smith, P., Hunjet, R.A., Aleti, A., Khan, A.: Swarm behaviour evolution via rule sharing and novelty search. CoRR **abs/1910.12412** (2019). http://arxiv.org/abs/1910.12412
753. Smith, P., Khan, A., Aleti, A., Lee, V.C., Hunjet, R.: Data communication assistance via swarm robotics: a behaviour creation comparison. In: Evaluation of Adaptive Systems for Human-Autonomy Teaming. IJCAI (2019)
754. Smith-Miles, K.A.: Towards insightful algorithm selection for optimisation using meta-learning concepts. In: Neural Networks, 2008. IJCNN 2008.(IEEE World Congress on Computational Intelligence). IEEE International Joint Conference on, pp. 4118–4124. IEEE (2008)
755. Solé, R.V., Corominas-Murtra, B., Valverde, S., Steels, L.: Language networks: Their structure, function, and evolution. Complexity **15**(6), 20–26 (2010)
756. Speith, T.: A review of taxonomies of explainable artificial intelligence (XAI) methods. In: Proceedings of the 2022 ACM conference on fairness, accountability, and transparency, pp. 2239–2250 (2022)
757. Spencer, J.: The eight rules of urban warfare and why we must work to change them (2021). https://mwi.usma.edu/the-eight-rules-of-urban-warfare-and-why-we-must-work-to-change-them/
758. Stafleu, M.: Philosophical ethics and the so-called ethical aspect. Philosophia Reformata **72**(1), 21–33 (2007). https://www.jstor.org/stable/24709625. Publisher: Brill
759. Stahl, R.: What the drone saw: the cultural optics of the unmanned war. Australian Journal of International Affairs **67**(5), 659–674. https://doi.org/10.1080/10357718.2013.817526
760. Steyvers, M., Tenenbaum, J.B.: The large-scale structure of semantic networks: Statistical analyses and a model of semantic growth. Cognitive science **29**(1), 41–78 (2005)
761. Stokey, R., Austin, T., Von Alt, C., Purcell, M., Forrester, N., Goldsborough, R., Allen, B.: AUV bloopers or why murphy must have been an optimist: a practical look at achieving mission level reliability in an autonomous underwater vehicle. In: Proceedings of the Eleventh International Symposium on Unmanned Untethered Submersible Technology, pp. 32–40 (1999)

762. Stommel, H.: The slocum mission. Oceanography **2**(1), 22–25 (1989)
763. Storms, R., Carere, C., Zoratto, F., Hemelrijk, C.: Complex patterns of collective escape in starling flocks under predation. Behavioral ecology and sociobiology **73**, 1–10 (2019)
764. Strachan, J.W.A., Albergo, D., Borghini, G., Pansardi, O., Scaliti, E., Gupta, S., Saxena, K., Rufo, A., Panzeri, S., Manzi, G., Graziano, M.S.A., Becchio, C.: Testing theory of mind in large language models and humans. Nat Hum Behav (2024)
765. Strauss, D.F.: The scope and limitations of von bertalanffy's systems theory. South African journal of philosophy **21**(3), 163–179 (2002). Publisher: Taylor & Francis
766. Strutt, J.E.: Report of the inquiry into the loss of autosub2 under the fimbulisen. Southampton Research and Consultancy Report 12, National Oceanography Centre (2006)
767. StudioDrift: Further dates of the light installation "breaking waves" have to be cancelled". Instagram.com https://www.instagram.com/tv/Cc8Q52_jyxH/. Retrieved from:
768. Sukkar, F., Best, G., Yoo, C., Fitch, R.: Multi-robot region-of-interest reconstruction with Dec-MCTS. In: 2019 International conference on robotics and automation (ICRA), pp. 9101–9107. IEEE (2019)
769. Sun, L.: Drones in 2016: 4 numbers everyone should know
770. Sun, R.: The CLARION cognitive architecture: Extending cognitive modeling to social simulation. In: R. Sun (ed.) Cognition and multi-agent interaction: From cognitive modeling to social simulation, pp. 79–99. Cambridge University Press (2006)
771. Sun, R.: Memory systems within a cognitive architecture. New Ideas in Psychology **30**, 227–240 (2012)
772. Sutton, R.S., Barto, A.G.: Reinforcement learning: An introduction. MIT press (2018)
773. Swidan, A., Joiner, K., Jewson, E., Carroll, N., Champ, D., Shpak, G.: A novel flying and diving wig craft for electronics intelligence-a conceptual design. In: 2022 International Telecommunications Conference (ITC-Egypt), pp. 1–5. IEEE (2022)
774. Taddeo, M., Blanchard, A.: A comparative analysis of the definitions of autonomous weapons systems. Science and Engineering Ethics **28**(5) (2022). https://doi.org/10.1007/s11948-022-00392-3
775. Talevski, J., Wong-Shee, A., Rasmussen, B., Kemp, G., Beauchamp, A.: Teach-back: A systematic review of implementation and impacts. PLOS ONE **15**, 1–18 (2020). https://doi.org/10.1371/journal.pone.0231350
776. Talha, M., Hussein, A., Hossny, M.: Autonomous UAV navigation in wilderness search-and-rescue operations using deep reinforcement learning. In: H. Aziz, D. Corrêa, T. French (eds.) AI 2022: Advances in Artificial Intelligence, pp. 733–746. Springer International Publishing, Cham (2022)
777. Taloni, A., Scorcia, V., Giannaccare, G.: Large Language Model Advanced Data Analysis Abuse to Create a Fake Data Set in Medical Research. JAMA Ophthalmology **141**(12), 1174–1175 (2023). https://doi.org/10.1001/jamaophthalmol.2023.5162
778. Tan, Y., Zhu, Y., Tan, Y., Shi, Y., Tan, K.C.: Fireworks algorithm for optimization. In: Y. Tan, Y. Shi, K.C. Tan (eds.) Advances in Swarm Intelligence, vol. 6145, pp. 355–364. Berlin, Heidelberg (2010). http://link.springer.com/10.1007/978-3-642-13495-1_44
779. Teh, Y., Bapst, V., Czarnecki, W.M., Quan, J., Kirkpatrick, J., Hadsell, R., Heess, N., Pascanu, R.: Distral: Robust multitask reinforcement learning. In: Advances in Neural Information Processing Systems, pp. 4496–4506 (2017)
780. Templeton, A., Conerly, T., Marcus, J., Lindsey, J., Bricken, T., Chen, B., Pearce, A., Citro, C., Ameisen, E., Jones, A., Cunningham, H., Turner, N.L., McDougall, C., MacDiarmid, M., Freeman, C.D., Sumers, T.R., Rees, E., Batson, J., Jermyn, A., Carter, S., Olah, C., Henighan, T.: Scaling monosemanticity: Extracting interpretable features from claude 3 sonnet. Transformer Circuits Thread (2024). https://transformer-circuits.pub/2024/scaling-monosemanticity/index.html
781. Thabtah, F., Cowling, P.: Mining the data from a hyperheuristic approach using associative classification. Expert Systems with Applications **34**(2), 1093–1101 (2008). https://doi.org/10.1016/j.eswa.2006.12.018
782. Thagard, P.: Conceptual Revolutions. Princeton University Press, Princeton, New Jersey (1992)

References

783. Thompson, F., Guihen, D.: Review of mission planning for autonomous marine vehicle fleets. Journal of Field Robotics **36**(2), 1–22 (2018)
784. Thrun, S., Mitchell, T.M.: Lifelong robot learning. Robotics and autonomous systems **15**(1–2), 25–46 (1995)
785. Tomasello, M.: Constructing a language: A usage-based theory of language acquisition. Harvard university press (2005)
786. Toolan, M.: Language. In: D. Herman (ed.) The Cambridge Companion to Narrative, pp. 231–244. Cambridge University Press (2007)
787. Trabasso, T., Sperry, L.: Causal Relatedness and Importance of Story Events. Journal of Memory and Language **24**, 595–611 (1985)
788. Trabasso, T., Sperry, L.L.: Causal relatedness and importance of story events. Journal of Memory and language **24**(5), 595–611 (1985)
789. Tranfield, D., Denyer, D., Smart, P.: Towards a methodology for developing evidence-informed management knowledge by means of systematic review. British Journal of Management **14**(3), 207–222 (2003). https://doi.org/10.1111/1467-8551.00375
790. Trentin, E., Gori, M.: A survey of hybrid ANN/HMM models for automatic speech recognition. Neurocomputing **37**(1–4), 91–126 (2001)
791. Trianni, V.: Evolutionary swarm robotics: evolving self-organising behaviours in groups of autonomous robots, vol. 108. Springer (2008)
792. Trinh, T.H., Wu, Y., Le, Q.V., He, H., Luong, T.: Solving olympiad geometry without human demonstrations. Nature **625**(7995), 476–482 (2024). https://doi.org/10.1038/s41586-023-06747-5
793. Tripathy, T., Sinha, A.: Generating patterns with a unicycle. IEEE Transactions on Automatic Control **61**(10), 3140–3145 (2016)
794. Tripathy, T., Sinha, A.: Unicycle with only range input: An array of patterns. IEEE Transactions on Automatic Control **63**(5), 1300–1312 (2017)
795. Trusted Autonomous Systems DCRC: Body of knowledge on the assurance and accreditation of autonomous systems: Key concepts: Assurance. https://www.rasgateway.com.au/resource-hub/assurance (2022). Accessed: 28 June 2022
796. Trusted Autonomous Systems DCRC: Primer on swarming technologies. https://www.rasgateway.com.au/resource-hub/introduction-to-swarming (2022). Accessed: 28 June 2022
797. Tsiotras, P., Castro, L.I.R.: The artistic geometry of consensus protocols. In: Controls and Art, pp. 129–153. Springer (2014)
798. Tsiotras, P., Castro, L.I.R.: Extended multi-agent consensus protocols for the generation of geometric patterns in the plane. In: Proc. American Control Conference, pp. 3850–3855. IEEE (Jun. 2011)
799. Turgut, A.E., Çelikkanat, H., Gökçe, F., Şahin, E.: Self-organized flocking in mobile robot swarms. Swarm Intelligence **2**(2), 97–120 (2008)
800. Tzu, S.: The Art of War by Sun Tzu. El Paso Norte Press
801. Uday, P.: System importance measures: A new approach to resilient systems-of-systems. Ph.D. thesis, Purdue University (2015)
802. Unit, D.I.: UAS solutions for the U.S. DoD https://www.diu.mil/blue-uas-cleared-list. Retrieved from
803. Unmanned boats demonstrate autonomous swarm: Office of Naval Research. https://www.youtube.com/watch?v=GGuMdJBpWdE (2014). Accessed: 30 May 2022
804. US Department of Commerce National Institute of Standards and Technology: Computer security resource centre glossary, 'deterministic algorithm'. https://csrc.nist.gov/glossary/term/deterministic_algorithm
805. Valentinov, V.: System-environment relations in the theories of open and autopoietic systems: Implications for critical systems thinking. Syst Pract Action Res **25**(6), 537–542 (2012). https://doi.org/10.1007/s11213-012-9241-0
806. Valentinov, V.: System-environment relations in the theories of open and autopoietic systems: Implications for critical systems thinking. Systemic Practice and Action Research **25**(6), 537–542 (2012). https://doi.org/10.1007/s11213-012-9241-0

807. Valianti, P., Kolios, P., Ellinas, G.: Energy-aware tracking and jamming rogue uavs using a swarm of pursuer UAV agents. IEEE Systems Journal pp. 1–12 (2022). https://doi.org/10.1109/JSYST.2022.3179632
808. Van Brussel, H., Bongaerts, L., Wyns, J., Valckenaers, P., Van Ginderachter, T.: A conceptual framework for holonic manufacturing: Identification of manufacturing holons. Journal of Manufacturing Systems **18**(1), 35–52 (1999)
809. van der Aalst, W.M., ter Hofstede, A.H.M., Weske, M.: Business process management: A survey. In: W.M. van der Aalst, M. Weske (eds.) Business Process Management, pp. 1–12. Springer, Berlin, Heidelberg (2003)
810. Vance, T.C., Doel, R.E.: Graphical methods and cold war scientific practice: The stommel diagram's intriguing journey from the physical to the biological environmental sciences. Historical Studies in the Natural Sciences **40**(1), 1–47 (2010)
811. Vaswani, A., Shazeer, N., Parmar, N., Uszkoreit, J., Jones, L., Gomez, A.N., Kaiser, Ł., Polosukhin, I.: Attention is all you need. Advances in neural information processing systems **30** (2017)
812. Vattimo, G., Zabala, S.: Hermeneutic Communism: from Heidegger to Marx. Columbia University Press, New York
813. Vegetabile, B.G., Stout-Oswald, S.A., Davis, E.P., Baram, T.Z., Stern, H.S.: Estimating the entropy rate of finite Markov chains with application to behavior studies. Journal of Educational and Behavioral Statistics **44**(3), 282–308 (2019)
814. Veitch, E., Alsos, O.A.: Human-centered explainable artificial intelligence for marine autonomous surface vehicles. Journal of Marine Science and Engineering **9**(11), 1227 (2021)
815. Verbruggen, M.: Beyond the buzz: A primer on swarms (2020). https://www.lowyinstitute.org/the-interpreter/beyond-buzz-primer-swarms
816. Verbruggen, M.: Beyond the buzz: A primer on swarms (2020). https://www.lowyinstitute.org/the-interpreter/beyond-buzz-primer-swarms. [Online; accessed 2022-01-31]
817. Verbruggen, M.: The question of swarms control: Challenges for ensuring human control over military swarms. Non-Proliferation and Disarmament Papers No 65, EU Non-Proliferation and Disarmament Consortium (2019)
818. Verkerk, M.J., Hoogland, J., Stoep, J., Vries, M.J.: Philosophy of Technology: an introduction for technology and business students. Routledge, Abingdon (2015)
819. Vinyals, O., Babuschkin, I., Czarnecki, W.M., Mathieu, M., Dudzik, A., Chung, J., Choi, D.H., Powell, R., Ewalds, T., Georgiev, P., et al.: Grandmaster level in starcraft ii using multi-agent reinforcement learning. Nature **575**(7782), 350–354 (2019)
820. Waltman, L., van Eck, N.J., Noyons, E.C.M.: A unified approach to mapping and clustering of bibliometric networks. Journal of Informetrics **4**(4), 629–635 (2010)
821. Wang, C., Xie, G.: Limit-cycle-based decoupled design of circle formation control with collision avoidance for anonymous agents in a plane. IEEE Transactions on Automatic Control **62**(12), 6560–6567 (2017)
822. Wang, W., Wang, Y., Huang, Q., Gao, W.: Measuring visual saliency by site entropy rate. In: 2010 IEEE Computer Society Conference on Computer Vision and Pattern Recognition, pp. 2368–2375 (2010)
823. Wannagat, W., Waizenegger, G., Nieding, G.: Multi-level mental representations of written, auditory, and audiovisual text in children and adults. Cognitive Processing **18**(4), 491–504 (2017)
824. Wasson, C.: Covert caution: Linguistic traces of organizational control. Yale University (1996)
825. Weber, J.: Artificial intelligence and the socio-technical imaginary. Manchester University Press, Manchester, England (2021). https://www.manchesterhive.com/view/9781526145949/9781526145949.000
826. Weber, J.: Autonomous drone swarms and the contested imaginaries of artificial intelligence. Digital War **5**(1), 146–149 (2024). https://doi.org/10.1057/s42984-023-00076-7
827. Weel, B., Crosato, E., Heinerman, J., Haasdijk, E., Eiben, A.: A robotic ecosystem with evolvable minds and bodies. In: Evolvable Systems (ICES), 2014 IEEE International Conference on, pp. 165–172. IEEE (2014)

828. Wei, D., Wang, Z., Si, L., Tan, C.: Preaching-inspired swarm intelligence algorithm and its applications. Knowledge-Based Systems **211**, 106,552 (2021). https://doi.org/10.1016/j.knosys.2020.106552
829. Weinberg, G.V.: Prediction of UAV swarm defeat with high-power radio frequency fields. IEEE Transactions on Electromagnetic Compatibility pp. 1–6 (2022). https://doi.org/10.1109/TEMC.2022.3193881
830. Whidden, C., Matsen IV, F.A.: Ricci-Ollivier curvature of the rooted phylogenetic subtree-prune-regraft graph. arXiv: 1504.00304 (2015)
831. White, S.: Political Theory and Postmodernism. Cambridge University Press, Cambridge
832. Williams, J.J.: Theory and the Novel: Narrative Reflexivity in the British Tradition. Cambridge University Press (1998)
833. Williams, S.M.: Swarm weapons: Demonstrating a swarm intelligent algorithm for parallel attack. US Army School for Advanced Military Studies Fort Leavenworth United States (2018)
834. Wilson, E.O.: Genesis: The deep origin of societies. Penguin UK (2019)
835. Winfield, A.F., Harper, C.J., Nembrini, J.: Towards dependable swarms and a new discipline of swarm engineering. In: International Workshop on Swarm Robotics, pp. 126–142. Springer (2004)
836. Winfield, A.F., Jirotka, M.: The case for an ethical black box. In: Towards Autonomous Robotic Systems: 18th Annual Conference, TAROS 2017, Guildford, UK, July 19–21, 2017, Proceedings 18, pp. 262–273. Springer (2017)
837. Winfree, A.T.: Varieties of spiral wave behavior: An experimentalist's approach to the theory of excitable media. Chaos: An Interdisciplinary Journal of Nonlinear Science **1**(3), 303–334 (1991)
838. Winner, L.: Do artifacts have politics? Dædalus **109**(1), 121–136
839. Woodworth, J., Gump, W.: Camelot: A Role Playing Simulation for Political Decision Making, 3rd edn. Wadsworth
840. Woolf, V.: To The Lighthouse. Oxford World's Classics, Oxford (2008)
841. Wu, B., Glesk, I., Prucnal, P.R., Narimanov, E.: Security analysis of stealth transmission over a public fiber-optical network. In: Conference on Lasers and Electro-Optics, p. CThBB3. Optica Publishing Group (2007)
842. Wu, G., Xu, T., Sun, Y., Zhang, J.: Review of multiple unmanned surface vessels collaborative search and hunting based on swarm intelligence. International Journal of Advanced Robotic Systems **19**(2), 17298806221091,885 (2022)
843. Xian, Y., Schiele, B., Akata, Z.: Zero-shot learning-the good, the bad and the ugly. In: Proceedings of the IEEE conference on computer vision and pattern recognition, pp. 4582–4591 (2017)
844. Xiao, J., Xu, H., Gao, H., Bian, M., Li, Y.: A weakly supervised semantic segmentation network by aggregating seed cues: the multi-object proposal generation perspective. ACM Transactions on Multimidia Computing Communications and Applications **17**(1s), 1–19 (2021)
845. Xue, J., Shen, B.: A novel swarm intelligence optimization approach: sparrow search algorithm. null **8**(1), 22–34 (2020). https://doi.org/10.1080/21642583.2019.1708830
846. Yamauchi, B., Beer, R.: Integrating reactive, sequential, and learning behavior using dynamical neural networks. From Animals to Animats **3**, 382–391 (1994)
847. Yang, J., Zhou, K., Li, Y., Liu, Z.: Generalized out-of-distribution detection: A survey (2024)
848. Yang, R., Han, B., Li, F., Zhou, Y., Xiang, J.: Nonlinear dynamic analysis of a trochoid cam gear. Journal of Mechanical Design **142**(9) (2020)
849. Yang, W., Green, A.E., Chen, Q., Kenett, Y.N., Sun, J., Wei, D., Qiu, J.: Creative problem solving in knowledge-rich contexts. Trends in Cognitive Sciences (2022)
850. Yang, W., Johnstone, M.N., Wang, S., Karie, N.M., Sahri, N.M.b., Kang, J.J.: Network forensics in the era of artificial intelligence. In: Explainable Artificial Intelligence for Cyber Security, pp. 171–190. Springer, Vol 1025, Issue January, . (2022)

851. Yang, X.: Nature-inspired algorithms for optimization. Springer Science & Business Media, Berlin (2010)
852. Yang, X.S., González, J.R., Pelta, D.A., Cruz, C., Terrazas, G., Krasnogor, N.: A new metaheuristic bat-inspired algorithm. In: J.R. González, D.A. Pelta, C. Cruz, G. Terrazas, N. Krasnogor (eds.) Nature Inspired Cooperative Strategies for Optimization (NICSO 2010), vol. 284, pp. 65–74. Berlin, Heidelberg (2010). http://link.springer.com/10.1007/978-3-642-12538-6_6
853. Yang, X.S., Watanabe, O., Zeugmann, T.: Firefly algorithms for multimodal optimization. In: O. Watanabe, T. Zeugmann (eds.) Stochastic Algorithms: Foundations and Applications, vol. 5792, pp. 169–178. Berlin, Heidelberg (2009). http://link.springer.com/10.1007/978-3-642-04944-6_14
854. Yaxley, K.J., Joiner, K.F., Abbass, H.: Drone approach parameters leading to lower stress sheep flocking and movement: sky shepherding. Scientific reports **11**(1), 1–9 (2021)
855. Yazdani, A.M., Sammut, K., Yakimenko, O., Lammas, A.: A survey of underwater docking guidance systems. Robotics and Autonomous systems **124**, 103,382 (2020)
856. Yee, E.: Abstraction and concepts: when, how, where, what and why? (2019)
857. Yee, E., Thompson-Schill, S.L.: Putting concepts into context. Psychonomic bulletin & review **23**(4), 1015–1027 (2016)
858. Young, G.F., Scardovi, L., Cavagna, A., Giardina, I., Leonard, N.E.: Starling flock networks manage uncertainty in consensus at low cost. PLOS Computational Biology **9**(1), 1–7 (2013). https://doi.org/10.1371/journal.pcbi.1002894
859. Yuan, Y., Yu, Z.L., Gu, Z., Deng, X., Li, Y.: A novel multi-step reinforcement learning method for solving reward hacking. Applied Intelligence **49**(8), 2874–2888 (2019)
860. Zakiev, A., Tsoy, T., Magid, E., Ronzhin, A., Rigoll, G., Meshcheryakov, R.: Swarm robotics: Remarks on terminology and classification. Lecture Notes in Computer Science, pp. 291–300. Cham (2018). https://doi.org/10.1007/978-3-319-99582-3_30
861. Zegzhda, P.D., Anisimov, V.G., Suprun, A.F., Anisimov, E.G., Saurenko, T.N., Los, V.P.: A model of optimal complexification of measures providing information security. Automatic Control and Computer Sciences **54**(8), 930–936 (2020)
862. Zhang, T., Tao, D., Qu, X., Zhang, X., Lin, R., Zhang, W.: The roles of initial trust and perceived risk in public's acceptance of automated vehicles. Transportation Research Part C: Emerging Technologies **98**, 207–220 (2019). https://doi.org/10.1016/j.trc.2018.11.018
863. Zhao, W., Ammar, M., Zegura, E.: A message ferrying approach for data delivery in sparse mobile ad hoc networks. In: Proceedings of the 5th ACM international symposium on Mobile ad hoc networking and computing, pp. 187–198. ACM (2004)
864. Zhao, Z.S., Granucci, F., Yeh, L., Schaffer, P.A., Cantor, H.: Molecular mimicry by herpes simplex virus-type 1: autoimmune disease after viral infection. Science **279**(5355), 1344–1347 (1998)
865. Zipf, G.: The psycho-biology of language: An introduction to dynamic philology. Houghton Mifflin, Cambridge, MA (1935)
866. Zipf, G., Thiele, L.: Human Behavior and the Principle of Least Effort. Addison Wesley, Cambridge (1949)
867. Zitkovich, B., Yu, T., Xu, S., Xu, P., Xiao, T., Xia, F., Wu, J., Wohlhart, P., Welker, S., Wahid, A., Vuong, Q., Vanhoucke, V., Tran, H., Soricut, R., Singh, A., Singh, J., Sermanet, P., Sanketi, P.R., Salazar, G., Ryoo, M.S., Reymann, K., Rao, K., Pertsch, K., Mordatch, I., Michalewski, H., Lu, Y., Levine, S., Lee, L., Lee, T.W.E., Leal, I., Kuang, Y., Kalashnikov, D., Julian, R., Joshi, N.J., Irpan, A., Ichter, B., Hsu, J., Herzog, A., Hausman, K., Gopalakrishnan, K., Fu, C., Florence, P., Finn, C., Dubey, K.A., Driess, D., Ding, T., Choromanski, K.M., Chen, X., Chebotar, Y., Carbajal, J., Brown, N., Brohan, A., Arenas, M.G., Han, K.: Rt-2: Vision-language-action models transfer web knowledge to robotic control. In: J. Tan, M. Toussaint, K. Darvish (eds.) Proceedings of The 7th Conference on Robot Learning, *Proceedings of Machine Learning Research*, vol. 229, pp. 2165–2183. PMLR (2023). https://proceedings.mlr.press/v229/zitkovich23a.html

868. Zuboff, S.: The age of surveillance capitalism: the fight for a human future at the new frontier of power, first edition. edn. Ingram Publishers, United States
869. Zwaan, R.: Situation models, mental simulations, and abstract concepts in discourse comprehension. Psychon Bulletin Review **23**, 1028–1034 (2016)
870. Zwaan, R.A.: Situation models, mental simulations, and abstract concepts in discourse comprehension. Psychonomic bulletin & review **23**(4), 1028–1034 (2016)
871. Zwaan, R.A., Magliano, J.P., Graesser, A.C.: Dimensions of situation model construction in narrative comprehension. Journal of Experimental Psychology: Learning, Memory, and Cognitionv **21**(2), 386–397 (1995)
872. Zylstra, U.: Living things as hierarchically organized structures. Synthese **91**, 111–133 (1992). Publisher: Springer

Index

A
Autonomous underwater vehicles (AUVs), 243–245, 247–252, 254, 257, 259, 261–263, 265, 266
Autonomy and society, 36, 155
Autopoiesis, 39, 41, 42, 47, 50
Awareness, 12, 14–16, 33, 99, 141, 146, 148, 156, 189, 194, 197, 198, 201–205, 243, 244, 248, 249, 257, 260, 266, 267, 272, 308, 314, 331

B
Behaviour selection, 272, 274–276, 282, 283

C
Communication, 10, 21, 41, 61, 97, 130, 161, 177, 189, 207, 243, 276, 291, 314, 337
Consensus, 12–14, 62, 207, 208, 210, 212–214, 223, 224, 227, 232, 241, 242, 276
Construct analysis, 5–11
Conversation, 61, 63, 72, 73, 80, 81, 88, 93, 105, 113, 189, 190, 194, 195, 197–205, 256, 257, 329
Cooperative control, 208
Curriculum design, 291–312
Curriculum learning, 301, 302, 308

D
Definitions, 4, 5, 35, 39–48, 51, 62, 148, 154, 155, 159, 161, 162, 165, 181, 214, 243, 246, 331, 334

Dooyeweerd, 49, 50, 53–58, 60
Drones, 7, 27, 59, 77, 127, 173, 202, 322, 331
Drone swarm, 68, 71, 73, 77, 78, 82, 83, 86, 133, 145, 147, 173, 176–178, 180, 184, 331
Dynamic systems architecture, 194

E
Emergent agents, 189, 196, 197, 201, 204, 205
Enkaptic, 49, 53, 55–60

F
Functional analysis, 11–15, 297

H
Hermeneutics, 77–93, 302, 332
Heterogeneous multi-robot systems, 314
Hidden swarm, 100, 117–119, 124

I
Infrastructure analysis, 15–19
Intelligence, 11, 21, 40, 53, 83, 96, 127, 153, 185, 189, 314, 330

L
Legal review, 164, 171–186
Lifelong learning, 271–289, 335

M

Machine education, 291, 293, 296–301
Machine learning (ML), 21, 25, 30, 32, 34, 36, 102, 110, 117, 124, 128, 142, 164, 167, 190, 204, 293, 295–297, 299–304, 307, 310, 311, 334, 335, 338
Manned-unmanned teaming, 317
Maritime autonomous swarms, 151–170
Multi-agent systems, 4, 40, 70, 156, 204, 208, 209, 219

N

Narrative, 7, 64, 69–71, 83, 84, 95–125, 127–148, 198, 200, 202, 203, 335
Navigation, 12, 13, 15, 16, 27, 29, 128, 159, 164, 243, 245, 246, 249, 252, 253, 257, 260–262, 266, 304
Neuro-symbolic systems, 330, 338
Noise, 17, 96, 117, 127–148, 261, 275–278, 307

P

Pattern formation, 12, 13, 27, 210, 212–214, 219–223, 227–232, 235, 241, 242
Philosophy, 49–56, 58, 62, 82, 277, 301, 330

R

Reconnaissance, 33, 82, 83, 185
Regulation, 9, 12, 13, 17, 18, 27, 70, 74, 81, 88, 93, 151, 153, 155–158, 160–162, 165–170, 173, 329–331, 333, 334
Robots, 7, 28, 39, 59, 66, 142, 152, 207, 243, 273, 291, 313, 331

S

Scenario-based simulation, 77, 78, 84, 90, 91
Shared awareness, 201
Simulation testing, 171
Surveillance, 27, 29, 33, 52, 82, 83, 89, 93, 151, 152, 177, 178, 208, 212, 239, 292, 314
Swarm coordination, 338
Swarming robots, 288, 329, 331–333, 336
Swarm robotics, 40, 155, 208, 271–289, 292, 294–296, 329, 330, 333–338
Swarms, 3, 21, 39, 49, 61, 77, 95, 127, 151, 171, 189, 208, 243, 271, 291, 313, 329
Swarm simulation, 39, 50, 77, 254, 336
Swarms of information, 96, 98, 99, 116, 127
Synthetic language, 95–125, 127, 142–146, 148, 194
Systems thinking, 23, 41–42, 46, 50–51

T

Thinking, 22, 39, 50, 67, 78, 102, 200, 321, 329
Thinking swarms, 3, 23, 39, 49, 63, 77, 185, 189, 191, 246, 313, 329
Trochoids, 207–242, 259

The manufacturer's authorised representative in the EU is Springer Nature Customer Service Centre GmbH, Europaplatz 3, 69115 Heidelberg, Germany. If you have any concerns regarding our products, please contact ProductSafety@springernature.com

Printed and bound by CPI Group (UK) Ltd, Croydon, CR0 4YY
26/03/2026
02078991-0002